To the memory of H.J. Bunker and K.R. Butlin

Microbes and Man

Microbes are everywhere. Normally invisible, they are abundant in the air we breathe, in soil, in water, on our skin and hair, in our mouths and intestines, on and in the food we eat. They make the soil fertile; they clean up the environment; they change, often improve, our food; some protect us from less desirable microbes. Yet most people are scarcely aware that they exist – except when they become ill. Microbes, as 'germs', are widely regarded as nasty, unpopular because a few can cause disease and a few can spoil food. Yet collectively microbes present a fascinating world of minuscule creatures, who together encompass all the processes of which terrestrial life is capable: creatures who have profound effects on our lives and surroundings. In this up-dated edition, the extraordinary impact which the microbial community has on our everyday lives is described in an accessible and easy to read style.

John Postgate FRS is Emeritus Professor of Microbiology at the University of Sussex and former Director of the AFRC Unit of Nitrogen Fixation. He was educated at Kingsbury County and other schools, and at Balliol College, Oxford, where he took a first degree in chemistry before turning to chemical microbiology. He then spent fifteen years in government research establishments – studying mainly the sulphur bacteria and bacterial death – before moving to the Unit at Sussex, where he spent the next twenty-two years. He has held visiting professorships at the University of Illinois and Oregon State University and has been President of the Institute of Biology and of the Society for General Microbiology. He is also the author of *The Sulphate-Reducing Bacteria, Nitrogen Fixation* and *The Outer Reaches of Life.*

He is the third Professor John Postgate: the first (his great-grandfather) taught medicine at Birmingham University, the second (his grandfather) taught classics at Cambridge and Liverpool Universities. His other grandfather was George Lansbury, the Socialist leader, and his father was Raymond Postgate, the historian and gourmet. Long ago John Postgate led the Oxford University Dixieland Bandits (on cornet) and he is known as a jazz writer. He and his wife, Mary, who read English at St Hilda's College, have three grown-up daughters and seven grandchildren.

University of Sussex

John Postgate

Microbes and Man

PUBLISHED BY THE PRESS SYNDICATE OF THE UNIVERSITY OF CAMBRIDGE
The Pitt Building, Trumpington Street, Cambridge, United Kingdom

CAMBRIDGE UNIVERSITY PRESS
The Edinburgh Building, Cambridge CB2 2RU, UK http://www.cup.cam.ac.uk
40 West 20th Street, New York, NY 10011–4211, USA http://www.cup.org
10 Stamford Road, Oakleigh, Melbourne 3166, Australia

First published by Penguin Books Ltd 1969
Reprinted with revisions and plates 1975
Reprinted 1976, 1979
Second edition 1986
Third edition published by the Cambridge University Press 1992
Reprinted 1992, 1996, 1997
Fourth edition 2000

Printed in the United Kingdom at the University Press, Cambridge

Typeface FF Scala 10.5/15pt. *System* QuarkXPress® [SE]

A catalogue record for this book is available from the British Library

Library of Congress Cataloguing in Publication data
Postgate, J. R. (John Raymond)
Microbes and man / John Postgate. – 4th ed.
p. cm.
Includes bibliographical references and index.
ISBN 0 521 66579 5
1. Microbiology–Popular works. I. Title.
QR56.P58.1999 579–dc21 99–13564 CIP

ISBN 0 521 66579 5 paperback

Contents

Illustrations *ix*

Preface *xi*

1 **Man and microbes** **1**

2 **Microbiology** **17**

3 **Microbes in society** **54**

4 **Interlude: how to handle microbes** **113**

5 **Microbes in nutrition** **133**

6 **Microbes in production** **171**

7 **Deterioration, decay and pollution** **243**

8 **Disposal and cleaning-up** **270**

9 **Second interlude: microbiologists and man** **291**

10 **Microbes in evolution** **308**

11 **Microbes in the future** **335**

Further reading *358*

Glossary *360*

Index *363*

Illustrations

A microscopic green alga 21
A protozoon 22
A filamentous microfungus 23
A virus 25
Some bacteria 29
Microbes grow in a hot spring 37
Life in the Galapagos Rift 41
Some sulphate-reducing bacteria 47
Electron micrograph of *Neisseria meningitidis* 59
A seal has died of phocine distemper 74
The hand of Sarah Nelmes 82
Bacteria on the surface of a tooth 99
The drastic effect of Dutch elm disease 102
Colonies of bacteria in a laboratory 117
A hoatzin or stinky cowbird 135
A case of BSE 140
Root nodules containing nitrogen-fixing bacteria 148
A market for green manure 151
Unseen friends in yoghurt 158
A sulphur-forming lake in Libya 177
A red, photosynthetic sulphur bacterium 178
An industrial fermenter 198
A plasmid 216
Diagram of a plasmid 218
DNR fingerprints resolve a question of paternity 232
A triumph of genetic engineering 237
Bacteria corrode iron pipes 259

Fish are killed by bacterial sulphate reduction 263
Genetic engineering – Shock! Horror! 294
Fossil microbes in ancient rock 318
A microbe's eye view of its family tree 331
Nanedi Vallis on Mars 350

Preface to the fourth edition

Nearly thirty years ago I wrote the first edition of *Microbes and Man*. It was an attempt to introduce educated readers, not necessarily scientists, to the world of microbes, and to survey the impact of these creatures on Mankind's economy and society. The book was a quiet success, I am happy to say: it has been translated into several languages and, because the study of microbes continues to advance in fascinating directions, this is its third major revision.

Microbes have a truly enormous influence on people's lives. They clean up the environment and they make soil fertile; they are all-important in food technology; they make vitamins within our bodies. They can live peacefully within, around and upon us, and some even protect us from other, harmful, microbes, Yet most people are hardly aware that they exist – until a microbial misfortune happens: perhaps a disease breaks out, some food is spoiled, or some valuable or treasured material is found to have rotted or otherwise deteriorated. Only then do most people remember microbes – so it is hardly surprising that they are widely regarded as nasty; invisible enemies that are to be feared or resented.

In my book I wished to change the microbe's unfortunate public image. Of course, it is true that some microbes, especially certain bacteria and viruses, cause illnesses; of course, too, microbes exist which spoil food and attack non-living material – even damaging such improbable objects as concrete and iron pipes. But most microbes are beneficial. Their value to the food, chemical and pharmaceutical industries is (I hope) well-known to students, and perhaps, since the flowering of genetic engineering, to a minority of laymen. But few non-specialists are aware of the crucially important part microbes play, for example, in sewage treatment and waste disposal. After all, who outside a sewage works wishes to know about sewage? Yet sewage treatment is absolutely fundamental to our social well-being. And microbes have effects of global importance, as when they clean up fouled lakes, rivers and beaches, or decompose plant material, animal corpses and excreta to renew soil fertility. By recycling the detritus of plants and animals they constantly renew the supplies of oxygen, carbon dioxide, nitrates and even water, on which all life on this planet depends. Microbes are an intimate and essential part of the lives of all higher organisms, including ourselves.

That, in outline, was – and still is – the principal message of my book. Since the days of my first edition I sense that awareness of microbes has become marginally more widespread among the general public. Perhaps my book helped a little. But awareness does not necessarily imply understanding. Thinking people all over the world have come to recognize the drastic impact the human population explosion is having on our planetary environment; in time, I hope, they will also realize that microbes can often provide the bases of economic and environmental remedies, whereby ecological disaster can be contained while Mankind gains time to cope with its own fecundity.

But as well as being economic and social, my message is intellectual. The world of microbes is intrinsically fascinating to the inquiring mind because, collectively, these seemingly primitive creatures encompass all the processes of which living cells are capable. Their range of biochemical and biophysical abilities is enormously wider than those that we encounter among higher organisms. I have developed this theme further in another book, *The Outer Reaches of Life*: as a group they define the limits of terrestrial life, and provide hints of how things may once have been on Earth – and how they could well be at present, elsewhere in the universe. Finally, as in earlier editions, I thank all those who drew my attention to occasional errors and misprints – I fondly hope that none has escaped the net this time. I am especially grateful to my wife, who has patiently read the revisions and, where necessary, sorted out my sometimes roundabout writing.

John Postgate

Lewes, UK

1 Man and microbes

This is a book about germs, known to scientists as microbes (or to some, who cannot use a short word where a long one exists, as micro-organisms). These creatures, which are largely invisible, inhabit every place on Earth where larger living creatures exist; they also inhabit many parts of the Earth where no other kinds of organism can survive for long. Wherever, in fact, terrestrial life exists there will be microbes. Conversely, the most extreme conditions that microbes can tolerate represent the limits within which life as we know it can exist.

The biosphere is the name biologists give to the sort of skin on the surface of this planet that is inhabitable by living organisms. Most land creatures occupy only the interface between the atmosphere and the land; birds extend their range for a couple of hundred metres into the atmosphere, burrowing invertebrates such as earthworms and nematodes may reach a few metres into the soil but rarely penetrate further unless it has been recently disturbed by man. Fish cover a wider range, from just beneath the surface of the sea to those depths of two or more kilometres inhabited by specialized, often luminous, creatures. Spores of fungi and bacteria are plentiful in the atmosphere to a height of about a kilometre, blown there by winds from the lower air. Balloon exploration of the stratosphere as long ago as 1936 indicated that moulds and bacteria could be found at greater heights; more recently the USA's National Aeronautics and Space Administration has detected them, in decreasing numbers, at heights up to thirty-two kilometres. They are sparse at such levels, about one for every fifty-five cubic metres, compared with 1,700 to 2,000 per cubic metre at three to twelve kilometres (the usual altitude reached by jet aircraft), and they are almost certainly in a dormant state. At the opposite extreme, marine microbes flourish at the bottom of the deep trenches of the Pacific Ocean, sometimes down as far as eleven kilometres; they are certainly not dormant. Living microbes have been found 750 metres deep in sediments beneath the sea bed, and in sedimentary rocks beneath the land surface down to 500 metres. They are abundant at comparable depths in oil formations, and highly specialized types of bacteria have been found much deeper: in samples of hot telluric water emerging from oil-bearing strata three kilometres beneath the bed of the North Sea. The current record depth for microbial life inside this planet is held by certain heat-tolerant bacteria living three and a half kilometres down a gold mine in South Africa, where the ambient temperature is about 65 °C. Thus one can say, disregarding the exploits of astronauts, that the biosphere has a maximum thickness of about forty kilometres. Active living processes occur only within a compass of about ten kilometres: in the sea, on and beneath the land, and in the lower atmosphere, but the majority of living creatures live within a zone of thirty metres or so. If this planet were scaled down to the size of an orange, the biosphere, at its extreme width, would occupy the thickness of the orange-coloured skin, excluding the pith.

In this tiny zone of our planet take place the multitude of chemical and biological activities that we call life. The way in which living creatures interact with each other, depend on each other or compete with each other, has fascinated thinkers since the beginning of recorded history. Living things exist in a fine balance, a balance often taken for granted because, from a practical point of view, things could not be otherwise. Yet it is a source of continual amazement to scientists, because of its intricacy and delicacy. The balance of nature is obvious most often when it is disturbed, yet even here it can seem remarkable how quietly nature readjusts itself to a new balance after a disturbance. The science of ecology – the study of the interaction of organisms with their environment – has grown up to deal with the minutiae of the balance of nature.

At the coarsest level, living creatures show a pattern of interdependence which goes something like this. Humans and other animals depend on plants for their existence (meat-eating animals do so at one remove, because they prey on herbivores, but basically they too could not exist without plants). Plants, in their turn, depend on sunlight, so the driving force that keeps life going on Earth is the Sun. So much every schoolchild knows. But there is a third class of organisms on which both plants and animals depend, and these are the microbes. I shall introduce these creatures more formally, as it were, in the next chapter, but I think it will be helpful to give here a sort of preview of what their importance in the terrestrial economy is, to show broadly how basic they are to the existence of higher organisms before going more deeply, in later chapters, into aspects that most influence mankind.

Microbes, then, are those microscopic creatures which some call germs, moulds, yeasts and algae – the bacteria, viruses, lower fungi and lower algae, to use their technical names. It will be instructive to give some idea of the abundance of microbes compared with other creatures.

In every gram of fertile soil there exist about 100 million living bacteria, of an average size of 1 or 2 μm (μm, a micrometre, is a thousandth of a millimetre; to use a familiar image, one thousand of them laid end to end would span the head of a pin). One can express this information in a form that is, to me, more impressive: there are 200 to 500 pounds of microbes to every acre of good agricultural soil. In world terms, this means that the total mass of microbial life on this planet is almost incalculably large – it has been estimated at five to twenty-five times the total mass of all animal life, both aquatic and terrestrial, and approaching the total mass of plant life. The actual global masses of any of these, plants, animals or microbes, are very difficult to estimate; it is easier to work out ratios in sample areas and use these to make an educated guess. No doubt that is why the estimates of total masses have been so vague – though the approach has established that animals (including ourselves) add up to but a trivial portion (perhaps a thousandth) of the Earth's biomass. However, the global census is improving. In 1998 a group of scientists at the University of Georgia, USA, were able to make a somewhat more precise estimate of the number of bacteria, a major class of microbes, in the world: it came out at between 4 and 6×10^{30} cells (that is, between 4 and 6 multiplied by 10 with 30 noughts after it!). It is an unimaginably huge number, and though each living cell is so tiny that it weighs only about a thousandth of a billionth of a gram (10^{-12} grams; three-quarters of that weight is water), taken together the world's population of bacteria adds up to around 5 thousand billion tonnes (or 5×10^{18} grams) of living matter. That figure, astronomical though it seems, is actually of a similar magnitude to the total amount of organic matter thought to constitute the world's plants. As I shall tell in the next chapter, there are lots of microbes that are not bacteria, so the global biomass of microbes probably exceeds that of all the plants and animals in the biosphere.

Microbes can multiply very rapidly when food and warmth are available. One type of bacterium divides in two every eleven minutes; many can double in twenty to thirty minutes; the slow ones double every two to twenty-four hours. This, of course, is a fantastic rate of multiplication compared with most organisms – one cell of the bacterium *Escherichia coli* could, if sufficient food were available, produce a mass of bacteria greater than the mass of the Earth in three days. Consequently, since microbes constitute well over half of the living material of this planet, and can multiply almost as fast as they can get suitable food, it follows that they are responsible for most of the chemical changes that living things bring about on this planet.

Now I must digress a moment. At intervals in this book I shall have to bring in a certain amount of chemistry, because it is in chemical terms that one can best understand most of the economic activities of microbes. I shall keep the chemistry as simple as possible, but I shall assume readers have at least some familiarity with chemical symbols: that they know, for example, that N symbolizes a nitrogen atom or Na a sodium atom; that free nitrogen gas occurs as molecules consisting of two atoms, formulated as N_2; that the formula of methane is CH_4 and signifies that its molecule consists of one carbon and four hydrogen atoms; that when one writes methane so:

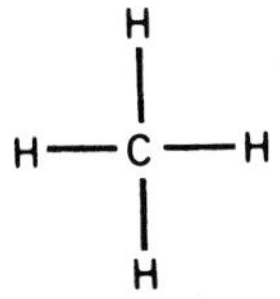

```
        H
        |
   H —— C —— H
        |
        H
```

it signifies that the hydrogen atoms are independently linked to the central carbon atom in the molecule and that they are symmetrically arranged around it.

I shall make use of the organic chemist's shorthand of:

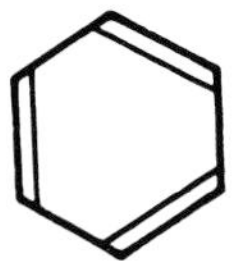

for six carbon atoms linked in a ring. Written out in full, the compound above (which is benzene) looks like this:

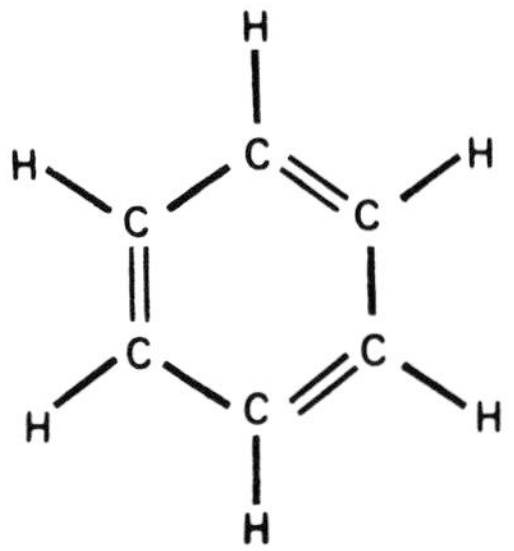

```
             H
             |
   H         C         H
     \     /   \\     /
       C           C
       ||          |
       C           C
     /   \      //   \
   H         C         H
             |
             H
```

but chemists learned long ago that writing out all those 'C's and 'H's was generally a waste of time.

I shall also assume an awareness, at least in principle, that dissolved salts dissociate into ions. That sodium nitrate, potassium nitrate and calcium nitrate, for example, all yield nitrate ions in water, so that when a plant uses nitrate from a fertilizer, it is largely irrelevant whether it arrived as sodium, potassium or calcium nitrate. Thus, for many purposes, it is legitimate to talk of nitrate (NO_3^-), sulphate (SO_4^{2-}) and so on, even though it would be impossible to obtain a bottle of sulphate.

Taking these principles for granted, I shall try to explain any more complex chemical concepts as they arise.

After that brief excursion into what the reader's homework should have covered, let me return to the question of the importance of microbes in the world's chemistry. My next thought on these matters is this: that nearly all the chemical changes that do take place on this planet are caused by living things. A few inanimate processes do occur: volcanoes bring about alterations in the neighbouring rocks and in the atmosphere; lightning causes oxides of nitrogen and ozone to be formed; ultraviolet light from the sun does so as well, and also causes a layer of ozone to exist in the upper atmosphere that protects us from some of the more harmful ultraviolet wavelengths. Rainstorms and erosion by the sea cause gradual chemical changes in rocks and minerals as they are exposed; radioactive minerals induce a certain amount of chemical change in the neighbouring rocks and keep the Earth's interior hot. But at the Earth's surface the purely chemical changes that now take place are trivial compared with those that took place in the infancy of this planet: the Earth's own chemistry has settled down, as it were, to a fairly quiescent state. The most obvious chemical changes are now brought about by plants, with animals as secondary agents, both on land and in the sea, and the energy needed to perform these chemical transformations comes from the Sun. The biosphere, therefore, is a dynamic system of chemical changes, brought about by biological agents, at the expense of solar energy.

I shall tell in Chapter 10 how the emergence of living things wrought dramatic changes many millions of years ago in the chemical composition of this planet's surface. The composition of the atmosphere, soil and rocks underwent gradual changes, often taking tens of millions of years, to yield the sort of biosphere we know today. No doubt that is still changing slowly, but as far as the last million or so years are concerned the average chemical composition of the biosphere has been constant. Another way of putting this point is that all gross chemical changes which occur on Earth, brought about by any one kind of biological activity, are reversed by some other activity. If one considers the elements that undergo chemical transformation on this planet, they are found to undergo cyclical changes, from biological (or organic) combination to non-biological (or inorganic) combination and back again.

Consider the element nitrogen, nowadays plentiful as the free molecules of nitrogen gas that comprise four-fifths of our atmosphere. Nitrogen gas, known to chemists as 'dinitrogen', is normally rather inert; it is harmless to living things, neither burning nor supporting combustion, and is generally reluctant to enter into spontaneous chemical combination. Yet all living things consist of proteins: their muscles, nerves, bones and hair, and the enzymes that manufacture these and everything else, that provide energy for growth, movement and so on, all consist of protein molecules. And something like 10 to 15 per cent of the atoms in every protein molecule are nitrogen atoms. The nitrogen atoms are combined with others: carbon, hydrogen, oxygen and sometimes sulphur. Compared with dinitrogen, N_2, molecules, protein molecules are huge and complicated, containing tens of thousands of atoms; this is why proteins can be so diverse in appearance and function. And since they constitute the major part of most living things, one can safely say that most living creatures consist of between 8 and 16 per cent of nitrogen, animals being on the high side, plants on the low side. The main exceptions are creatures that form thick chalky or siliceous shells: they seem to have low nitrogen contents, but even they have the usual chemical composition if one regards the shell as a non-living appendage and excludes its composition from one's calculations.

Living things therefore need nitrogen atoms to grow. When they die, they rot and decompose, and their nitrogen becomes available for other living things. Rotting and decomposition are largely the result of the action of microbes on the organism and, of course, microbes die too, either naturally or by being consumed by protozoa, nematode worms and so on. Gradually the nitrogen is assimilated by larger living things – plants, worms, birds, etc. – and so it becomes part of new creatures. (A process dramatically enshrined in that essentially macabre Yorkshire song *On Ilkley Moor baht 'at*: 'Then shall ducks have eaten thee . . .'.) So a process of constant transformation of the state in which nitrogen atoms are combined takes place, which is known to biologists as 'the nitrogen cycle'. In this cycle certain microbes return nitrogen as N_2 to the atmosphere (the denitrifying bacteria) and others bring it back to organic combination (the nitrogen-fixing bacteria). One can write the biological nitrogen cycle schematically as below.

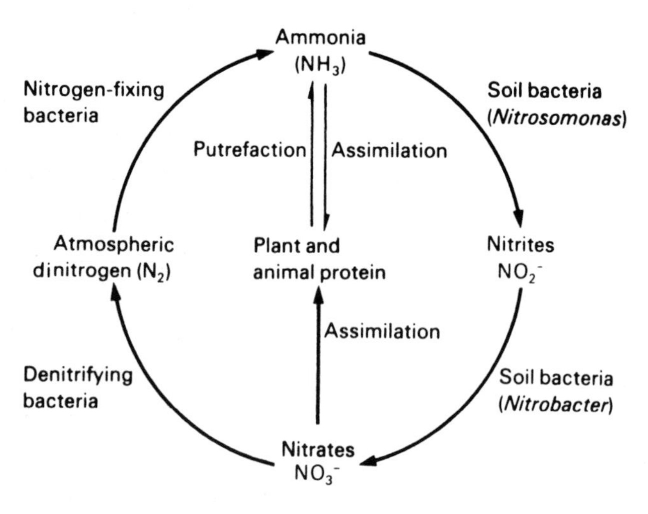

In this scheme nitrates in the soil are used by plants for growth and become plant and animal protein. Later these decompose through the action of microbes, releasing ammonia. Plants can recycle this, but they prefer nitrates, and two groups of soil bacteria convert ammonia back to nitrate by way of nitrite. Denitrifying bacteria, found in soil, compost heaps and so on, can release the nitrogen of nitrates as free nitrogen molecules, and this loss of biological nitrogen to the atmosphere is compensated for by the activities of the nitrogen-fixing bacteria. Some of these live in association with the roots or leaves of plants, other live freely in soils and water. I shall return to them in Chapter 5, but for present purposes the important point is that, in many soils, particularly agricultural soils, the supply of fixed nitrogen (ammonia or nitrate) determines the productivity of that soil. Hence the number of animals, or people, that can feed from that soil depends on how rapidly the nitrogen cycle is turning, on how actively nitrogen-fixing bacteria are performing.

Of course, the cycle may be bypassed to a limited extent. Artificial nitrogen fertilizers made industrially from atmospheric dinitrogen increase soil productivity by bringing chemically fixed nitrogen to the soil. Thunderstorms, and ultraviolet light from the sun, generate oxides of nitrogen in the atmosphere without the intervention of living things, and rain washes these into the soil as nitrates. These processes have been left out of the scheme above because, although together they account for over a third of the newly fixed nitrogen in soils, on a world scale the Earth's productivity of vegetation, and hence of food for man and animals, still depends primarily on the activity of the nitrogen-fixing bacteria. In a year, something in the region of 3 thousand million tonnes of nitrogen as N pass through the cycle, and nearly 10 per cent of this turnover involves loss of N to the atmosphere as dinitrogen and its return to the biosphere by nitrogen fixation. C. C. Delwiche has calculated that every nitrogen atom in the atmosphere passes through organic combination on an average once in a million years. Obviously the microbes are of crucial importance to the economy of living things on this planet.

The nitrogen-fixing bacteria are of basic importance to the nitrogen cycle, but one should not underestimate the rest. The putrefying microbes return protein nitrogen to circulation by forming ammonia and, since most plants prefer to assimilate their nitrogen as nitrate rather than ammonia, the two groups of bacteria which convert ammonia to nitrate (collectively called nitrifying bacteria) perform an economically useful function. This is not an unqualified virtue, however, because nitrates are washed out of soils by rain more easily than ammonia; to avoid such waste, agricultural chemists sometimes advise the use of ammonia fertilizers, which most plants can manage with perfectly well, together with chemicals that inhibit multiplication of nitrifying bacteria.

Another biological cycle, of equally basic importance to the biosphere, is the carbon cycle. This, as far as higher organisms are concerned, is intimately involved with the cycle of changes undergone by oxygen. All living things respire; in effect, respiration is the transformation of the carbon and hydrogen compounds that constitute food into carbon dioxide (CO_2) and water, usually with the aid of the oxygen of air. Thus living things tend to remove oxygen from air and replace it by carbon dioxide. The reverse process, that of fixing carbon dioxide as organic carbon and of replenishing the oxygen of air, is conducted by green plants: they absorb CO_2 to form the constituents of their own substance with the aid of energy derived from sunlight and in so doing they release the O of H_2O as oxygen (O_2). Today, on a world scale, these processes are more or less in balance, such that the atmosphere consistently contains about 21 per cent of oxygen and just over 0.03 per cent of CO_2. The main contribution of microbes to this cycle is in decay and putrefaction, whereby they break down residual organic matter such as wood, faeces and so on, and thus return carbon dioxide to the cycle. In so doing, they often provide an important diversion of the carbon cycle, because their carbon turnover need not necessarily be tied to the oxygen cycle. I shall introduce in Chapter 2 the anaerobic bacteria, which have no need of oxygen for their respiration and which can produce such materials as methane, hydrogen or butyric acid from organic matter. They are most important in deposits of organic matter to which oxygen does not readily penetrate, such as vegetation decaying deep in a pond, and methane in particular is important in the carbon cycle, because it is a gas and, by diffusing away from the deposits, it transposes carbon from air-free zones to aerated zones. Here the methane is oxidized; indeed, on a world scale, most of the products formed by anaerobic bacteria are oxidized by other microbes, using oxygen, to yield finally CO_2. Thus the carbon is returned to circulation and the cycle proceeds. The turnover rate of the carbon cycle overall is about 10 thousand million tonnes of carbon a year. On land, most of the CO_2 fixation is conducted by higher plants, but in the sea microbes are still the most important CO_2 fixers: microscopic cyanobacteria, algae and diatoms, microbes that float in the plankton layer of the sea surface, together with more dispersed microbes called picoplankton, form the bulk of the organic matter that fish feed upon. One can present the carbon cycle so:

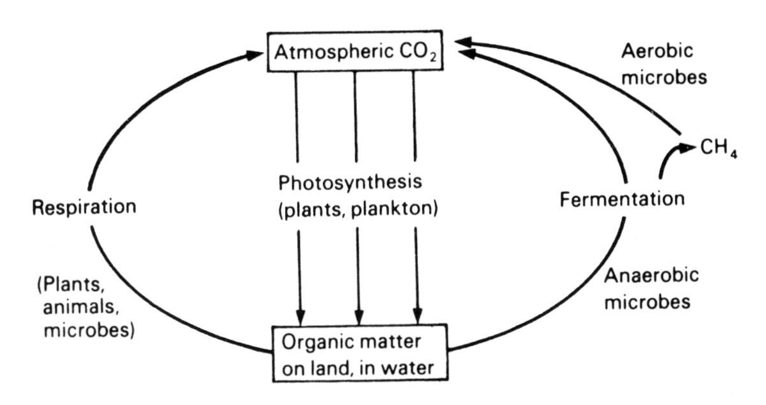

The microbes of plankton use sunlight, as land plants do. I shall introduce in later chapters several groups of microbes that can fix CO_2 using chemical reactions, not sunlight, but, though they may have been important during the early history of life on this planet, they now contribute little to the carbon cycle except in certain very special environments.

Today environmentalists are rightly worrying about how balanced the carbon cycle really is. The proportion of CO_2 in the atmosphere has risen during the latter part of the twentieth century and continues to do so, which means that more CO_2 is appearing than plant and microbial photosynthesis can cope with. It appears that mankind is responsible: by burning fuel, especially coal, oil and natural gas, but in some other ways too, we are adding significantly to the amounts of CO_2 reaching the atmosphere. Because CO_2 traps heat from the sun and so helps to keep our planet warm (it is a so-called 'greenhouse gas'), the fear is that the extra CO_2 will gradually make the world warmer. The consequent disturbances in weather patterns may not be as pleasant as one might at first imagine, but as they are still being debated I must refer readers to current magazines and quality newspapers for details.

Elements such as hydrogen, iron, magnesium, silicon and phosphorus are all part of the structures of biological molecules and undergo comparable cyclical changes. The phosphorus cycle is also worrying, because it involves a net transfer of something like 13 million tonnes of phosphorus a year from the land to the sea with no obvious return process. Microbes play a certain part in this and in the other cycles just mentioned, but their part is not a major one and I shall not discuss them further. However, there is one cycle of great importance that I must not neglect, if only because it depends almost exclusively on microbes: the sulphur cycle. The element sulphur is a component of protein and of certain vitamins – living creatures contain between 0.5 and 1.5 per cent sulphur – and the biological sulphur cycle is of critical importance in maintaining supplies of that element. But before I discuss it I must introduce a technicality that will be important here and later in this book: the concepts of oxidation and reduction.

Coal, which is carbon, becomes oxidized when it is burned, and the chemical energy of this reaction is dissipated as heat. The process is called oxidation because oxygen atoms are added to the carbon atoms to give carbon dioxide. Using chemical notation one can represent it as a sort of equation so:

$$C + O_2 \rightarrow CO_2$$

If insufficient oxygen is available, some carbon monoxide is formed:

$$2C + O_2 \rightarrow 2CO$$

(This, incidentally, is the poisonous component of motor exhaust fumes.) Thus there are degrees of oxidation in the sense that carbon can be partly or wholly oxidized. In a similar way, other elements may form stable compounds in more than one degree of oxidation.

Food consists of carbon compounds which, when used by the body, are oxidized to form carbon dioxide and water. A typical example is glucose, which has the formula $C_6H_{12}O_6$:

$$C_6H_{12}O_6 + 6O_2 \rightarrow 6CO_2 + 6H_2O$$

Some of the energy of such a reaction appears as heat; much of it goes to drive the various chemical reactions which keep the body functioning.

All microbes live by comparable oxidative reactions, but there are some that can conduct such processes without using oxygen gas. The sulphate-reducing bacteria, for example, use sulphate:

$$\text{Carbon compound} + CaSO_4 \rightarrow CO_2 + H_2O + CaS$$

They steal oxygen atoms from sulphate and use them to oxidize carbon compounds. In that reaction, calcium sulphate becomes converted to calcium sulphide. The calcium sulphate undergoes a process called a reduction and, generally speaking, if some chemical is being oxidized, another is being reduced. (In burning, for example, the carbon is oxidized while the oxygen is reduced.) The denitrifying bacteria reduce nitrates in a comparable way:

$$\text{C-compound} + NaNO_3 \rightarrow \text{Na carbonate} + N_2$$

In nitrate or sulphate reduction, the oxygen atoms of nitrate or sulphate are used to oxidize the carbon source, so the ion is said to be reduced.

So far the concepts of oxidation and reduction are easy to follow. Things get a mite complicated when chemists refer to reactions that do not involve oxygen at all as oxidations and reductions, but this only means that the reactions in question have the same general character as those that concern oxygen. Compounds of iron, for example, can exist as ferrous salts (ferrous sulphate, nitrate and so on) or ferric salts; the ferric group are all more oxidized than the ferrous ones from the chemist's point of view, though they need not necessarily contain more oxygen (or, indeed, any oxygen at all: ferric chloride – $FeCl_3$ – is more oxidized than ferrous chloride – $FeCl_2$).

Microbes can make use of all sorts of oxidative reactions to obtain chemical energy for growth, movement and multiplication, including the conversion of ferrous compounds to ferric. Since oxidations are coupled with reductions, they bring about interesting reductions too, and in appropriate circumstances one can find one group of microbes conducting reductions and others oxidizing whatever they have reduced. This occurs particularly clearly in the biological sulphur cycle, which is turned by a group of soil and water bacteria called the sulphur bacteria. (In fact, they have little or no biological relationship: the main thing they have in common is that their metabolism is based on the sulphur atom.) Here is the sulphur cycle:

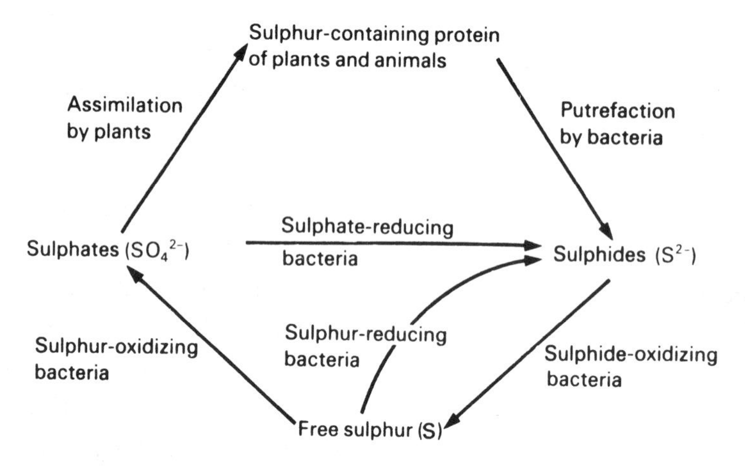

(Notice that sulphur appears in two oxidation states: sulphur itself is more oxidized than sulphide, though containing no oxygen, and sulphate is even more oxidized.)

In this cycle the sulphur of animal protein comes from plants, which get it from sulphates in soil. In decomposition and putrefaction of dead material, bacteria release the sulphur as sulphide, which is a reduced material. Other bacteria can oxidize this to sulphur, which some can reduce to sulphide again; yet others oxidize sulphide or sulphur to sulphate, which plants can re-use. The sulphate-reducing bacteria can bypass the top part of the cycle, reducing the sulphate straight back to sulphide, obtaining energy to do this by oxidizing organic matter, and thus a microbial sulphur cycle can go on without involving higher organisms at all. Such microcosms of sulphur bacteria are often encountered in nature, in sulphur springs, in polluted waters and so on, and, as I shall tell in Chapter 9, they may have been the dominant living systems during the early history of this planet. They are called sulfureta (singular: sulfuretum) and are responsible for a variety of economic phenomena that will appear in later chapters of this book. The individual bacteria of the sulphur cycle will also appear again later.

Microbes, then, play an important part in the cyclical changes that the biological elements undergo on Earth. In this sense they are of transcendental importance in the terrestrial economy, because without them higher organisms would rapidly cease to exist. Yet they couple these fundamental activities with a number of other functions which may be valuable, trivial or a thorough nuisance to mankind. Most diseases, for example, are caused by microbes. From a biological point of view disease is valuable in that it limits excessive animal or plant populations, but the reader need hardly be told how thoroughly inconvenient it can be in the civilized world today. Pollution and putrefaction are all very well in their place – our sewerage systems depend on them – but out of control they can be disagreeable and destructive. Microbes ferment foods, yielding delicious delicacies and wines, but tainted food is dangerous. Microbes aid our digestion and nutrition, but upset our stomachs in a strange land. Over geological time microbes formed several of the world's most valuable mineral deposits, but when they corrode steel and concrete we do not welcome their peculiar propensities. And so it goes on. Microbes are neither generally good nor generally bad; they can be either. The important thing, which is not widely realized, is that they have an enormous effect on the economy and well-being of mankind. That, in fact, is what this book is about. How do microbes come into our lives? What do they do? And why? These are far-ranging questions because, as the patient reader will learn, it might be more pertinent to ask whether there are any aspects of our daily lives in which microbes are not involved. I shall have to skip and skim in places, but in a book intended to introduce readers to an unfamiliar subject this is, I think, excusable. Let me start, therefore, by introducing that huge group of invisible or scarcely visible creatures we call the microbes.

2 Microbiology

In the early 1950s I was involved in the foundation of the National Collection of Industrial Bacteria (NCIB), a sort of bank established in Britain from which strains of industrially significant microbes could be obtained. Today it has grown into the National Collection of Industrial and Marine Bacteria (NCIMB) in Aberdeen and it is part of a valuable network of collections of microbes. The NCIMB has an important function: not only does it act as a reserve of organisms used in industry and non-medical research, but it also keeps typical bacteria involved in spoilage and deterioration, so that technologists can obtain reference strains to compare with those which may be causing trouble. In the early days of the NCIB's existence, parties of visitors used to come to see it. On one occasion a small party of civic dignitaries and their wives visiting the locality from France came round. I never clearly understood why, as it seemed a rather soggy sort of entertainment for the local municipality to arrange. However, I well recall the alarm shown by the wives when, not having at first understood the word *bactéries*, they suddenly realized they were amid a collection of *germes*. As one woman they pulled out handkerchiefs, covered their noses and left as soon as they politely could.

Laymen always associate bacteria, microbes and germs with disease. Microbes seem to have a faintly alarming or disgusting aura, and the fact that by far the majority are nugatory or even beneficial is rarely understood. Yet it is so. One's hands, hair, mouth, skin and intestines are teeming with bacteria; all but freshly cooked or sterilized foods are contaminated with living bacteria and their spores; drinks, soil, dust and air have populations of microbes, the majority of which are harmless and many of which are beneficial. Disease-causing (pathogenic) microbes are the minority, except where sickness is prevalent.

The fact that we eat, sleep, live and breathe microbes has only slowly been realized during the last hundred and fifty or so years and has led to the enormous advances in hygiene and medicine of the twentieth century. I mentioned some of the impacts of microbes on man's existence in the first chapter and I shall look at these in greater detail later on. In this chapter I propose, as it were, formally to introduce some of the microbes readers will encounter later, to familiarize them with the way in which they are classified and what generally they do, and to show how their study has crystallized into the branch of biology known as microbiology.

Microbes were first described by a famous Dutch scientist, Antoni van Leeuwenhoek,* in the seventeenth century, in a fascinating sequence of letters to the Royal Society of London. He had constructed for himself a primitive but very effective microscope, and in his letters described the extraordinary menagerie of 'animalcules' he had observed in samples of canal water, broth, vinegar, saliva and so on. Leeuwenhoek's drawings leave no doubt that, among tiny worms, water fleas, particulate matter etc., he saw normally invisible creatures: the bacteria, yeasts and protozoa that we now call microbes.

However, the subject of microbiology can fairly be said to have been created in the late nineteenth century by Louis Pasteur. He was a French chemist, who proved that fermentation and putrefaction, hitherto believed to be purely chemical processes, were due to microbes. The manner of his proof is now a matter of history, with which I shall not be concerned; from the point of view of the development of microbiology the important point was Pasteur's realization that the air contained a variety of microbes likely to fall randomly on any susceptible material and to putrefy or ferment it. Consequently, scientists who wish to study and understand these microbes need to

* A terrible name for anglophones to spell or pronounce; Dutch friends tell me that 'layvenhook' approximates the correct pronunciation.

develop special methods for sorting out, conserving and keeping separate the different types of microbe. Since the individual microbes are invisible to the naked eye and are very numerous, it has rarely been practicable to take one microbe and study it. For, in the first place, it is inconveniently small, and secondly, if it does not die, it turns into two new ones, then to four and so on. The microbiologist, generally speaking, is obliged to study great numbers of microbes at once and deduce an average behaviour for the whole lot. It is therefore necessary to be at pains to see that they are all as nearly the same as possible and, above all things, that the pure family (usually called a species or strain) is not contaminated with little strangers from elsewhere, be it hair, skin or the air.

How this is done I shall discuss in Chapter 4. For present purposes the essential point is that the techniques of microbiology are, on the whole, very different from those of the rest of biology. You can take a dog, dogfish or plant and observe it in a variety of ways, doing a variety of things. While you *can* do this with a microbe, it would not, at present, get you very far. Just as chemists deal in the behaviour of millions upon millions of molecules, and rarely derive information from the study of single molecules, so microbiologists study microbes in thousands of millions, and rarely have recourse to the individual germ. This is not a matter of choice in either instance: easy techniques are, at present, just not available for fruitful study of the unitary bodies of the two sciences. For this reason, microbiology is a science defined more by the techniques it uses than by the subjects it covers. Indeed, when macrobiologists come to study the component cells of multicellular organisms, when they use tissue cultures, for example, they adopt many of the techniques of microbiology.

But tiny multicellular creatures such as nematode worms or water fleas do not count as microbes in the language of microbiologists. As a general rule, microbiology is concerned with organisms that consist of one cell, or perhaps a cluster of very few similar cells, and which do not develop into more complex aggregations of dissimilar cells. And since cells are very small, a microscope is almost always needed to see microbes. The borderlines of the subject are fuzzy, however: in three of the major groups of microbes there are a few creatures which at some stage in their lives form clusters including two or even three types of cell, or which reach a size just large enough to be visible with the naked eye; in such instances the subject overlaps into the provinces of botany and zoology. However, it is fair to say that microbiology is generally concerned with the following five great groups of unicellular or non-cellular living things – and one group of oddball entities, the prions, of which more in due course.

ALGAE (pronounced with a hard 'g'; singular: alga)

These are unicellular plants of the kind that one sees on the walls of goldfish aquaria and which turn ponds and waterbutts green. Seaweeds and many pond weeds are in fact multicellular algae, but these are normally the province of the botanist. Typical unicellular green algae are *Scenedesmus*, *Chlorella* and *Chlamydomonas*. The latter is a common inhabitant of green water, and consists of single egg-shaped cells, about 10 μm (0.01 mm) long, capable of swimming around (motile, in biologists' jargon) with the aid of two hair-like appendages, the flagella (singular: flagellum). The cells are green, and the green colour is due to chlorophyll, contained in a portion of the cell called the chloroplast (in *Chlamydomonas* the chloroplast occupies almost all of the cell). There is a nucleus, as in the cells of higher organisms, and a cell wall composed of cellulose. Like plants, the green algae need light to grow, and with it they reduce carbon dioxide to sugars and starch and so they multiply. They do not use organic food at all: light, CO_2 and certain minerals are all they need for growth. Creatures that use exclusively mineral matter for growth are known to microbiologists as autotrophs, and the green algae come in a particular class of autotrophs called photo-autotrophs, because of their need for light. The antonym of autotroph, heterotroph, pertains to organisms that require organic food (as you and I do). I shall need these words in later sections.

A group of microbes exists which was earlier called the blue-green

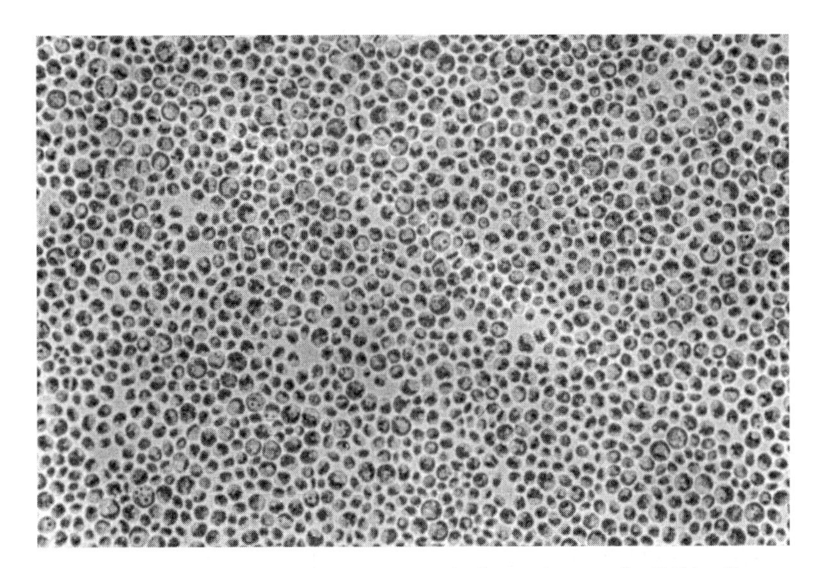

A MICROSCOPIC GREEN ALGA. A photomicrograph of *Chlorella*, which grows as a mass of round, green cells. It is common in stationary fresh water. It is a source of food for fish and other water fauna and has actually been considered as a foodstuff for people (see Chapter 5). The magnification is about 60-fold. (Courtesy of Dr H. Canter-Lund, Freshwater Biological Association)

algae; they are now classified among the bacteria and will be dealt with later.

PROTOZOA (singular: protozoon)

These are single-celled creatures of which the schoolchild's *Amoeba* is a typical example. They are heterotrophs and are in fact the most complex of the microbes. For some reason they tend to be neglected by microbiologists (possibly because specialized types of zoologists, protozoologists, exist and regard them as their special province), but they have in fact been extremely valuable in nutritional and genetical research. *Paramecium*, the slipper animalcule, was probably one of the first microbes to be observed by Antoni van Leeuwenhoek. *Astasia*, a motile ovoid protozoon, is interesting, because it has a nearly identical cousin, *Euglena*, which possesses a chloroplast. This

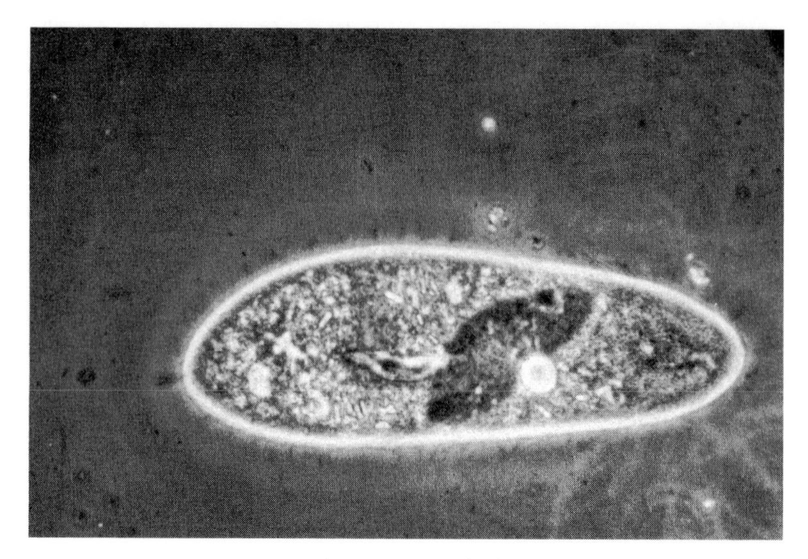

A PROTOZOON. A photomicrograph of *Paramecium*, a single-celled protozoon. It is about 250 μm long and little hairs (cilia), which enable it to move about, can be seen, as well as its quite complex internal structure. (Courtesy of Dr B. J. Findlay, Freshwater Biological Association)

creature thus bridges the gap between algae and protozoa. Protozoa cause one or two diseases in plants, animals and man, but as far as we know they have a relatively small impact on people compared with other microbes. Therefore, though they will crop up occasionally in later chapters, I shall say no more about their classification here.

FUNGI (pronounced with a hard 'g'; singular: fungus)
Mushrooms and toadstools are familiar to botanists, both amateur and professional, but rarely provide material for study by microbiologists. Moulds, mildews, rusts and yeasts, however, are very important and, because of their simple structure and metabolism, have become honorary microbes to the microbiologist, despite the fact that many of them are not unicellular. A common bread mould, *Neurospora*, forms red spores which give the characteristic colour to mouldy bread (though a relative, *Aspergillus*, is often present too).

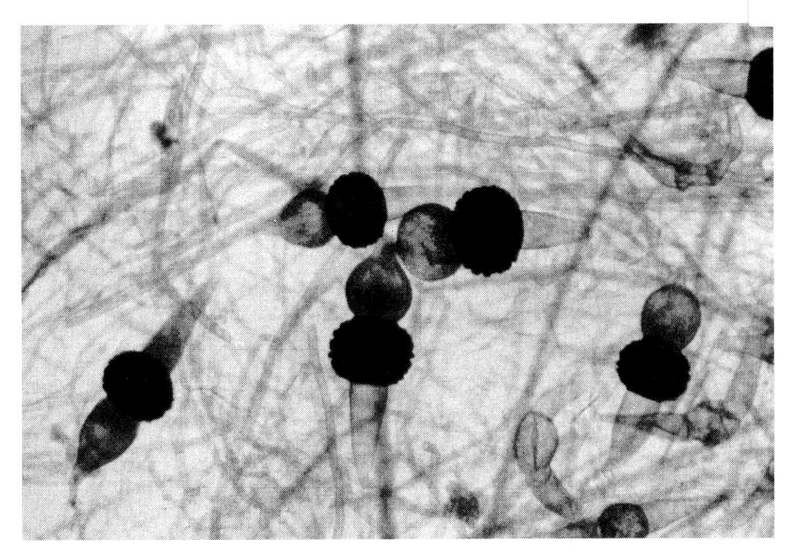

A FILAMENTOUS MICROFUNGUS. A photomicrograph of a species of *Rhizopus*. It is one of several common bread moulds but it is important in several other areas of economic microbiology and crops up in Chapters 5, 6 and 7. The pale filaments are the actual mould and the dark, round bodies are its coloured spores. The magnification is about 200-fold. (Bruce Iverson/Science Photo Library)

Bluish colours are often due to the justly famous *Penicillium*; the grey bread mould is *Mucor*. Mouldly cheese often features *Penicillium* too. Ordinary soil is rich in small, thread-like fungi, which are in many ways plant-like: they grow as threads that sometimes branch and they spread by forming spores (analogous, in a general sense, to forming seeds). However, they lack chlorophyll, so they cannot photosynthesize. They are heterotrophs: they need organic material in order to grow and are therefore normally found on decaying organic matter of almost all kinds. They are particularly versatile at breaking down such resistant materials as wood, leather and so on. Yeasts, used in baking and brewing, are more orthodox microbes, being single-celled fungi.

Certain fungi live in association with special algae, forming the composite creatures called lichens. Sometimes the so-called alga is a cyanobacterium (see 'bacteria' below). In these circumstances, aided by the autotrophic abilities of the algal partner, they can grow in extremely barren environments such as the roofs of houses, bare rocks and so on. Quite what benefit, if any, the alga receives from this partnership is obscure.

VIRUS (singular: virus)

Though the correct plural name of this group should be virus, I am going to be like almost everyone else and call them viruses. These creatures are between ten and a hundred times smaller than bacteria, from 0.2 to 0.02 μm long. They have become important in recent years as the major causative agents of disease: as I shall tell in the next chapter, most of the bacterial diseases are now under control, but the viruses remain largely unconquered. Viruses are responsible for many plant diseases (wilts, scabs and so on); diseases such as poliomyelitis and the common cold in humans; foot and mouth, among other diseases, in cattle; diseases of fish and doubtless of other organisms. They also attack bacteria, and the viruses responsible for diseases among bacteria have been given the special name of bacteriophages by microbiologists. There is a class of bacterial viruses called temperate phages which seem to live harmlessly in their host until some stress causes the infection to develop.

Viruses lie on the borderline of living things. They have, for example, no metabolism of their own: they do not respire, break down carbon compounds, fix CO_2 or do anything like that. When they infect a creature, they pervert its own metabolism so that it synthesizes more of the virus. When the host dies or, in the case of higher organisms, when the infected cells die and break up, many hundreds of virus particles are liberated and can spread the infection further. When they are not infecting a host, some viruses behave like stable chemical molecules. They do not die and, in fact, certain plant viruses have been concentrated, crystallized and stored for many years in the laboratory. If you can take a crystalline substance from a bottle, infect an organism with a trace of it and later harvest relatively

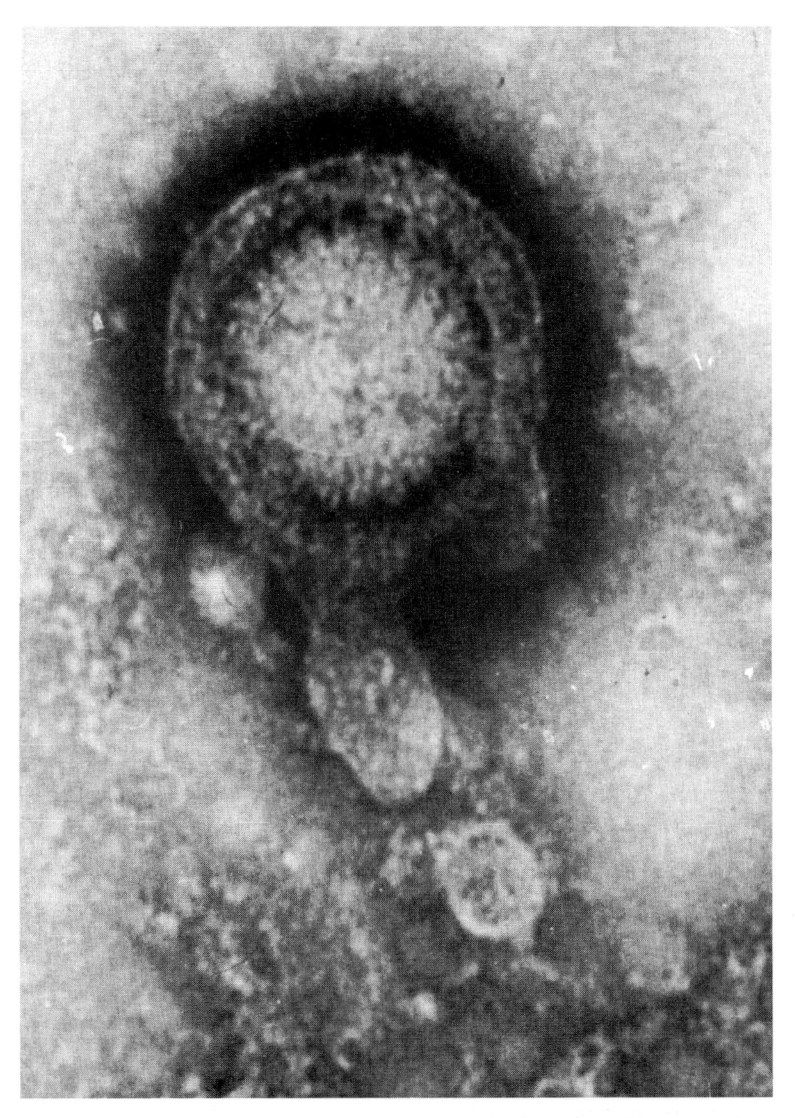

A VIRUS. The herpes simplex virus (which causes cold sores and, rarely, some nastier infections) in its protein envelope. It is too small to be seen even with the most powerful optical microscopes; this is an electron micrograph at about a half-million-fold magnification. (Courtesy of Professor D. H. Watson)

vast quantities of those crystals from the infected creature, are those crystals living or dead? It is an interesting question to which there is no straight answer. I shall return to it in the next section. As far as their impact on man is concerned, viruses are only too alive and, in this book, I shall treat them as microbes.

PRIONS

Classification is even more of a problem when it comes to the agents known as prions (earlier 'sub-viral particles' or 'slow viruses'). These appear to be the cause of a group of infectious diseases of brain and nervous tissue known as transmissable spongiform encephalopathies – TSEs for short. Scrapie, a disease of sheep, is a typical TSE: it develops slowly and causes a spongy degeneration of the brain (the spongiform encephalopathy) and the animal becomes un-coordinated and dies. Scrapie has been present as a rare disease among sheep herds for over two centuries, during which period it seems to have remained restricted to sheep and caused no harm to other creatures. Two comparable TSEs have been known in man for most of the twentieth century: a tropical disease called kuru and an inheritable form of human dementia known as Creutzfeldt–Jakob disease. Both make the brain spongy and both are slow but lethal. In the 1980s, as I shall tell later, a form of scrapie became established in cattle.

What causes these spongy brain diseases? Research has necessarily been slow-moving because the diseases progress very slowly and can only be studied in animals, not in laboratory cultures. As in ordinary microbial infections, the agents are transmissible: within a species they can pass from one animal to another. But research soon eliminated bacteria, and cellular agents of any kind; and, though viruses were long suspected of being the culprits, they too were finally excluded on the grounds that the scrapie agent is amazingly resistant to sterilizing agents. For example, preparations from scrapie-infected tissue were immune to ultraviolet light, survived boiling in water, and were actually toughened by disinfectants such as chlorine or formaldehyde – all sure-fire ways of killing viruses.

Moreover, evidence has gradually accumulated that the agents do not possess the ordinary hereditary materials of living things, DNA and RNA (see pp. 209 ff.). By the 1990s the majority of microbiologists agreed that they were something new to microbiology.

So what are they? They seem to be very special proteins. According to Dr Stanley Prusiner of California, who received a Nobel Prize in 1997 for his work in this area, they are aberrant forms of perfectly natural proteins, called prions, which are present in healthy brain and nerve tissue, in all mammals. Nobody yet knows what exactly the 'healthy' prion protein does, but to cause the disease it simply changes its shape, in quite a small way, and it then becomes able to convert more of the normal protein into the aberrant form. A sort of chain reaction starts; the body replaces the normal protein, only for more and more of the new material to be converted. As the disease progresses, the body can no longer cope and the encephalopathy advances.

Are prions microbes, then? Personally I think not. The consensus – and lest I give a false impression of certainty, let me add that a few scientists still remain unhappy with it – is that they are ordinary protein molecules which sometimes, for reasons not yet understood, become aberrant and pathogenic: entities which cause disease in a way hitherto unknown to biology. Viruses or more complex microbes seem to play no part in either their formation or their transmission. Prions are infectious because they can be absorbed from the gut when contaminated tissue is eaten, and when they reach their new host's nerve tissue they start generating more aberrant protein. However, whether we call them microbes or not, the techniques needed to study prions are those of microbiology, and that is why their investigation is the province of microbiologists – and why they feature in this book.

Nevertheless, the purist will ask, are they alive? I pointed out earlier that the status of the viruses is dubious enough in this context, but at least they possess the nucleic acids, DNA or RNA, that are found in all living organisms (some viruses consist of nothing else), and they evince heredity in that they reproduce themselves exactly, subject to the occasional genetic variations (mutations) which all living things show. Prions do not have nucleic acids, yet they still display something resembling heredity in that different strains can be distinguished which 'breed' true. For some biologists, such as the distinguished evolutionist John Maynard Smith, anything which shows multiplication, heredity and variation is alive, a view which would certainly encompass both viruses and prions. Yet to many biologists the definitive criteria of living things also include the ability of each organism to react to and adjust to environmental changes: the quality charmingly named 'irritability'. (A mental image of an irascible *Amoeba* has remained with me since my school biology lessons.) Prion molecules do not display irritability; nor, probably, do virus particles. Well, I must leave the question to examiners and linguistic philosophers; my own prejudice is that viruses and prions lie on either side of a border separating living and non-living matter.

BACTERIA (singular: bacterium)

I have left this group until last because it is the one which will feature most prominently in this book. 'Bacteria' is a collective name for the traditional germs. They are microscopically small creatures, usually 1 to 2 μm in length or diameter. They have almost no visible internal structure and, particularly important, they lack the nucleus which is an essential feature of the algae, protozoa and fungi, as well as of the cells of higher organisms. They are thus a distinct group of living things, called prokaryotes to separate them from all nucleate organisms, which are called eukaryotes. Bacteria were the first disease-causing microbes to be identified, though microbiologists now recognize pathogens among the fungi, viruses and protozoa as well. They are generally so small that they can only be seen clearly with the most powerful of optical microscopes and, though there are many thousands of species and strains known, they tend all to look much the same. Three main shapes are known: rods (bacilli), spheres

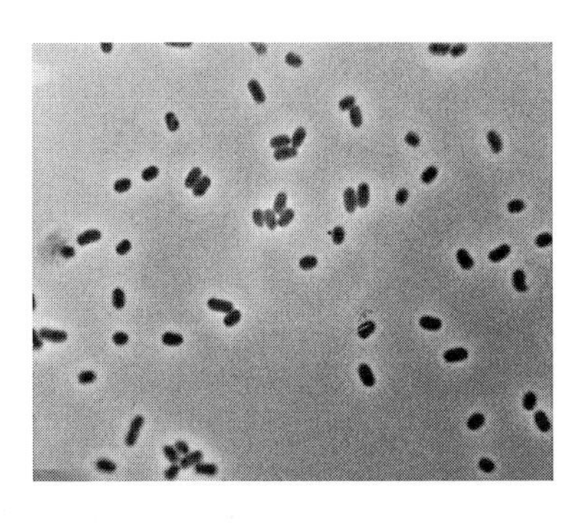
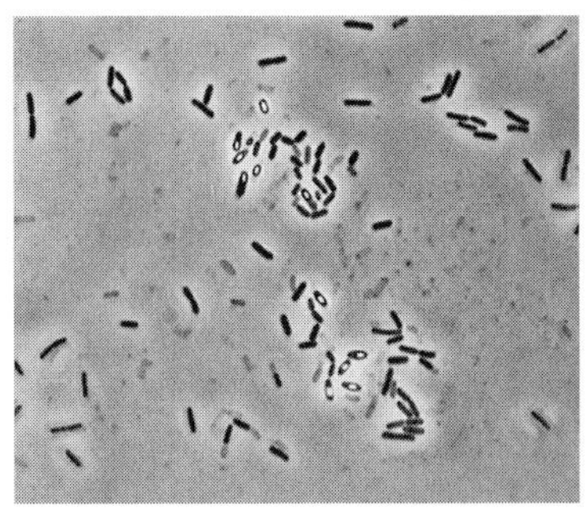
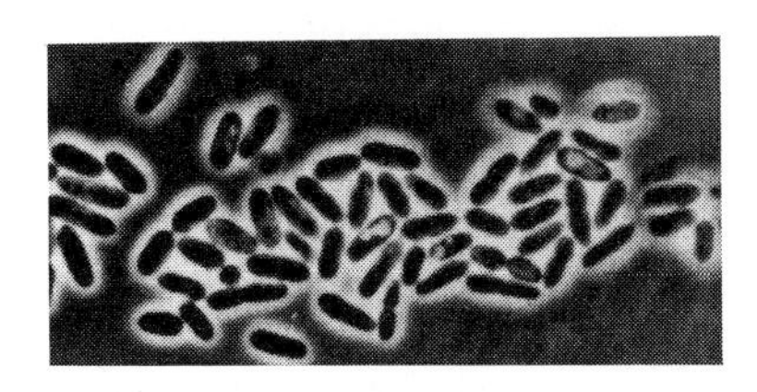
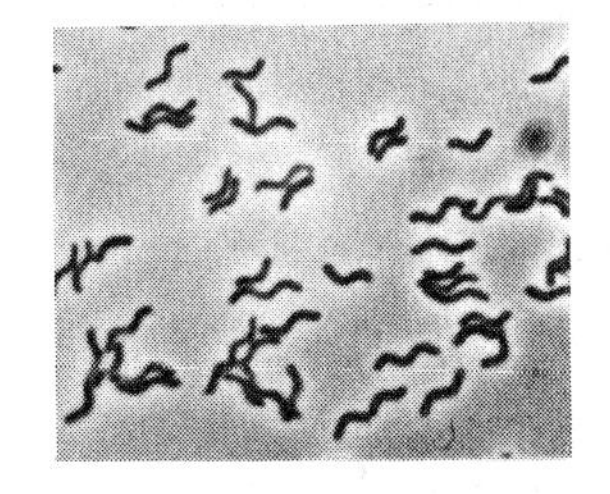

SOME BACTERIA. The photomicrographs show stubby rods (*Klebsiella*); longer rods, some forming glistening spores in their middles (*Bacillus*); egg-shaped bacteria (*Azotobacter*), some dividing; curly bacteria (*Rhodospirillum*). The magnifications are 500 to 800-fold. (Courtesy of Dr Crawford Dow, Dr Peter Dart) Photomicrographs of other bacteria are on pages 47, 99, 158 and 178.

(cocci) and commas (vibrios). Some of the rods are filamentous; rather less common shapes are S-shaped forms (spirilla) and cork-screw or wavy forms (spirochaetes); rare are Y- and V-shaped bacteria; very rare are lemon-shaped, pear-shaped and even square bacteria. When bacteria multiply, they mostly do so by simply growing to a maximum size and splitting into two; sometimes the two daughter cells fail to separate and they grow in pairs, clusters of chains. Some are motile; some form spores and can then resist heating or drying. Sexual reproduction does not occur among bacteria, though certain strains are now known to undergo a primitive kind of sexual congress.

The majority of bacteria are heterotrophic: dependent on pre-formed organic matter for their food. Certain bacteria can conduct photosynthesis and grow autotrophically, and among these the cyanobacteria (earlier known as blue-green algae) evolve oxygen like plants. Other photosynthetic bacteria do not; a group which form sulphur from sulphides will feature in Chapter 8. Some bacteria are distinctly fungoid in appearance, forming branched filaments, and these are called actinomycetes. Some filamentous bacteria are almost as large as true algae, other bacteria are so small that they are practically invisible even under the most powerful optical microscope. Mollicutes, earlier called mycoplasmas, are fragile, shapeless specks of protoplasm that fall between bacteria and viruses in size. They resemble certain forms (L-forms) that true bacteria may occasionally take up and some mollicutes cause diseases among plants and animals. Bdellovibrios are true bacteria, but they are about an order of magnitude smaller: they are tiny comma-shaped creatures, 0.1 to 0.3 μm long, that exist in soil and are parasitic on normal soil bacteria. One can culture them on larger bacteria. Even smaller are the *Rickettsiae*, round particles of about 0.2 μm in diameter that cause diseases such as scrub typhus or trench fever in both man and animals. They are right on the borderline of visibility even with the best optical microscopes. Some ordinary bacteria of the kind which inhabit soil or marine sediments become equally small when starved, and yet they remain alive; they have been called 'ultramicrobacteria' or sometimes 'nanobacteria'. A few types have been found living in such harsh environments which are permanently 'ultramicro': their size does not increase even when they are well fed.

The bacteria, being prokaryotes, represent a distinct division of living things, separate from the eukaryotic animals and plants. In the late 1970s a group of scientists working in the USA realized that the numerous genera and species of bacteria included types which were very different from the rest in their biochemical behaviour and their genetic structure, though they did not look unusual under the microscope. The details of those differences are too specialized to be of concern just now; for present purpose I record that they proved to be so different that microbiologists soon agreed that they represented a wholly new group of microbes within the prokaryotes. In the belief that they were present-day representatives of very primitive bacteria, members of the new group were given the collective name 'archaeobacteria'; more recently a change to the 'archaea' (singular: archaeon) has been proposed. The formal name 'eubacteria' has been adopted to distinguish conventional bacteria from archaeobacteria, as I shall call them here; and I shall use the term bacteria to cover both types together. The recognition of the archaeobacteria had a dramatic impact on general evolutionary biology, as I shall tell in Chapter 10.

Archaeobacteria are most abundant in conditions which are unsuitable for most forms of life, such as saturated salt lakes, hot sulphur springs or totally oxygen-free sediments. The best known of the archaeobacteria are the methanogens, microbes which are responsible for the formation of methane in pond and river sediments, or in the intestinal tracts of humans and other animals.

The classification of living things into their broadest categories still presents problems. For several centuries, biologists divided terrestrial life into animals and plants, called (perhaps as an echo of Aesop's Fables) the animal and plant Kingdoms. But even 100 years ago biologists recognized that simple organisms such as protozoa (classed as microscopic animals), fungi and bacteria (classed as primitive, or perhaps degenerate, plants) fitted into those Kingdoms only clumsily. Many revisions have been proposed; most popular recently has been a five-Kingdom system of animals (*Animalia*), plants (*Plantae*), fungi (*Fungi*), protozoa (*Protista*) and bacteria (*Monera*). Of these, the Monera are prokaryotes (the eubacteria and archaeobacteria) and the rest are eukaryotes. The way in which microbiologists classify microbes is extremely important to the specialist, for the obvious reason that one cannot study microbes usefully until one can pinpoint what one is talking about and dealing with. Yet for a survey such as this, I must disregard the intricacies of microbial classification, because I am more concerned with what microbes do. That is why the foregoing treatment of microbes has been cursory and restricted to outlining the major types that will appear later on. Microbial classification is a difficult and changing science; I shall return to it in Chapter 10.

Microbes have a remarkable capacity to exist in association with other organisms. I have already mentioned the lichens, combinations of algae and fungi, and in Chapter 5 I shall indicate the importance of bacteria that live in the intestines of animals, and also write about the special groups of nitrogen-fixing bacteria which live in the roots of plants and fix atmospheric nitrogen, thus providing the plant with an essential nutrient. Even more intimate associations may exist: bacteria may form an integral part of the protoplasm of protozoa and harmless viruses may become part of the genetic apparatus of bacteria. Ultimately it becomes possible to imagine that many of the characters of highly developed organisms were derived from associated microbes at earlier stages of their evolution – I shall return to this matter towards the end of this book.

One of the most important properties of microbes, as far as their impact on mankind is concerned, is their adaptability. If you take a microbe which, for example, cannot grow with the milk sugar called lactose, and then grow a culture of, say, a few thousand million progeny from it, about one in a hundred thousand of those progeny is likely to be able to use lactose. Suppose, to take a different example, one has a population whose growth is stopped by a certain amount of penicillin. If one gives the population a little penicillin, but insufficient to prevent them all from growing, the few persistent organisms will multiply and their progeny will be found to be resistant to much more penicillin than was the parent population. If one performs the process again and yet again, stepping up the penicillin concentration each time, one can breed strains of microbes with enormous drug resistances. Finally, as the third example, if one takes bacteria that are normally simple, discrete rods and grows them in an environment that they can manage with, but which is not the best for them (starve them of magnesium, for example, or have a trace of disinfectant present), their appearance will be quite altered: they may form long, snaky filaments, develop weird protuberances and even make the environment coloured; their chemical composition will also change and in many ways they will seem to be quite different organisms. These examples should be sufficient to show how the properties of microbes can often depend on how they have been treated or where they came from. The mutability of microbes is so great that a central problem of microbial classification is less that of giving the little beasts names than that of discovering what characters are truly immutable and truly distinctive.

From a practical point of view the adaptability of microbes means that, in almost any terrestrial environment, one will find living or dormant microbes capable of all kinds of biochemical activity. The chemical versatility of microbes, as a group of organisms, is possibly their most impressive feature, and for the rest of this chapter I shall look at the range of these abilities.

I outlined briefly a biological classification of microbes. The classification of plants and animals has what biologists call phylogenetic significance: creatures closely related in the evolutionary sequence of living things are classified close together, more divergent types further apart, so that one is made aware at a glance that cows, for example, are more closely related to buffaloes than to horses, but that all three are closer together than they are to dogs, yet all four form a group separate from frogs, and so on. With microbes the groupings are far less well defined. One can say that algae and viruses, for example, are far apart but, within the bacteria, for instance, even organisms that look and behave in similar manners may have evolved from different ancestors. Phylogenetic classification of microbes has always been difficult and, with bacteria, it has become possible only quite recently. I shall give some account of the position in Chapter 10, but for much of this book it will be more useful to classify microbes according to what they do, not according to some hypothesis about how they came to do it. In a similar way, there are other ways than the conventional one of classifying animals and these can be very useful in special contexts. An obvious way is to do so according to environment: there are arctic, temperate and tropical animals acclimatized to diverse ranges of temperature, desert animals accustomed to extreme dryness and aquatic animals enjoying extreme wetness. One useful classification of animals makes use of their eating habits, so one has carnivorous, herbivorous and omnivorous animals. Yet another system uses the duration of their activity, so one has nocturnal or diurnal creatures and those that become dormant – hibernate – for cold or dry seasons. Other minor classifications include parasitic or non-parasitic, wild or tame, fierce or timid. These classifications all have their uses, and they cut completely across the natural or biological system. In general biology they are usually of secondary importance, but in microbiology, because of the deficiencies of the more formal system, they are often the more important. Let me start by classifying microbes according to the sort of environment in which they flourish.

Mankind and other mammals, as most people know, require very precise physical and chemical conditions to live at all. Their temperatures must lie within the range 35 to 40 °C and they must breathe an atmosphere of about 21 per cent oxygen with about 0.03 per cent CO_2 at a pressure around 760 mm of mercury. Temporary deviations from these conditions can be tolerated, but their bodies in fact have built-in mechanisms to maintain such an environment: temperature regulation and ventilation rate adjust automatically to cold, heat or carbon dioxide changes. The salt concentration (salinity) of the blood is also closely regulated by the kidneys; the breathing rate and kidney function control the acidity of the blood. Mammals, in fact, only withstand the fluctuations of the terrestrial environment by controlling their internal environments very closely. Cold-blooded creatures have wider tolerances of temperature, but are otherwise pretty exacting; plants tolerate, and indeed flourish in, atmospheres with excessive carbon dioxide and, in some instances, they tolerate salty, acid or very dry soils. But again they need air, light and an equable temperature to do well. Plants and animals have learned to grow and multiply on dry land, but they do this by controlling the state of their interiors minutely so as to guard their cells against external fluctuations.

Microbes are mainly aquatic. Some filamentous fungi grow in air, on damp materials such as decaying bread, but all bacteria, protozoa, algae and viruses, all the truly unicellular microbes, require a watery environment in which to grow. A microscopic film of water on a leaf, on skin, in soil or on a jelly is sufficiently aquatic for most of them, but they never grow actually out of water. However, though they may not grow out of water, they do not necessarily die when dried. Many bacteria and moulds form spores, resistant bodies which will withstand desiccation not just for years but for decades. Professor Peter Sneath produced the most impressive death curve known to microbiology when he examined soil samples attached to ancient pressed plants kept at Kew: he found live bacterial spores in specimens dating back to the seventeenth century (but no earlier). Live bacterial spores have been found in sediment strata several thousand years old, but the tomb of Tutankhamun was sterile as regards bacteria when, in 1923, it was opened for the first time in 3,000 years. So it seems that, though bacterial spores last a great many years, they do not last for ever. Viruses do not form spores, but they survive drying if there is a little protein around in the fluid from which they dried (as there always is in droplets from a sneeze, for example). Once dried, they last indefinitely, as far as we know, and it is a pity that techniques for recognizing unknown viruses were not available (indeed, still are not) when Tutankhamun's tomb was opened.

Most microbes that do not form spores die if they get too dry, but even some of those can be protected if there is some protein around: a few bacteria will survive, for instance, in mucus dried on a handkerchief. (Handkerchiefs, like drying-up cloths, are among the most infectious of civilized appurtenances from the microbiological point of view; more of that in Chapter 3.)

Spore formation by moulds and bacteria, dormancy as a result of drying or starvation, and the property some bacteria and protozoa possess of forming relatively resistant bodies called cysts together comprise an important general property of microbes: that of going into a state of suspended animation when conditions become adverse. In Chapter 1 I noted that microbes could be detected many miles up in the stratosphere. They are, in fact, nearly all spores of the moulds *Cladosporium* and *Alternaria*, with dormant bacteria of the *Micrococcus* group, blown there by winds from the lower atmosphere. It is improbable that microbes actually multiply in the airborne state, though such a thing could conceivably happen on a wet dust particle of suitable character and buoyancy. Dormant microbes are very important in dispersal, being the main form in which microbes become spread around everywhere.

Many microbes are killed by freezing but, here again, protein can protect them and so, it seems, can soil. The permafrost zones of the Arctic and Antarctic contain viable bacteria and fungi, and a most curious fact about them is that many of the bacteria are thermophilic. This means that they require an uncommonly high temperature to grow: 55 to 70 °C is commonly provided in laboratories. The hottest hot bath that one can comfortably bear is between 45 and 50 °C; at the temperatures at which thermophilic bacteria grow, normal creatures would rapidly scald and die. Thermophilic bacteria are normal inhabitants of hot springs, hot artesian wells and other geothermal environments; this is comprehensible enough, but why should they be present in ordinary temperate soils and even in the permafrost? This is still one of the more baffling problems of microbial ecology. In practice it means that hot environments such as central-heating systems, cooling towers and so on are as liable to microbial contamination as any other environment.

Thermophilic creatures are restricted to the microbes, and some exceptional bacteria show quite extraordinary heat tolerance: there is a methane-forming archaeobacterium which can grow in water super-heated under pressure to 112 °C. At the elevation of Yellowstone National Park, USA, water boils at 92 °C; in the boiling water of some of the springs there are dark deposits of small, rod-

MICROBES GROW IN A HOT SPRING. The 'Morning Glory Pool' at Yellowstone National Park, Wyoming, USA, is not one of the hottest but, at about 70 °C, it is nevertheless lethal to all higher organisms. But some bacteria flourish in it. Its romantic name arises because it is coloured bright blue-green by thermophilic cyanobacteria and edged by brown filaments of thermophilic flexibacteria. (J. H. Robinson/Science Photo Library)

shaped bacteria multiplying at that temperature. Professor T. D. Brock in the USA has studied the microflora of that area: a filamentous bacterium called *Flexibacterium* flourishes up to 83 °C; the temperature limit for cyanobacteria seems to be 75 °C; fungi and true algae are found up to 60 °C; protozoa to 50 °C; insects to 40 °C. Obviously the more complex the creature, the lower its maximum temperature.

Heat above 80 °C is lethal to most microbes, even most thermophiles. However, the spores of certain bacteria called clostridia are very resistant even to heat. Boiling water, for instance, will kill spores of the bread mould in about ten minutes, whereas spores of some clostridia will stand six hours' boiling; some will even survive five minutes' pressure-cooking in live steam at 120 °C.

Salty water tends to kill microbes, which is why foods such as bacon and fish can be preserved by pickling in brine. Strong syrups have similar effects. The sea is sufficiently salty to kill most (but by no means all) fresh-water bacteria and viruses (a circumstance for which the inhabitants of Great Britain have cause to be grateful, because their islands are situated in what is now a sea of dilute sewage). Yet there exists a whole microbial flora adapted to life in the sea, and a branch of microbiology (marine microbiology) has grown up around their study. Even pickling brines and preserving syrups become infected with specialized bacteria, moulds and yeasts that can grow in such strange environments. A familiar example in the home is jam that has gone mouldy and begun to ferment. Microbes that withstand strong salt or sugar solutions are called strong halophiles by microbiologists; as well as being found in foods and brines, halophilic archaeobacteria crop up in natural salt pans, and the red-brown colour of many brackish lakes in the Middle East is due to a halophilic alga called *Dunaliella*.

I wrote just now of bacteria in the seas. Most of the sea is very cold and only the upper layers (the thermosphere) change temperature according to the seasons. Below a certain level (called the thermocline; its depth depends somewhat on latitude and season) exists the psychrosphere, where the temperature lies below 5 °C at all times. More than 90 per cent of the volume of the Earth's oceans is psychrosphere. In addition, for every ten metres, approximately, the pressure in the sea increases by about one atmosphere. What, then, of the microbes which inhabit this zone? As the reader may guess, they are peculiar, and in particular, most are psychrophilic and some are thought to be barophilic. These words mean that they grow only at low temperatures and high pressures. For many years the existence of true psychrophilic (low temperature) bacteria was doubted, and the main reason was that microbiologists simply failed to refrigerate their samples while transferring them from the sea to the laboratory. So these bacteria, which die fairly rapidly at temperatures above 20 °C, had mostly died by the time the samples could be examined properly and only their hardier neighbours survived. Barophilic microbes are even more difficult to study: specially strong apparatus is necessary to reproduce the pressures of 500 to 1,000 atmospheres encountered in, for example, the deep Pacific trenches, and it is almost certain that many of the microbes that inhabit such deep sedimentary oozes have never been detected. Only those that tolerate a brief exposure to low pressure can readily be cultured, even in specialized laboratories. An interesting feature of growth at high pressures is that, at such pressures, bacteria show increased heat resistance: thermophilic, barophilic bacteria from an oil well have been grown by Professor ZoBell under a pressure of 1,000 atmospheres at 104 °C, well above the boiling point of water at ordinary pressures.

At the opposite extreme of pressure, few normal microbes mind a near-vacuum, provided it is wet. Most anaerobic bacteria (see below) can be cultured in vessels that have been evacuated and contain nothing but a little water vapour over the culture fluid, a property which can be convenient in a laboratory but a thorough nuisance when they get into vacuum-packed foodstuffs.

Bacteria and most viruses do not tolerate acids. Even acidity as weak as that of vinegar prevents the growth of most bacteria. This is why pickling works: the normal putrefactive bacteria cease growing, though certain acid-forming bacteria survive and, in fact, can aid the pickling process by forming acid. Yeasts and moulds, on the other hand, somewhat prefer mild acidity, doubtless because their normal habitat is fruit juices, plant exudates and fermented matter. Strong mineral acids such as sulphuric or hydrochloric are, however, lethal to most microbes and it is therefore curious to discover that there exist micro-organisms that can not only tolerate sulphuric acid solutions but actually generate them. The sulphur bacteria called thiobacilli oxidize sulphur or pyritic ores to form sulphuric acid, and couple these reactions to the fixation of carbon dioxide – much as green plants couple the trapped energy of sunlight to a similar process of CO_2-fixation. They are autotrophs, but different from those that I introduced under the heading 'algae' earlier in this chapter, because they couple a purely chemical, not a photochemical, reaction to biological syntheses and growth. Such creatures are called chemo-autotrophs, to distinguish them from photo-autotrophs, which, like plants, use sunlight.

Thus one can classify microbes according to their temperature relationships into thermophiles, mesophiles and psychrophiles, according to salinity into normal organisms and halophiles, according to pressure into normal and barophilic, according to resistance to heat and drying in terms of whether they form spores or not, according to their tolerance of acidity (the acid-tolerant microbes are called acidophiles; those preferring alkaline environments, alkalophiles). These classes have obvious analogies to the arctic–temperate–tropical, the desert–aquatic and other classifications of animals mentioned earlier. Let me now describe a classification according to nutritional habits, because this is much more important among microbes than is the carnivore–herbivore division among animals.

From the nutritional point of view microbes span the gap that distinguishes plants from animals and include categories of nutrition that do not occur at all among higher organisms. Most notable of these are the autotrophic types of metabolism such as that of the thiobacilli just mentioned. I shall say more about chemo-autotrophy later. To return to this matter of acidity: when thiobacilli generate acid, in sulphur springs or in the seepage waters of mines containing pyrites, they prevent the growth of common microbes. However, there develops in such waters a whole microflora of acid-tolerant creatures: yeasts, actinomycetes, bacteria and even some protozoa, all depending primarily on the CO_2 fixed by the chemo-autotrophs that grew there in the first place. Just as heterotrophs such as humans and animals depend on plants to live in the neutral (or, strictly, very faintly acid) conditions of this planet, so, in acid environments, a specialized microbial microcosm can develop analogous to ours.

An even more exotic community was discovered in the 1970s associated with hydrothermal vents in the deep trenches of the

LIFE IN THE GALAPAGOS RIFT. The photograph was taken by the *Alvin* submersible. It shows clams and sessile worms living in the neighbourhood of a hydrothermal vent, 2.5 km deep in the Pacific Ocean. Both depend on the local sulphur bacteria for their nutrition. (Courtesy of Professor H. Jannasch)

Pacific Ocean. The sea, like the land, has its volcanic zones and one is to be found near the Galapagos Islands in the so-called Galapagos Rift, at a depth of some 2.5 km. Here scientists in the research submersible *Alvin* discovered submarine hot springs from which spouted hot water, sometimes above boiling point, rich in hydrogen sulphide. In cooler zones, this sulphide is used by sulphur bacteria for growth and CO_2-fixation. Some of the bacteria live in the gills of clams or the intestines of worms in the locality, which use them as symbiotic sources of food; a local variety of crab parasitizes the clams and worms; occasional deep-sea fish consume both. The zone round the hydrothermal vent, for about 50 metres, is teeming with life in an otherwise dark, dead seascape, for no sunlight penetrates to such depths. The whole community depends on the sulphur bacteria which, in turn, depend on the sulphide in the water of the volcanic vent. The species of creature which live there are mostly unique, though they have their counterparts in more normal habitats, and to biologists they are fascinating examples of how living systems *can* manage remote from sunlight.

The thiobacilli are part of a fairly small group of chemotrophic microbes. As I wrote just now, they couple the oxidation of sulphur or pyrites to CO_2-fixation. Other reactions that can be used by bacteria for chemo-autotrophic growth are the following:

Oxidation of hydrogen to water (by *Pseudomonas facilis*).
Oxidation of ammonia to nitrite (by *Nitrosomonas*).
Oxidation of nitrite to nitrate (by *Nitrobacter*).
Oxidation of ferrous ions to ferric (by *Thiobacillus ferro-oxidans*).
Oxidation of methane to water and CO_2 (by *Methanomonas*).
Oxidation of sulphide to sulphur (by *Thiovulum* and some other sulphur bacteria).

If readers are dismayed by the chemistry implied by these reactions, they need not be. There is nothing to be gained by writing out correct chemical equations for these processes (they are mostly obvious and elementary anyway, provided one remembers they are taking place in water), because the essential point is that a number of purely chemical reactions exist which microbes can use as alternative energy sources to sunlight for the primary synthesis of biological material. Chemotrophic ways of life such as these are only found in bacteria; they are not encountered among the fungi or protozoa nor, of course, among the viruses.

All the processes listed above assume the presence of air, and autotrophic reactions that need neither air nor light are, as far as one knows, rare. However, a few well-established instances exist. The bacterium *Thiobacillus denitrificans*, while able to grow at the expense of sulphur oxidation in air, can, if no air is available, oxidize sulphur while reducing the nitrate ion. The sulphur goes to sulphuric acid and the nitrate to nitrogen gas; the organism couples this interaction to the reduction of CO_2. An organism called *Paracoccus denitrificans* can conduct a similar process using hydrogen to reduce nitrate to nitrogen gas; an organism called *Desulfonema* can reduce sulphate with hydrogen and couple this to CO_2-fixation in the total absence of air; an archaeobacterium called *Thermoproteus* can reduce sulphur in hydrogen, at about 80 °C, and couple this to CO_2-fixation.

Microbes which grow without air are called anaerobes and they are quite common. Some anaerobes need both light and a chemical reaction for autotrophic growth. The coloured sulphur bacteria, which will reappear later, oxidize sulphide to sulphur provided they are illuminated, and in these circumstances they fix CO_2 and grow. The importance of this process is that air is unnecessary: provided both light and sulphide are present, they grow in the total absence of air. A comparable group of anaerobes exists whose members require light to grow autotrophically but instead of sulphide they need some organic matter to oxidize while reducing CO_2; they do not grow photosynthetically in air. (One has to make the reservation that they do not grow photosynthetically in air, because they can usually grow in air if there is no light.) These bacteria are always coloured, though they are not necessarily (or even commonly) green. Red colours (due to relatively large amounts of carotenes, which swamp the green chlorophyll) are very common among such bacteria.

The photosynthetic anaerobes are a little like green plants in that they use light to fix CO_2 but, unlike plants, they need an oxidizable substance – sulphide or organic matter – to couple to CO_2 reduction. They also differ fundamentally from plants in that their photosynthesis does not yield oxygen. Looking at the matter another way, one can say that plants can use water in place of sulphide or organic matter for photosynthesis: they split the water molecule (H_2O) to H_2, which they use to fix CO_2, and oxygen atoms, which they release as O_2. A large group of bacteria, the cyanobacteria (once called blue-green algae) can conduct a similar oxygen-producing photosynthesis (it is interesting that certain strains can also use sulphide and photosynthesize without forming oxygen – an evolutionary link here, I expect). Like the algae proper and higher plants, they grow readily in air and oxygenate their environment when illuminated.

While I am dealing with the chemical versatility of microbes, I should mention the iron bacteria. This is a group of bacteria, often filamentous or in the form of twisted, branching stalks, that occur in iron-rich waters. The brown 'iron' deposits on rocks and stones in mountain streams are often formed by these bacteria, which oxidize dissolved ferrous ions to the ferric form. At one time this reaction was thought to permit chemo-autotrophic growth, but the evidence for this seems now to be unsound. Yet microbiologists may well be mistaken, because the process seems to be of little use unless, as has been suggested, the precipitate tends to concentrate organic matter by adsorption (just as charcoal adsorbs smells) and so enables the bacteria to feed more easily.

One aspect of the metabolism of microbes which has no analogy among higher organisms yet which provides an important way of categorizing them is their respiration – or lack of it. I mentioned a while ago the class of bacteria called anaerobes, which live without air. All higher organisms require air (or at least an inert gas with 21 per cent of oxygen) or else they die. Certain plant tissues (e.g. seeds) can respire for a while without air and some primitive animals (nematodes, insect larvae) seem able to tolerate considerable oxygen starvation. But their metabolism remains based on the use of oxygen to oxidize foodstuffs. Yeasts, some moulds and many bacteria (but not, generally speaking, protozoa and algae) can grow anaerobically, which means without air. (Viruses, of course, don't care: they use their hosts' metabolisms anyway.) Anaerobic bacteria are quite as common and widespread in nature as aerobic ones. They grow anaerobically in one of two ways, either by splitting the food molecules into smaller fragments, so as to yield energy without the participation of oxygen, or by using an alternative oxidizing agent to oxygen. These processes are called fermentative or oxidative respectively.

A typical fermentative reaction occurs in the alcoholic fermentations brought about by yeasts: fruit sugars (mainly glucose) are decomposed by the yeasts to alcohol and carbon dioxide, a sequence of reactions that provides enough energy for the yeasts to grow and multiply yet which involves no air. Many moulds and bacteria, if deprived of air, can conduct such fermentations, forming CO_2 together with products such as lactic acid, succinic acid and butyl alcohol in addition to ethyl alcohol. Most of these microbes can use air if it is present, in which case the glucose and other products become oxidized completely to CO_2. Other bacteria exist, however, that can grow only if air is absent, and these are called obligate anaerobes, to distinguish them from the optional facultative anaerobes. Members of the genus *Clostridium* (the clostridia mentioned earlier) do not grow unless air is absent and thus they flourish in polluted or putrescent environments where other microbes have used up all the available oxygen. They ferment sugars or amino-acids, derived from carbohydrates and proteins and, particularly when using protein, produce some evil-smelling and even poisonous by-products such as the ptomaines (amines formed from protein). Clostridia are characterized by forming spores, and this property confers on them resistance to heat and desiccation, which makes them dangerous in food technology. But obligate anaerobes that do not form spores are known, such as the *Bacteroides* found in the intestines of mammals and in milk. The rumen or first stomach of ruminant mammals contains little if any air normally and is rich in fermentative bacteria; it is one of the few environments in which anaerobic protozoa may be found.

The second class of anaerobes, the oxidative anaerobes, functions rather differently, and are exclusively bacteria. Instead of oxygen, they use an ion such as nitrate, sulphate or carbonate to oxidize organic food, and this material becomes reduced. I introduced two nitrate-reducing bacteria earlier: *Thiobacillus denitrificans* and *Paracoccus denitrificans*, both authenticated anaerobic chemoautotrophs. Quite a number of ordinary bacteria can use nitrate in place of oxygen for respiration, generally reducing it only to nitrite and not to dinitrogen gas but, in dung heaps, compost heaps and polluted muds, bacteria which reduce nitrate to dinitrogen flourish. They are important in the nitrogen cycle as denitrifying bacteria. Most denitrifying bacteria are facultative: they can use oxygen if it is available. This is not true of the bacteria that use sulphate or carbonate. The sulphate-reducing bacteria, the microbes we first met in Chapter 1, are very strict anaerobes which, while they are not killed by air, cannot grow in its presence. What they do is reduce sulphate to sulphide while oxidizing organic matter to acetic acid and CO_2; they thus produce a horrible smell and, since sulphide reacts fairly rapidly with oxygen, they remove oxygen from any neighbourhood in which they get established. This property is at the bottom of the extraordinary variety of economic nuisances they cause – and the few benefits they have engendered – as I shall tell in Chapters 6 and 7.

Let me pause a moment here and consider the sulphate-reducing bacteria in slightly more detail, because they will appear often in this book and they illustrate well the way in which the classifications I have been discussing cut across each other. They are, as I said, strict anaerobes, because they 'breathe' sulphate instead of oxygen, but within this restriction they have representatives in several of the categories I have discussed. One of the main groups (called *Desulfotomaculum*) forms spores, the others (*Desulfovibrio, Desulfobacter, Desulfonema*, etc.) do not. One species of *Desulfotomaculum* can be thermophilic; there is also a genus of thermophilic sulphate-reducing archaeobacteria called *Archaeoglobus*; some of the *Desulfovibrio* and other groups

SOME SULPHATE-REDUCING BACTERIA. These bacteria, which are of considerable economic importance, come in a variety of shapes and sizes, most often curvy, but round, straight, lemon-shaped and fat filamentous types exist. The desulfotomaculum shown here is forming spores; only part of the relatively large *Desulfonema* can be seen. Desulfovibrios are the first to grow in enrichment cultures and are responsible for most bursts of sulphate reduction in nature, but they are not the most abundant. For scale, the little bar on each micrograph is 10 μm long. (Courtesy of Dr F. Widdel)

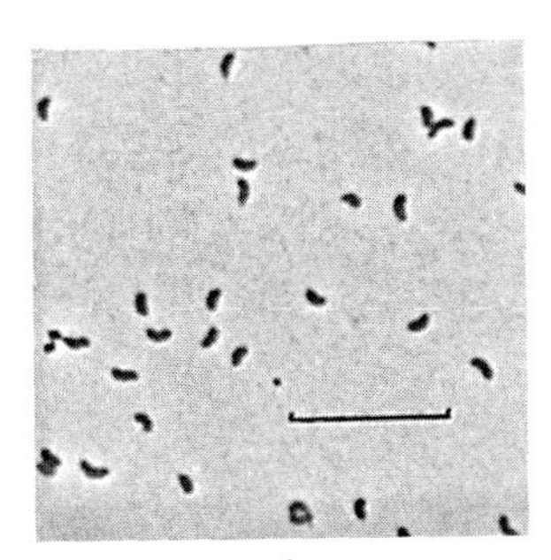
Desulfovibrio vulgaris

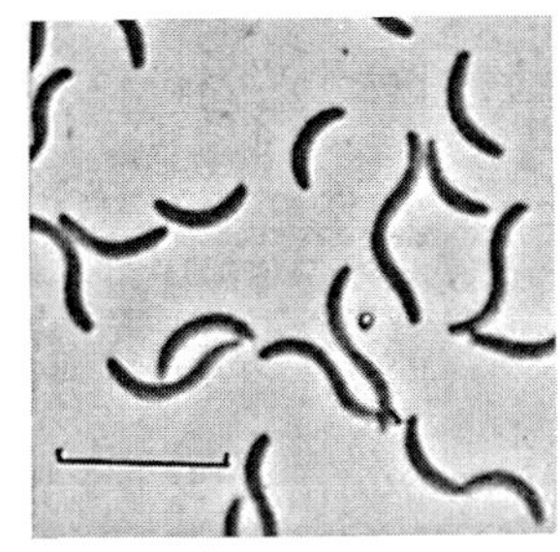
Desulfovibrio gigas

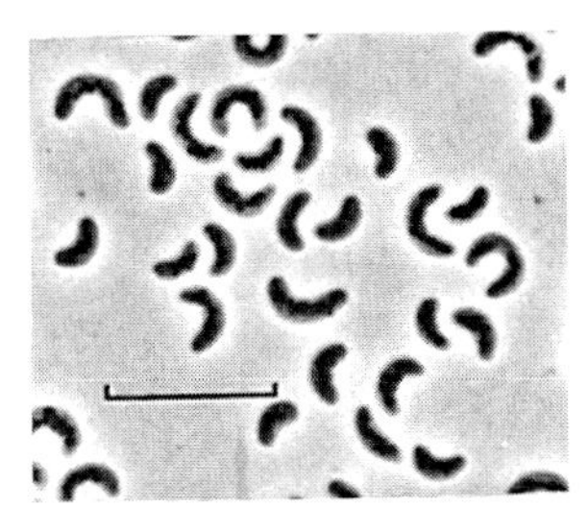
Desulfovibrio sapovorans

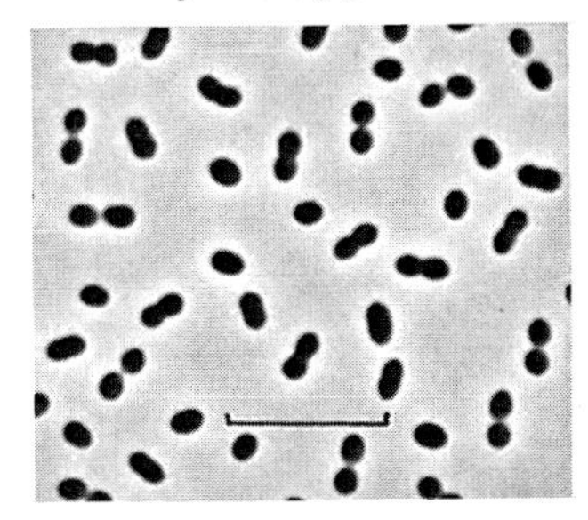
Desulfobacter postgatei

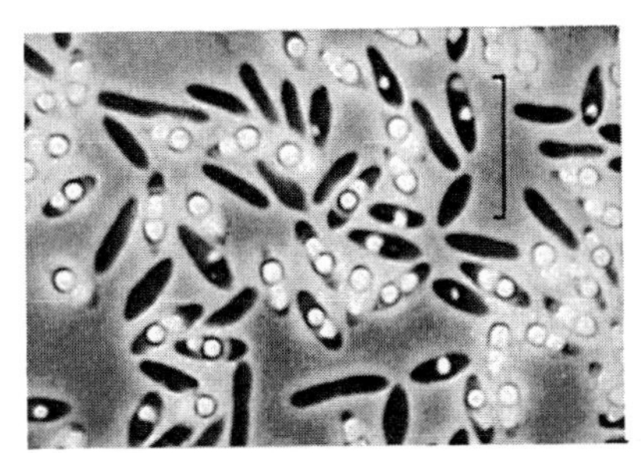
Desulfotomaculum acetoxidans

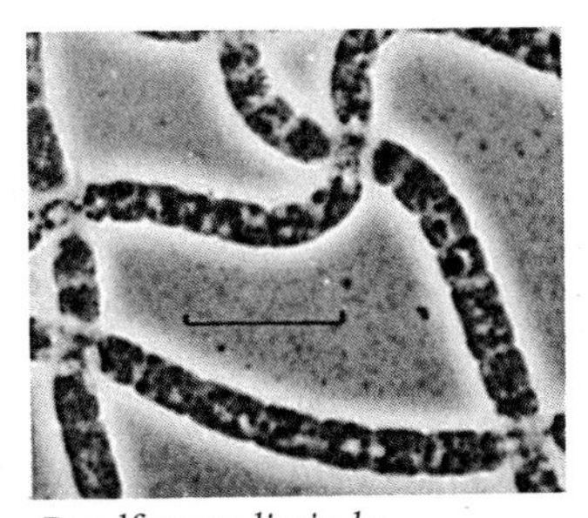
Desulfonema limicola

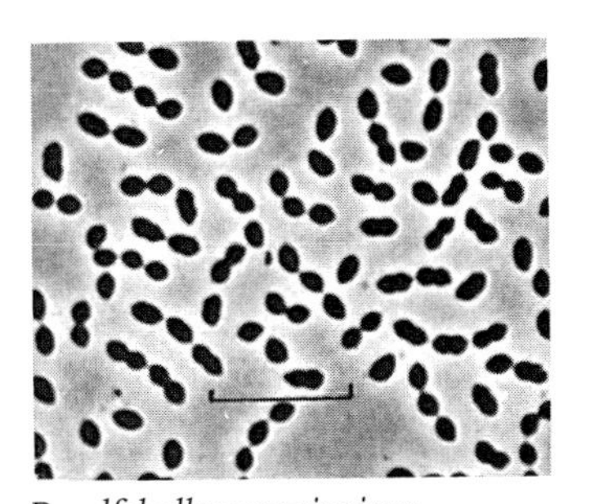
Desulfobulbus propionicus

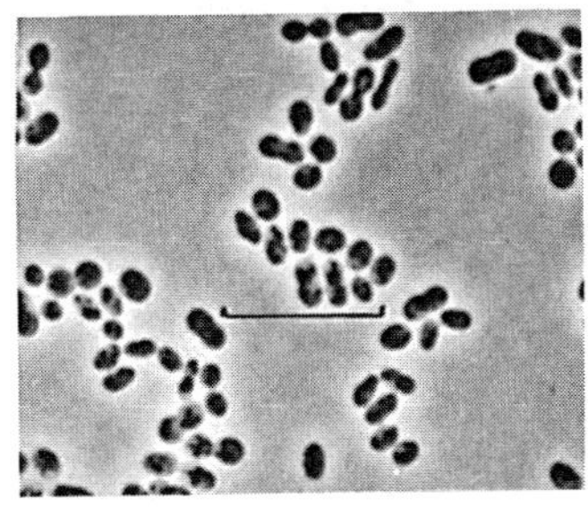
Desulfosarcina variabilis

are halophilic. Representatives of the sulphate-reducing bacteria are found in environments ranging from brackish, super-cooled Antarctic waters to hot artesian springs; barophilic types have been found in deep Pacific sediments. Bacterial sulphate reduction is believed to be one of the commonest biological processes on earth, although, because of its anaerobic nature, it mostly occurs out of sight: deep in seas, in soils and in polluted waters.

Carbonate reduction is also an extremely anaerobic process. The marsh gas that bubbles up when polluted mud at the bottom of a stagnant pool is disturbed is methane and it is formed by the action of certain archaeobacteria collectively called methanogens or methane bacteria. Some of these organisms form methane from other carbon compounds by fermentative reactions, but most couple the oxidation of such compounds to a reduction of carbonate to methane. The methane bacteria, like the sulphate-reducing bacteria, are very widespread and are even more sensitive to oxygen. (They find a warm, anaerobic habitat in the stomachs of ruminant mammals.) They are difficult to culture in the laboratory in the absence of other bacteria. They should not be confused with the methane-oxidizing bacteria, mentioned earlier among the chemo-autotrophs, which are quite different: they need air and couple the oxidation of methane, in air, to CO_2-fixation and growth.

The categories I have used so far to discuss microbes – I have assembled them for convenience in the box opposite – are, as I said earlier, often more useful to microbiologists than the 'biological' classification into algae, fungi, bacteria and so on. This is particularly true when one is discussing their impact on mankind – as, rest assured, I shall shortly start doing – because, though one would ideally wish to know exactly what all the types of microbe in a certain environment are, one can often say useful things about the microbiology of an environment without such detailed knowledge. To give a simple example: if sulphate-reducing bacteria are becoming dominant in a certain environment, then a fairly predictable sequence of changes in the chemistry and microbiology of that environment is

Ways of categorizing microbes

As well as classifying microbes in the traditional biological way, in genera, species, varieties etc., it is often useful to categorize them ac ing to a particular character as, for example, when one divides microbe into pathogenic or non-pathogenic organisms. Here are more examples of such groupings which appear in this book – mostly in this chapter.

Categories of habitat:

Thermophiles like it hot.
Mesophiles like it middling.
Psychrophiles like it cold.
Halophiles like it salty.
Acidophiles like it acid.
Alkalophiles like it alkaline.
Barophiles like high pressures.

Nutritional categories:

Heterotrophs need organic matter as food.
Autotrophs make their food from CO_2.
  Photo-autotrophs use light energy to do so.
  Chemo-autotrophs use chemical energy to do so.

Metabolic categories:

Aerobes have to breathe air.
Facultative anaerobes can breathe air but do not have to.
Anaerobes do not breathe air.

going to take place no matter to what genus and species the actual bacteria belong. Information about genera and species becomes important at a more detailed level – when one has to make a choice among possible counter-measures, for example. The variety of categories used by microbiologists may seem confusing at first, though I have tried to show that they are in principle no different from categories of animals, such as herbivorous versus carnivorous, or tropical versus temperate. However, I undertake to remind readers of the details of this chapter when they arise elsewhere in this book (reminders are also available in the glossary) and I shall bring it to a close by drawing one important moral about the behaviour of microbes.

ing, and what makes these various clas-
portant, is the extraordinary chemical
ave told how microbes can utilize quite
is for growth and multiplication, and so far
ly reactions involving primarily mineral or
such as sulphur and iron. I have been entirely
ajor nutrients, materials that provide energy, and
bout the fixation of nitrogen, for example, which is a
cess in the biological economy of this planet (I shall return to it in Chapter 4). Nor can I say more of such processes here, except to point out that, where higher organisms other than plants need fixed nitrogen, amino-acids, vitamins, fats and so on in their diet, microbes range from those that can synthesize all of these things from mineral sources to those that are so exacting that one wonders that they survive at all. *Mycobacterium leprae*, the causative organism of leprosy, has still never been cultivated away from living tissues, and many disease-causing organisms need quite complex brews if they are to be cultivated in the laboratory. One can almost say that, whatever minor nutrient one mentions, there will be microbes that need it and others that do not; the only exceptions seem to be vitamin C (ascorbic acid) and the fat-soluble vitamins which, as far as scientists know, are not *required* by any known bacteria, fungi or algae. (The position regarding protozoa is uncertain: representatives of the group may need steroid materials, which are analogous to fat-soluble vitamins, and some mollicutes certainly need them.) Viruses, of course, fall outside this discussion, because, in one sense, they require no nourishment at all.

The basic foodstuffs of animals are carbohydrates, proteins and fats, but many organic materials having comparable compositions cannot readily be used. The cellulose of plants, the lignin of wood, the chitin of crustacean shells, the keratin of hair and processed materials such as leather, paper and so on are either not eaten or are discharged undigested (except, as happens quite often, when the animal carries commensal microbes within itself to effect their digestion). The range of organic matter suitable as food for animals is in fact rather limited. This is not so for microbes. Many fungi utilize the cellulose and lignin of wood and break down paints, leather and paper. There are bacteria that decompose cellulose, and bacteria or yeasts that metabolize waxes, hydrocarbons such as petroleum, kerosene and petroleum grease. Asphalt, coal and road-building materials are slowly attacked by certain bacteria. The gases hydrogen and methane may be utilized by certain bacteria; even polythene, unknown to the living world before this century, is attacked by some soil microbes, though not very effectively; plastics such as nylon, polystyrene and polyurethane are, it seems, attacked by bacteria, but again only slowly. The odd thing is that they are attacked at all. Equally curious are the organisms that attack strong poisons. Phenol, for example, is a powerful disinfectant, yet there exist strains of bacteria that grow readily with it. Others can metabolize and grow with antibiotics and, though cyanides are possibly the most universal poisons for terrestrial living things, a mould exists that can grow with cyanide as its main source of carbon. Fluoracetamide, which is also a powerful, universal poison used sometimes as an insecticide, caused deaths of cattle in 1963 when a field in Smarden, Kent, was accidentally contaminated with it; microbes have since been obtained from soil that decompose fluoracetamide and actually grow at its expense. Most of these microbes only tolerate small concentrations of the toxic substances – if they get too much phenol, say, or cyanide, then they too are killed. But provided they are not over-fed, as it were, they convert these materials to harmless waste products. This property is vitally important in the disposal of certain industrial wastes.

After some acquaintance with the chemical versatility of microbes, the microbiologist tends to be more surprised by organic materials that are not attacked by one or another creature: at present certain plastics and detergents, plus pure carbon in the form of graphite or diamond, seem definitely immune, but few other substances are. I shall tell in Chapter 7 how such materials as iron, steel, concrete, stone, glass or rubber, while not necessarily consumed by bacteria, may be corroded or decomposed as a result of their activities.

Finally, though I have mentioned bacteria that consume and detoxify poisons such as phenol, I should also refer again to the general ability of bacteria to acquire resistance to toxic substances. I mentioned earlier in this chapter how one could train bacteria to resist, for example, penicillin, and this provides quite a good general illustration. There exist in nature bacteria which can grow in the presence of penicillin because they contain an enzyme, penicillinase, which enables them to destroy penicillin. Such organisms have caused trouble in hospitals. Bacteria trained to resist penicillin, however, do not necessarily make penicillinase. They adjust their metabolism in such a way as to avoid the damage penicillin would have done, and by similar adaptive processes one can get microbes adapted to sulphonamide drugs, flavin disinfectants (such as acriflavin) and various antibiotics. Copper sulphate is a powerful general poison for living things, yet at Rutgers University, New Jersey, USA, I was shown a strain of mould that grows in 20 per cent copper sulphate in weak sulphuric acid provided a little sugar is present. The creature has developed a mechanism for keeping the copper outside its cell walls.

I could continue listing the variety of chemical activities that microbes are capable of, and the toxic environments they will tolerate, almost indefinitely. However, not only would such a list become wearisome, but also, I now realize, a quite detailed acquaintance with the biochemistry of ordinary multicellular living things is necessary to appreciate quite how impressive this diversity of microbial behaviour is. Yet, essentially, all terrestrial living things have similar biochemistries. The chemical mechanisms whereby they build up and break down proteins, carbohydrates and fats, the ways in which they control these processes, even the ways in which they use or store energy, all these are similar in most of their chemical details. Microbes are no exception (leaving out, as usual, the viruses, which get a host organism to do these things for them). The distinctive features of microbes are that they can make use of unusual and ingenious methods of obtaining energy in order to drive a fairly conventional metabolism, and that they can adjust themselves to run such a metabolism in circumstances that would be lethal to higher organisms. They can be found all over this planet, can grow and perform chemical transformations in what seem, from an anthropomorphic point of view, the most unlikely places. For this reason they have a profound, and often unrealized, effect, not only on the balance of nature but on the existence of mankind and on our economy. Microbiology is the study of microbes as a pure science; the study of their impact on man and other higher organisms, with which this book is concerned, has been variously called applied or industrial microbiology, economic microbiology and biotechnology. I rather prefer the term economic microbiology, because it brings out the diverse effects microbes have on our economy, but biotechnology is the 'OK' word at the time of writing, although it largely concerns productive processes and does not necessarily involve microbes. Call it what you will, the facets of microbiology with which I shall be preoccupied are good examples of the sort of hybrid of pure science with technology that impinges on all aspects of our daily lives.

3 Microbes in society

I ended the last chapter with a comment hinting at the relationship between pure science and its applications. A clear example is that most personal and important technology, the one called medicine. For medicine is not a science, despite the fact that its practitioners are called doctors. It is a model example of a technology, the application of various branches of science to one facet of the human condition. It is, historically, one of the most admirable examples of science and its application progressing hand-in-hand; even today it is unique among technologies in that a fundamental discovery in, say, a biochemical or physical laboratory may find application in medical practice within weeks instead of years.

The reason is simple to see. We all hate being ill, whether we are doctors, scientists, treasury officials or laymen, and we will support with positive enthusiasm research intended to cure or alleviate this condition, whereas the study of quasars or the ecology of plankton might cause us reservations. Microbiologists have particular cause to be grateful for the self-interest that underlies the relative affluence accorded to medical research because, since microbes are the cause of most illnesses, microbiology has progressed very rapidly in the twentieth century, particularly in its medical aspects. Naturally, this imbalance has left non-medical microbiology in a somewhat neglected state, as will become obvious in later chapters of this book, but though pure microbiologists have at times been critical of the narrowness of their more medical colleagues, this fact should not blind one to the enormous contribution the traditional Path. and Bact. type of scientist has made to the science as a whole.

In this chapter I cannot hope to survey medical microbiology even superficially. Microbes cause disease in man, animals, plants and each other. They do not, of course, cause all known diseases. Some, such as schistosomiasis, are caused by higher organisms (a worm in that case); others, such as lung cancer, have environmental origins (e.g. in tobacco smoke); yet others, such as haemophilia, are hereditary: due to genetic defects peculiar to certain families. (I shall tell in Chapter 6 how microbes can today be exploited to detect genetic diseases and may one day be used in the same way to cure them.) But microbes cause most of our day-to-day ailments, and most of our more serious ones, too. For a catalogue of which microbes cause which diseases the reader must look elsewhere, in the specialized literature of medicine, veterinary medicine, agriculture and pure microbiology. There is a selected list of diseases and the microbes which cause them in the box overleaf. Though a few illnesses, such as tuberculosis or mumps, are attributable to a single type of microbe, an important message which this list conveys is that different microbes can cause similar symptoms when they establish an infection. A very clear example is diarrhoea, which can be caused by bacteria, viruses, protozoa or occasionally even fungi – not to mention something inanimate, such as castor oil. The names given to diseases, especially those which have long been recognized, are more often derived from their symptoms than from the microbes that cause them. So pneumonia, which means an inflammation of the lungs, can be due to one of a variety of microbes: bacteria, viruses or fungi; meningitis, an inflammation of the brain lining, can be bacterial or viral. In all instances, the first step in medical diagnosis is to spot the likely cause of the disease from details of the intensity, complexity and provenance of the symptoms, and then to follow up with whatever treatment and/or further tests may be indicated. However, I shall make no attempt to provide a diagnostic handbook here; I shall naturally discuss specific diseases and the microbes responsible for them, but in general I shall be more concerned with why microbial diseases happen at all, how they are spread and how they may be avoided or treated.

Disease is a sort of parasitism, but an inefficient sort. The

Microbial diseases and their agents

Here are some examples of familiar microbial diseases and their agents. All feature in this book, most of them in this chapter. The microbes listed all attack people and some attack other animals too. The abbreviation 'spp.' means that the genus quoted includes more than one, and sometimes several, species which cause similar diseases, as well as species harmless to man.

Diseases caused by bacteria:

Diarrhoea. This condition is a symptom of many diseases, including serious ones such as typhoid, dysentery and cholera. Transient attacks may be caused by *Salmonella* spp., *Campylobacter* spp., *Listeria* spp. or certain strains of *Escherichia coli*.

Pneumonia. This, too, is a symptom of many disorders. The two most common bacterial forms are 'classical' pneumonia due to *Streptococcus pneumoniae* and legionellosis caused by *Legionella* spp.

Tuberculosis. Caused by *Mycobacterium tuberculosis*.

Leprosy. Caused by its relative *Mycobacterium leprae*.

Whooping cough. Caused by *Bordetella pertussis*.

Plague. Caused by *Yersinia pestis*.

Syphilis. Caused by *Treponema pallidum*.

Gonorrhoea. Caused by *Neisseria gonorrhoeae*.

Tonsillitis. Caused by *Streptococcus pyogenes*, which can also cause sore throat, scarlet fever or erysipelas.

Boils, spots and pimples. Caused by *Staphylococcus aureus*. Its relative, *Staphylococcus albus*, lives harmlessly on the skin.

microbes which live on the human skin, in the mouth and in the intestines are mostly parasites; they make use of their host as a supplier of food and warmth and provide little or nothing in return. (Some intestinal bacteria actually contribute B vitamins, as I shall tell in Chapter 5, thereby graduating to the status of symbionts.) Though unusually heavy blooms of skin, mouth and intestinal bacteria can cause rashes or discomfort, these parasitic microbes generally cause the host no harm at all, and they probably do some good by consuming materials that would otherwise support the growth of more damaging microbes. As parasites, these microbes are perfectly adapted to

Diseases caused by viruses:

Diarrhoea. Caused by a variety of viruses loosely called enteroviruses.
Pneumonia. Caused by members of several virus groups.
Mumps. Caused by a type of myxovirus.
Measles. Caused by a morbillivirus.
Influenza. Caused by groups of myxoviruses.
Common cold. Caused principally by rhinoviruses.
Smallpox. Caused by *Variola*.
HIV. Caused by human immunodeficiency virus, leads to AIDS (see p. 71).

Diseases caused by fungi:

Thrush. Caused by *Candida* spp.
Ringworm. Caused by a group of fungi called *Microsporon*.
Pneumonia. Caused by *Pneumocystis carinii*; can be a symptom of AIDS.

Diseases caused by protozoa:

Diarrhoea. Caused by several types including *Cryptosporidium*; amoebic dysentery is a serious hazard in the tropics.
Malaria. Caused by *Plasmodium* spp.
Sleeping sickness. Caused by *Trypanosoma* spp.

their hosts: they live in their little microcosms peacefully causing no one any harm, and this is the case with the normal commensal microbes found in association with all living creatures. Disease occurs when a microbe finds its way into a host, or some part of a host, to which it is imperfectly adapted yet within which it finds it can grow and flourish. When this happens, the biological defence processes of the host are brought into play and, if these are overstrained or unsuccessful, the host sickens and may die. Now, it is obviously a poor sort of parasite that kills its host. From an evolutionary point of view, when the host dies, the parasite's own microcosm is destroyed and all the parasites dependent on it are likely to die too. Thus the most perfectly adapted parasites, as I said before, cause little or no damage and it is the poorly adapted and inadvertent parasites that are dangerous and sometimes lethal.

If diseases are due to parasitic microbes growing in the wrong host, or in the wrong part of a host, then it should be possible to find a host, or part of a host, where they are harmless. In many instances this is true. The bacterium called *Bordetella pertussis*, which causes whooping cough, can sometimes be isolated from most healthy throats, and so can the *Streptococcus* species that cause sore throats and tonsillitis. They seem normally to be in a sort of balance with the hosts' defence mechanisms: seemingly without effort, the host keeps them at bay and only when some variation in the condition of the host occurs will disease strike. In the case of whooping cough, for instance, children who have this disease sometime in early life are subsequently immune, or so lightly susceptible that subsequent infections pass unnoticed, because at the time of the first infection the body developed defence mechanisms against *Bordetella* which it can return to for the rest of its life. What precisely those mechanisms are I shall discuss later.

Streptococcus pneumoniae, the bacterium which causes 'classical' pneumonia, looks like a string of spherical beads under the microscope. It can regularly be isolated from the throats of healthy people; related strains just give people sore throats in winter which, albeit unpleasant, are rarely life-threatening. The related species, *Streptococcus pyogenes*, also inhabits the throats of healthy people, yet it can cause the range of disorders noted in the box on p. 56 and, very rarely, a horrible but exceedingly rare disease called necrotizing fasciitis (malignant ulcer), in which breakdown of the patient's flesh spreads in a matter of hours. Happily, streptococcal infections generally respond well to appropriate antibiotics or drugs. Yet another common bead-like bacterium which occasionally turns nasty is the agent of bacterial meningitis. Called *Neisseria meningitidis*, or more colloquially 'the meningococcus', it seems also to cohabit with healthy people, inhabiting the nose or throat, without doing them any obvious harm. However, on occasions, again very rarely, it will cause its rapid and usually lethal disease. (Most often in infants, by the way, though occasional outbreaks among students have recently

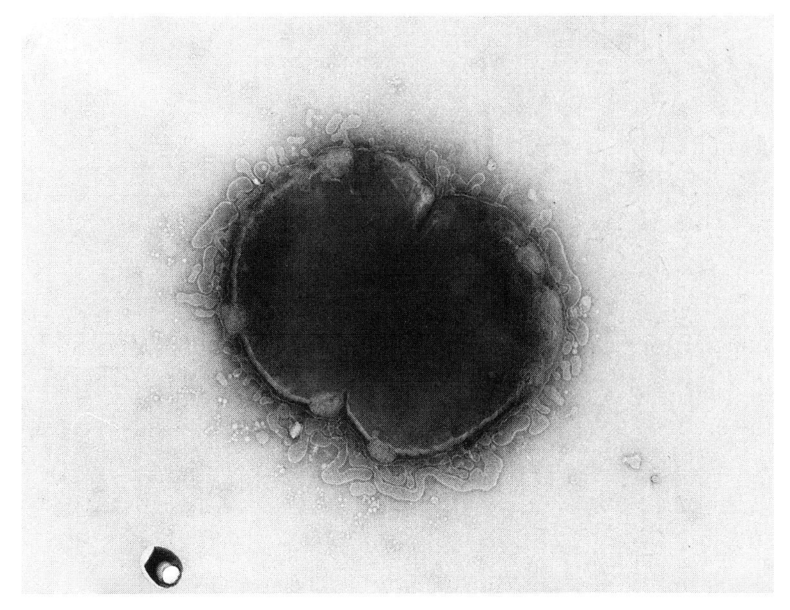

ELECTRON MICROGRAPH OF *Neisseria meningitidis*. The bacteria, which cause meningitis in susceptible people, exist as pairs of spheres – called diplococci – of about 7 μm diameter. The protuberances or blebs in the membrane surrounding the cells contain the toxins which cause the disease's symptoms. (Courtesy of Professors C. A. Hart and J. R. Saunders)

attracted press publicity.) It is the speed with which bacterial meningitis progresses which presents the most serious medical problem: as with streptococcal infections, drugs are available which can cure it provided the infection is diagnosed quickly, but since the early symptoms are rather like mild 'flu this is not easy. Vaccines can be made which prevent most forms of bacterial meningitis, at least for a few years, and vaccination is usually offered to those who have been in contact with a patient – though in fact the disease seems not to spread very readily from person to person. Why these normally tame, or at worst mildly irritant, bacteria should occasionally become virulent in certain unfortunate people is still obscure. Outbreaks of necrotizing fasciitis and bacterial meningitis are still, I repeat, very rare, but they understandably cause widespread alarm.

As a much milder instance of a comparable phenomenon, the skin and the inside of the nose are populated by staphylococci, tiny spherical germs which can be seen under the microscope to grow in clusters. The majority are *Staphylococcus albus* (the white staphylococcus), which is quite harmless, yet among them we will often find a few of its yellow brethren (*Staphylococcus aureus*), which can cause pimples, boils and more drastic skin conditions. For some reason that is not understood – and, since this type of skin infection is prevalent during adolescence, it is probably a change in the host and not the microbe that allows it – hair follicles or sweat glands can become infected with the yellow pyogenic (pus-forming) cocci and the familiar, unpleasant sequence of rash formation, inflammation and eventual exudation of a mass of pus and debris takes place.

In our intestines we have quite a balanced microbial flora and, despite modern hygiene in handling foods and general cleanliness of habits, there is little doubt that these bacteria get passed round from individual to individual in families and communities. So our bodies develop immunity to the local intestinal microbes and we all live together in relative harmony. One does not always have the same fellow travellers. A newly-born baby is almost sterile, carrying (both inside and out) the bacteria derived from its mother's vagina. Very soon it picks up lactobacilli – doubtless *via* its mother's milk – and only gradually is the adult population of mixed bacteria – *Escherichia coli*, clostridia, *Bacteroides*, lactic acid bacteria, yeasts of the genus *Candida*, and so on – established. That population, incidentally, seems to be beneficial in that it prevents less desirable microbes from gaining a foothold, so to speak. Some authorities advocate the 'probiotic' use of cultures of appropriate bacteria to alleviate gastrointestinal disorders in convalescent patients – and to control epidemics in agricultural flocks.

Normally we live in peaceful equilibrium with our personal intestinal microbes. Travel abruptly to another country, however, and be a little incautious in eating or drinking over the first few days, and a sometimes catastrophic rearrangement of the intestinal flora may take place. How many gastronomic delights have been spurned after one tasting by travellers who did not realize that *E. coli* from, say, Paris, Rome, Cairo or Bombay was not quite the same, as far as their bodily immunity was concerned, as *E. coli* from, say, Finchley, London? There are, of course, troublesome intestinal bacteria to be found abroad – those causing dysentery, typhoid and paratyphoid, even cholera – but it is almost certain that the majority of 'Gyppy tummy'-like diseases suffered by travellers are not caused by virulent pathogens; they are mostly due to ordinary, locally harmless, microbes that have suddenly found a host whose immunity to them is faulty.

If I may digress into quackery for a moment – and I must assure the reader that it is quackery, for I have no medical qualification – there is a simple routine for travellers which will considerably lower their chances of getting this kind of vague intestinal infection. There is no real chance of avoiding these microbes, so the thing to do is to admit them in as small a dose as possible, to avoid non-microbial disturbances of the gut, and build up a normal immunity painlessly. In practice this means that one should avoid gastronomic excess during the first few days. Drink the local water, for instance, but in small amounts at first; choose freshly cooked dishes; wash fruit and so on. In a few days you should be able to gorge yourself gluttonously on the local delicacies with no more disastrous consequences than you could reasonably expect in your own home.

Escherichia coli is a major and perfectly normal inhabitant of the lower intestines of ourselves and most mammals: we deposit billions upon billions of them into our sewage systems daily, where they die fairly rapidly. Their presence in rivers, soil and foods is the basis of a microbiological test for faecal contamination which is widely used by public health laboratories. The vast majority of strains of *E. coli* do us no harm, but just as unfamiliar *E. coli* strains, encountered in remote countries for example, can be troublesome, so *E. coli* from other mammals can sometimes do us harm. An example, which has become a particular problem in the last twenty years or so is a strain called *E. coli* O157. There are hints that it originated in Latin America, but it is now endemic in cattle all over the world. When slaughtered animals are being cut up *E. coli* O157 can easily get from the animal's intestines on to the surfaces of butchered meat and, if it somehow survives subsequent handling and is eaten, it can give one acute, bloody diarrhoea, and even cause kidney failure and death in a few patients. Several serious outbreaks occurred during the 1990s; one was traced to undercooked hamburgers in the USA (1992–3), another to centrally distributed school meals in Japan (1996), a third to cross-contamination of cooked meat dishes from fresh meat in Scotland (1996–7). In the Japanese outbreak some 10,000 people became seriously ill and thirteen died; in Scotland only 500 were ill but eighteen died. Rigid attention to hygiene when handling farm produce in bulk, especially fresh meat, is an important preventative precaution; so, of course, is thorough cooking of meat – more about that later on in this chapter.

Our relations with our personal intestinal microbes can sometimes become touchy. Duodenal ulcers were for many years thought to be stress disorders: self-inflicted physical damage brought about by anxiety, pressure of work, irregular meals and generally trying to do too much, and it is true that they are very common in conscientious and ambitious people. There was no real cure: diet and sedatives, or radical surgery for the less fortunate, were the only available remedies. However, in the last fifteen or so years medical scientists have realized, with some disbelief at first, that the usual cause is a bacterium now called *Helicobacter pylori*. It is acquired in youth (house-flies have been suggested as a likely source) and it settles within the lining of the duodenum, or sometimes the stomach, becoming almost dormant and quite difficult to detect. But it remains alive and produces an irritant (probably ammonia) that inflames that lining. Most infected people seem to adjust to the situation and suffer little more than chronic indigestion, but in a minority the inflammation leads to ulcers and their further consequences. Infection by *Helicobacter* is quite common – hence the profitability of marketing indigestion remedies to the general populace – and the question of why some people develop a serious and debilitating condition while others only burp rather often is still debated by experts. The good news is that a fortnight or so's treatment with a cocktail of antibiotics can nowadays cure most sufferers.

The story of *Helicobacter* illustrates a situation in which pathogenic bacteria settle into some special niche within us and, instead of causing acute illness, bring about a long-term, often slow-moving, malaise. The *Mycobacterium* which causes tuberculosis has long been known to be capable of doing this: mass X-ray surveys in the mid-century showed that some people had had a mild infection for several years, and recovered spontaneously, without ever knowing it. However, in the majority of patients tuberculosis 'flares up' quite quickly. Another microbe which is now suspected of causing a slow but widespread 'stress ailment' belongs to a genus known as *Chlamydia*, bacteria so small that they were once thought to be big viruses. They are curious pathogens which have an infectious free-living form but which invade the cells of a host and then change their form, remaining live and active therein. Chlamydiae of differing species may invade the cells of the eyes, lungs or urinary system, ordinarily causing acute diseases, but there is evidence that one species colonizes the walls of the arteries, promoting gradual fat deposition and consequent heart attacks. In this instance, the case is not proven, but the evidence that least a few chronic diseases are caused by bacteria makes one wonder how many other bewildering long-term disorders, such as arthritis and chronic fatigue syndrome, have long-term microbial origins.

I have become side-tracked from my discussion of where pathogenic bacteria go when they are not being pathogenic. With some, then, we know they are carried around by healthy people who have become immune. With a killer disease such as typhoid, immune carriers can be extremely dangerous and antisocial. The case of Typhoid Mary is celebrated in medical history: she was a New York cook, who, though immune to typhoid herself, managed to infect numerous

Americans in the 1920s and who, refusing to believe that she was the source of the trouble, insisted on returning to her former trade under assumed names, with disastrous results. She was at large for twenty-three years after her recognition as a typhoid carrier, but was eventually detained. Nowadays, with the aid of antibiotics, carriers of typhoid can usually be cured, but even this is a long and tedious process, most unwelcome to the unfortunate carrier, who feels perfectly well throughout.

Animals are quite often the reservoirs of human diseases. *Brucella abortus*, which causes undulant fever in humans, is a bacterial pathogen of cattle, and *Francisella* (earlier *Pasteurella*) *tularense*, the cause of a rare but highly lethal disease called tularaemia, is endemic among certain rodents (e.g. ground squirrels in California) and transmitted to humans by tick bites; a spirochaete called *Borrelia burgdorferi*, which causes a rare arthritic condition called Lyme disease, is also transmitted by the bites of ticks, which are spread by red deer, mice or birds. Sleeping sickness, an African disease of cattle, is caused by one of two species of parasitic protozoa called *Trypanosoma* and is transmitted to man by the bite of the tsetse fly. *Plasmodium*, a genus of protozoa which includes a couple of species responsible for malaria, is by now well known to be transferred from patient to patient by mosquitoes, and recently virus diseases, forms of dengue, have been found to be transmitted by mosquitoes, too.

Rodents such as mice and rats can harbour and transmit microbes that cause gastro-enteritis, and the association of rats with bubonic plague (caused by the bacterium *Yersinia pestis*, earlier called *Pasteurella pestis*) is now a matter of history. In 1665 London suffered from the Great Plague, during which the bulk of its population died of the Black Death – the medieval name for bubonic plague. Some believe that the nursery rhyme *Ring-a-ring o' roses* enshrines the folk-medicine myth that a posy of aromatic herbs and flowers, by disguising the stench of death and decay, somehow protected against the plague. In fact, rats only carry plague from place to place and the infective agents are rat fleas, which transmit the disease from infected rodents to man. Even today bubonic plague persists in parts of Asia and, during the Vietnam War, it became a serious problem, with over 2,000 cases diagnosed in South Vietnam in the first six months of 1965. War, strife and the plague go together: in 1947, 57,000 people died of plague in one Indian state during upheavals following independence.

Animals can, of course, be carriers of virus diseases. Foot and mouth disease, a virus infection of cattle, may occasionally infect man. Rabies (earlier known as hydrophobia or mad dog's disease) is particularly dangerous to man but endemic in some mammals such as squirrels and, it is said, vampire bats. It was all but eliminated from Western Europe in the first half of the twentieth century but then began to creep back from the East, carried principally by foxes. Thanks to vigorous quarantine regulations it was kept out of the UK. More recently, deliberate spreading of fox baits laced with an anti-rabies vaccine which is effective when eaten has been encouragingly successful in controlling, and in some regions eliminating, rabid foxes in mainland Europe, and it might soon be safe to ease the UK's quarantine measures.

Though animals, insects and carriers can act as reservoirs of infection in many instances, such reservoirs need not be living things. Soil and waters contain pathogenic microbes: cholera is caused by a virulent form of a bacterium which inhabits estuaries. They may also contain what microbiologists call opportunist pathogens: microbes that are not ordinarily pathogenic but which, if they find a damaged host (such as a sore or a recent cut on a person) can establish themselves there and delay healing or, at worst, cause disease. Tetanus and gas gangrene are diseases of this kind, due to microbes which normally live in quite different habitats and which would not affect healthy tissue, but which can establish themselves in wounds and cause dire consequences. I shall say more about them shortly. Many milder opportunist pathogens are known; usually our natural immunity copes with microbial opportunists so quickly that we do not notice them. A remarkable example of an unexpected reservoir of infection arose in 1976 when, after a convention of US Legionnaires in Philadelphia, a number of participants succumbed to a mysterious pneumonia which was given the name Legionnaire's disease (legionellosis). The organism was very difficult to culture in the laboratory and corresponded to no known pathogenic microbe. However, by 1979 it had been identified, given the name *Legionella* and cases were being discovered all over the world. Two or three in a hundred cases of pneumonia in the UK are apparently legionellosis: mostly the disease is mild, though people over 50 tend to be more susceptible. Probably many cases go undiagnosed as vague kinds of 'flu in younger people.

Where did this apparently new pathogen come from? Microbiological detective work established that its natural habitat is water and it is often found, though in small numbers, in domestic, hotel and hospital water supplies – particularly warm water. It is killed by chlorination and by really hot water, and so is absent from good tap water and hot tank water. But cool or tepid tanks from which the chlorine can evaporate, particularly if they have a sludge, are likely to contain it. The water reserves of air-conditioning systems are one such site; the water tanks of communal showers and washing facilities are another, particularly after a period of disuse. It now seems likely that the original outbreak in Philadelphia occurred because the water in the air-conditioning system became infected, so a fine spray of droplets (an aerosol) carrying *Legionella* was blown steadily into the conference hall. Many participants must have been immune, but a high proportion of the legionnaires were elderly and the incidence of the diseases was especially high. Since about 1979, several comparable examples have been identified – tanks feeding showers in Mediterranean hotels, air-conditioning systems in hospitals and offices, humidifiers and cooling systems. Fortunately, even susceptible people need to inhale quite a heavy dose of *Legionella* to get the disease and the situation is readily cured once it is detected. The conclusion? *Legionella* has probably been around as long as man has, probably longer, but modern life has given it new opportunities for pathogenic effect.

Not all diseases have such clear reservoirs. Venereal diseases, virus infections such as the common cold, poliomyelitis, mumps and so on seem to have no clear origin or reservoir. In these cases it is probable that, in ordinary communities, there are at all times a number of people with clinical infections who act as reservoirs of disease. Some authorities, but not all, believe that syphilis, a venereal disease caused by a spirochaete, was unknown in Europe until Columbus's crew brought it back from Haiti in the late fifteenth century. It remained unknown among the South Sea Islanders until the visit of the *Endeavour*, captained by Captain Cook, in 1769. The disease is only, or almost only, transmitted by sexual intercourse and, having been introduced to Polynesia by the Europeans, it rapidly became endemic in that part of the world, with the ghastly degenerative consequences, both physical and mental, that characterize its later stages. The common cold, too, is probably kept in circulation by persons who contract mild infections during the summer and, in special environments, it can die out altogether. Persons who spend a year or two in the Antarctic research stations usually cease having colds within a few weeks, despite the climate, but the arrival of a supply ship can trigger off a new round of colds throughout the whole community. The islanders of Tristan da Cunha proved remarkably susceptible to colds and bronchial disorders after their transfer to Britain in 1961, when their local volcano became overactive; this probably contributed as much as 'beat' music and the stresses of our society to their understandable desire to return to their island.

The reserves of infective microbes provided by any densely populated community are quite dismaying when one considers even familiar virus diseases. Scientists can recognize the influenza virus, for instance, by the kind of immune reaction that patients who have recovered from it have developed, and over the last few decades it has become clear that not one type but a huge variety of viruses cause respiratory disorders, ranging from colds to influenza. Such afflictions are subjects near to the hearts of most of us, especially in winter, so I shall attempt some indication of the problems they present.

The throat can harbour a great number of viruses, which fall into three main groups. Adenoviruses, particularly prevalent in the tonsils, can cause sore throats, and by 1975 thirty-one different types were known. In the mucus normally coating the tissue of the nose and throat are found myxoviruses, viruses that may be recognized relatively easily in the laboratory, because they cause blood cells to clump; among them are the influenza and mumps viruses, as well as organisms that cause mild, influenza-like diseases. Then there are innumerable types of very small virus that have been called picornaviruses and a subgroup of these, the rhinoviruses, includes some of the causative organisms of the common cold. Unhappily, there are at least 90 types of rhinovirus. A second group of picornaviruses is the enteroviruses, also found in the throat but distinguished by being found in the intestinal tract as well. Some of these cause sore throats and chest infections (e.g. the 30 known types of Coxsackie virus, named after the town of Coxsackie, USA); others, the echoviruses, either cause respiratory infections or seem to have no harmful function at all. One is forced to the conclusion that among viruses there exists an enormous variety of types of which only a few are pathogenic but that, unlike most of the larger microbes where one kind of organism generally causes one kind of disease, many different types of viruses can cause similar diseases. This conclusion is rather depressing from the point of view of developing immune reactions: if there are 30 kinds of common cold one can have, and we know that immunity to colds does not last long, what prospect is there for control of this trivial scourge? No doubt an answer will be found, but the reason why progress is slow should now be fairly obvious.

As to influenza, the situation is different but quite as troublesome. An especially virulent strain of influenza virus spread all round the world just after the First World War, causing the death of some 20 million people in 1918–19. It originated in China; apparently pigs were the reservoir of the disease and somehow it became transmitted to humans, and then it spread like wildfire because no-one had an immunity to it. The same strain still causes disease occasionally, but it has never caused so drastic a pandemic since. (A pandemic, incidentally, is simply the name for a worldwide epidemic.) Other 'flu pandemics have also spread from the far East; the 'Asian 'flu' pandemic of 1956–7, an altogether milder disease, probably had a similar source. However, pigs are not the sole reservoirs of 'flu viruses; in 1997 an influenza of chickens, which die rapidly if they catch it, infected about eighteen humans in Hong Kong, six of whom died. Happily containment measures, including a chicken cull, stopped its spread.

Avian 'flu actually kills chickens. However, despite the scary death tolls recorded in the famous world pandemics, it is probable that influenza in humans has rarely, by itself, killed anyone. What the infection usually does is pave the way for other more lethal infections by stressing our general immunity. Then so-called secondary infections join in; over the century secondary pneumonia, either bacterial or viral, has been the most usual complication to lead to death. This is the reason why some doctors prescribe antibiotics for patients with 'flu: they know well that viruses, including 'flu, are unaffected by antibiotics, but their medication is intended to keep any bacterial secondary infection at bay while the body copes with the viral infection in its own way. So do not expect a miracle cure from antibiotics if you simply have 'flu!

Many familiar virus diseases such as measles and 'flu attack within days and are largely delocalized: their effects are felt all over the body as a cluster of symptoms – headaches, snuffles, fever and so on. But some virus infections are slow and localized. The most familiar instance, I imagine, is the common wart. Caused by a so-called papillomavirus, warts can be spread by contact and will take six months or so to generate a visible lump. Common skin warts are not in fact very contagious: the virus has to reach a layer of tissue beneath the outermost surface of the skin to establish itself, and more often than not it will be sloughed off, or even washed off, before it gets there. Human papillomaviruses mostly cause harmless, if unsightly, lesions, which sometimes appear and disappear capriciously, but a subgroup which infects the genitalia is dangerous: it is now known to be associated with cervical cancer, which is still, despite widespread screening, a worldwide cause of premature death in women. There are numerous different papillomaviruses capable of infecting diverse species of mammals, birds and even reptiles; they are not generally transmissible between host species.

Turning to more exotic diseases, I have already indicated that international travel is a great mixer, microbiologically as well as socially. The local pathogens find new and susceptible hosts among the visitors. Many cause disorders which, albeit troublesome at the time, are only transient: they yield sooner or later to their unwilling host's natural resistance, sometimes helped by modern pharmaceuticals. But others, such as malaria and some tropical diarrhoeas, can be persistent and dangerous. It was always so, but in recent decades the risks have become more widespread because remote and exotic places have been opened up to tourism, and the speed of air travel has much increased the risk of local diseases being imported or exported by travellers who are incubating an infection.

In the 1950s and 1960s I was caused to travel rather widely and my passport contained a little wad of certificates of immunization against regional diseases – smallpox, typhoid, cholera, yellow fever; these were common-sense precautions against foreseeable illnesses. Unforeseen hazards remain, however. One which generates great concern when it breaks out is ebola disease. Caused by a filamentous virus, it has been known since 1976 when, appearing first among nurses at a mission in Zaire, it ultimately infected at least 278 people, killing about 90 per cent of them; it also appeared in the Sudan in that year. Early in 1995 a new outbreak occurred, in Zaire again, which killed over 170 people before being contained. No reliable therapies or vaccines exist, and ebola kills very rapidly in a gruesome manner. Its lethal character usually means that an ebola outbreak limits itself before it can spread far among the sparse or relatively isolated populations where it has occurred. However, one case from the 1995 ebola outbreak had travelled by air to South Africa before being diagnosed and quarantined, an event which understandably caused great public disquiet, because the consequences of a disease as lethal as Zairian ebola reaching a densely populated community will hardly need spelling out to readers. (Those consequences formed the theme of at least two widely-read works of fiction published in 1994–5.) Containment procedures exist, but spotting the disease in time to mobilize them still causes anxiety among experts. Needless to say, research on such microbes is difficult and dangerous, and requires the most rigid of containment precautions.

Where does the ebola virus come from? There is another highly lethal African virus, called the Lassa fever virus, which is carried by the local rodents, and yet another lethal virus, the Marburg virus, came from African green monkeys – they were imported into Germany for making poliomyelitis vaccine in 1967. But the natural reservoir of ebola is not known. Monkeys have come under suspicion but it is almost as lethal to them as to humans, so they are unlikely to be primary sources. *Ex Africa semper aliquid novum* (loosely: 'there is always something new from Africa') – but not only Africa. In 1989 a variety of ebola, fortunately less pathogenic to humans, killed a group of macaque monkeys in the USA, which had been imported from Asia.

Viruses can even disrupt the immune system. In 1981 a disease was discovered among homosexual men on the West Coast of the USA, which causes a breakdown of the host's immunity so that, sooner or later, he dies of an infection by another microbe. More than 95 kinds of infection capable of causing death in such patients are now known; the most common are 'opportunist' infections, such as a kind of fungal pneumonia (p. 57) which healthy people resist. The microbe responsible for the breakdown in immunity was identified as a virus, now called human immunodeficiency virus (HIV), by Professor Luc Montagnier in Paris and Dr Robert Gallo in the USA, more or less simultaneously. When HIV infects someone it may remain apparently dormant for a few years, but ultimately the great majority of patients develop AIDS (acquired immunodeficiency syndrome): the infection which leads to death. HIV is present in the blood, semen and saliva of its host, but it is not infectious or contagious in the usual sense; it is transmitted only when infected body fluids reach the blood of a new host, as when semen and blood mingle as a result of anal intercourse between male homosexuals. By the mid-1980s, AIDS had reached the heterosexual population of North America and Europe, though cases were very few, brought in by bisexual men, who transmitted it to women, notably prostitutes, and also by addicts who shared needles for intravenous drug abuse. At the same time as AIDS was spreading over the northern hemisphere, a similar disease was found to be reaching epidemic proportions in sub-Saharan Africa, infecting heterosexual men and women as well. The African HIV virus is detectably different from the American one, but HIV, like the influenza virus, can change readily. Such evidence as there is suggests that HIV/AIDS is a truly new disease which originated in the 1950s, in Africa, by transmission from a primate.

During the 1980s and 1990s HIV infection leading to AIDS became the worldwide public health threat with which most readers must be familiar. It has continued to spread, especially in the heterosexual population, among whom, curiously, it is the African variety of HIV that has become the more common: the inter-group transmission of HIV noted during the first decade of the epidemic has apparently become rare. It has generated considerable public anxiety, not only because it impinges on basic human urges, morality and taboos, but because it is still incurable and generally fatal. The outlook improved appreciably in the mid-1990s, at least for patients in affluent countries, when combinations of drugs were developed which delay the onset of AIDS in HIV-positive patients for several more years, but in the longer term drug intolerance and serious side-effects remain a problem. In world terms HIV/AIDS remained a relatively minor epidemic for about a dozen years, but as the 1990s progressed it has exploded in Africa, causing nine-tenths of the 2.3 million deaths from AIDS recorded by the United Nations in 1997. In Botswana and Zimbabwe a quarter of the population is said to be HIV-positive, and the African disease seems set to rival malaria and tuberculosis as a major global scourge. Medication is expensive and transient and the disease remains incurable; prevention is still the only remedy for HIV infection, which raises the intractable problem of changing people's behaviour and attitudes.

Several milder virus diseases tend to leave patients susceptible to secondary infections. The name post-viral fatigue syndrome has been given to a cluster of disorders, of which an example is ME (myalgic encephalomyelitis, misleadingly called 'yuppies disease'); in that instance patients are especially prone to develop thrush, in the mouth or vagina, caused by the yeast (*Candida albicans*) which has hitherto lived harmlessly about their persons. Such immune deficiencies occur in animals, too. In 1988 a remarkable epidemic afflicted North Sea seals, both grey and common varieties, killing some 14,000 in all. The primary cause proved to be a hitherto unrecognized virus resembling one which causes distemper in dogs (it was named phocine distemper virus) but in fact death was caused by a variety of secondary bacterial infections: the virus had weakened the seals' immunity. The catastrophe caught the public's imagination and the press aired a variety of rather wild theories: that sewage pollution, over-fishing, organo-chlorine residues or global warming had precipitated the epidemic. However, by 1989 it had all but disappeared, and the seal population was recovering, without any of these factors changing. Its origin remains a mystery.

Why do microbes grow in the throat, intestines or wherever we find them? In general, we do not know the answer to this question, but there are one or two instances in which we do. I mentioned earlier the bacterium *Brucella abortus*, which can cause undulant fever in man. This microbe, however, more usually causes contagious abortion in cattle, a disease in which the pregnant cow aborts a

A SEAL HAS DIED OF PHOCINE DISTEMPER. A dead seal being collected during the 1988 epidemic of this virus disease. It led to the death of some 3,000 seals around Britain's North Sea coastline alone. (Courtesy of the Sea Mammal Research Unit)

stillborn calf. When this happens, the *Brucella* is found to inhabit almost exclusively the placenta, the organ attaching the embryo calf to its mother's uterus; the rest of the cow and the calf are relatively free from the pathogen. In the 1960s Professor Harry Smith and his colleagues discovered the reason for this. *Brucella* normally requires a number of vitamins and such trace materials in order to grow properly, and among these is a sugar-like substance called erythritol. Erythritol is fairly rare in animal tissues but, for reasons that we do not wholly understand, it is plentiful in the calf's placenta. Hence the infection flourishes there but not elsewhere, and since the placenta is, as it were, the embryo's lifeline, it dies, and duly the mother's uterus expels it.

The reason for the action of *Brucella* in contagious abortion is a particularly clear example of what is called the specificity of infections: the fact that microbial diseases are often localized. By injecting erythritol artificially into experimental animals, for example, Professor Smith and his colleagues were able to induce a generalized brucellosis. Microbes grow in a special location, and perhaps cause disease, because some nutrient they need can be found only there or because something they dislike is absent. Another example of the former case is *Corynebacterium renale*, which causes kidney disease in cattle. It grows only in kidneys because it has a particular affinity for urea, which, as a component of urine, is primarily concentrated in the kidney. An example of the case where microbes flourish because something they dislike is absent is gas gangrene, mentioned earlier in this chapter. It is a putrefactive disease of wounds which can occur after serious injury. The microbe responsible, *Clostridium welchii*, is fairly common in polluted waters and soils, but is normally harmless because it is an anaerobe: it does not grow if air is present (see Chapter 2). However, it forms spores which survive in air. Wounded tissue usually has its blood supply interfered with, if only because the small blood vessels are damaged and inflammation and swelling tend to squeeze them tight so that the flow of blood is restricted. Consequently, the supply of oxygen brought by the blood to wounded tissue may be low and, if the wound is extensive, conditions in the damaged tissue may become quite anaerobic. If, now, spores of *Clostridium welchii* have got in accidentally from external contamination, they find the situation much to their liking and grow. Quite incidentally these bacteria make a substance (a toxin) which is highly poisonous to tissue and which extends the area of damage rapidly.

A curious application of this principle – the ability of normally harmless anaerobes to grow in tissue that is deficient in oxygen – was tried in the mid-1960s for the treatment of cancer. Cancerous tissue can be deficient in oxygen because it grows relatively rapidly, and regression was induced by deliberately infecting certain cancers with the normally harmless *Clostridium butyricum*. The cure was unhappily incomplete and impermanent. However, the idea was revived in the 1990s with a new slant: it is possible, by genetic manipulation, to graft genes into bacteria which cause them to make substances which are toxic to cancer cells. A normally harmless microbe could thus be made into a selective anti-cancer agent. Experiments along these lines have apparently been successful in laboratory tissue cultures – but it is a long haul from these to real patients.

Returning to gangrene. In this disease the microbe is an opportunist pathogen: the fact that a product of the microbe's growth is toxic to the host is unfortunate for that host but irrelevant to the microbe, which is normally non-parasitic and, so to speak, uninterested in finding a host. Tetanus, caused by *Clostridium tetani*, has similar origin: a soil clostridium grows in wounded tissue because it becomes anaerobic and, quite incidentally, forms the powerful poison, called a toxin, that causes lockjaw. In this case the patient will almost always die by the time the symptoms of tetanus are detectable. This is why anyone suffering a deep wound in country or farmyard areas should immediately have prophylactic treatment against tetanus unless they have been properly immunized already – as most country children are these days.

An extreme case of this kind is the disease known as botulism. Here the microbe, an anaerobe called *Clostridium botulinum*, does not grow in the host at all but in infected canned or preserved meat or fish. But in growing, it produces a toxin which is one of the most powerful poisons known to man, which rapidly kills anyone who eats the food. The organism itself does not grow at all when eaten. Fortunately modern methods of food preservation are such that botulism is rare, otherwise we should all have to be immunized against botulinus toxin.

I am conscious that I have alluded rather glibly to the body's defence mechanisms against microbes and immunity to infection. What does this mean? The answer is fairly complex; in fact the body has at least four lines of defence. The first is an enzyme called lysozyme, which is found in saliva, tears and nose mucus and has the property of dissolving many bacteria. The second is a group of substances called interferons, proteins produced by virus-infected cells which interfere with the further growth of viruses. The body's third line of defence is based on the fact that the blood contains certain white corpuscles (leucocytes) which are rather like domesticated protozoa and live in the bloodstream. There are several varieties of white corpuscle, involved at various stages of our immune responses. The so-called phagocytes actually ingest alien particles and are able to kill and digest any extraneous microbes that get into the blood. The way ingested microbes are killed is interesting: macrophages, a type of phagocyte, make a simple but very unstable chemical, nitric oxide, which is a powerful bactericide. (Nitric oxide also proves to be a very important substance in general human physiology, relaxing the walls of blood vessels, promoting blood flow and decreasing blood pressure; perhaps destroying cancerous cells.) If a slight wound occurs, the damaged tissue causes these phagocytes to congregate near the site of damage and thus be ready to forestall infection. The body also has a system of cells, centred on the liver, called the reticulo-endothelial system, from which it can generate reserves of phagocytes if need be.

This is all very well, but a bacterial infection of the blood, for example, once well established, involves billions upon billions of microbes, far more than the phagocytes could possibly cope with. How, in such circumstances, does the body cope? The short answer is, of course, that it does not, at least at first. Massive microbial growth only occurs if the body's initial defences have been broken down, and then one is very ill and, if the bacteria produce particularly nasty toxins, one may die. If one recovers, the reason is that the fourth defence mechanism has been successful: the body has made certain proteins called antibodies which, dissolved in the blood stream, react with the invading microbes and cause them to coagulate in lumps. In this condition they do less harm and are more easily ingested by the phagocytes. The serum of the blood is now immune to the particular microbe and this immunity can be retained, sometimes only for a few months, sometimes for many years, even a whole lifetime. Colds and influenza, for example, seem to generate rather short-lived immunities; mumps, measles and such childhood ailments seem to confer lifelong immunity. Immunity is very specific: immunity to a virus such as that of mumps confers no immunity at all to that of poliomyelitis, though both diseases are due to myxoviruses. One of the few exceptions to this is the cross-immunity that exists between cowpox and smallpox: vaccination was originally the practice of deliberately infecting people with the almost harmless cowpox virus, to which they develop immunity and which also renders them resistant to the far more dangerous smallpox – more about that story shortly. BCG vaccination against tuberculosis makes use of a live but harmless culture of the tubercle bacillus to induce immunity against natural, virulent tuberculosis; the Sabin poliomyelitis vaccine is a live, non-virulent strain of the virus. But generally the medical profession, quite reasonably, prefers to induce immunity to disease by injecting microbes that have been killed in such a way that they can still provoke the immune reaction. Injections against bacterial diseases such as typhoid or diphtheria are of this kind.

Immunity can be developed against the toxins formed by microbes as well as against the microbes themselves, and a serum that has developed such an immunity is said to contain antitoxins or to be an antiserum against such a toxin. Antisera against tetanus and botulism are induced in horses and used in emergencies where there is a risk of these diseases; in such cases the patient acquires no permanent immunity to the disease, but in an emergency this does not matter.

For much of the twentieth century diseases such as mumps, measles and influenza had no reliable antisera, but, since most of the population is immune to these diseases most of the time, pooled sera from numbers of people would, if injected into a sufferer, alleviate the disease to a large extent by providing temporary immunity. This is the logic of the use of gamma-globulin, a form of pooled serum obtainable from blood banks, to treat such diseases in patients (such as adults with mumps or pregnant women with German measles) where it could be dangerous to let the disease take its natural course. However, in the last decade or two, good vaccines for forestalling such diseases have become available and pooled gamma-globulin is rarely needed.

Immunity is our major defence against most diseases, but even immunity carries its hazards. Allergies arise when, for a variety of reasons which are not well understood, some component of the environment sets up an immune response in which the system over-reacts. Mosquito bites and bee stings are of this character: the dogma is that the very first bite or sting passes unnoticed but generates immunity to substances injected along with the bite or sting, such that, on the next occasion, local over-reaction (e.g. swelling, inflammation, elevated temperature) causes irritation. One can over-react to microbes, and several illnesses are caused less by the products of the alien organisms than by an over-reaction of the patient's immune system to them. Tuberculosis has something of this character, so has pneumonic plague – both happily rare these days. But cases of the kind re-emerge: toxic shock syndrome is an acute allergy to *Staphylococcus aureus* (the yellow staphylococcus) which occurs mainly in women and is especially associated with the use of tampons. It is rare but devastating, and before its character became understood in the late 1970s, it caused great consternation in medical circles.

The specificity of immune reactions is extremely valuable to microbiologists and, in fact, sometimes provides them with the only available method of recognizing organisms. In 1964 there was a frightful outbreak of typhoid in Aberdeen, apparently because, when some infected canned corned beef had been cut on a mechanical slicer, the slicer had become infected and contaminated other cooked meat that was sliced on it. As a result, the organisms became spread widely among the customers of one particular food store, causing an explosive epidemic that reached over five hundred cases before it was contained. The whole story was a fantastic sequence of mishaps. It seems almost unbelievable, for example, that the original beef could have been as heavily infected as it was, yet not have been obviously bad, but later experiments, in which tins of meat were deliberately infected with typhoid bacteria alone, looked perfectly wholesome for three months. The manner in which the source of infection, a cooked-meat counter, was tracked down was also an impressive piece of detective work. (In one family, for example, everyone was infected except one person who, it transpired, hated corned beef and had eaten none.) But most impressive to non-scientists was the identification of the infective organism as a South American strain, and the consequent discovery that the original reason for infection of the beef was failure to use chlorinated water to cool the cans at the original South American factory. This identification was done partly by means of antisera: different strains of typhoid bacteria (*Salmonella typhi*) generate appreciably different antibodies, and a collection of antisera to various known strains is kept at the Government's Enteric Reference Laboratory in North London. Once a culture from the Aberdeen outbreak was available, its identification as a South American strain was a matter of routine (though a second character, the strain's susceptibility to a bacterial virus, was also used).

The way in which microbes react with antisera, and the sort of antibodies they generate, is called by microbiologists their antigenic pattern, and collections of antisera to both medical and non-medical bacteria exist which are used entirely for identifying and typing various microbes. The antigenic pattern of a strain of microbe may be said to have something of the quality of a fingerprint in man for purposes of identification: if one has it on a file somewhere, the chances of recognizing the culprit are extremely high.

Once the sources of the Aberdeen typhoid outbreak was tracked down, the means by which it spread was obvious. People actually ate the bacteria along with cold meats which had become contaminated by the slicer. Once the organisms got inside a patient, they multiplied and the disease took its normal course, leading to fever, vomiting, diarrhoea and so on.

Despite tremendous advances in chemotherapy, of which more later, stimulation of the immune response by vaccination or comparable immunization remains medicine's most powerful preventive weapon in the battle against pathogenic microbes.

The story of smallpox, though it has received considerable publicity over the last decade, is worth outlining again because it illustrates the point so impressively. Until the end of the eighteenth century the disease was so common in Europe, including Britain, that at least one child in ten would die of it, and in some parts of Europe it was apparently customary to delay naming a child until after he or she had recovered from smallpox. One in four people, child or adult, would die of it; most people would catch it at some time, many would be blinded, and all survivors would be in some degree disfigured by pock-marks. It was a major scourge which was always there, unlike the equally feared plague, which came and went. In June of 1796, long before microbes were suspected of playing any part in disease, a country doctor, Edward Jenner, showed that deliberate infection of an uninfected boy, James Phipps, with matter taken from a cowpox pustule on the hand of a milkmaid, Sarah Nelmes, protected the boy against smallpox. It was an experiment based largely on rumour and intuition (one which no laboratory safety or medical ethics

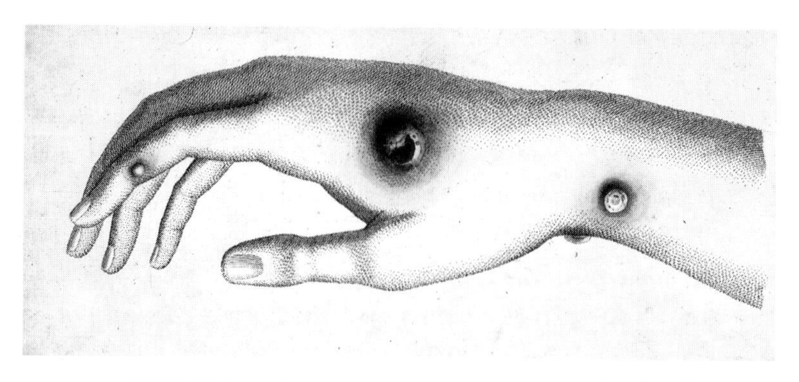

ENGRAVING OF THE HAND OF SARAH NELMES, showing the cowpox pustules from which Edward Jenner vaccinated James Phipps in 1796. From Jenner's report of 1798. (Courtesy of Dr Baxby)

committee would countenance today!), but it was repeatable. Very soon vaccination became widely practised and, though it had its ups and downs during the nineteenth century, it was by and large a success. During the first half of the twentieth century, as a result of improved methods of preparing and handling the cowpox virus, it became even more reliable. Smallpox was eliminated from Britain between 1920 and 1950 by a programme of almost mandatory vaccination at a young age, coupled with a legal obligation on travellers abroad to have regular booster vaccinations. It was re-introduced accidentally on occasions, often by visitors from the Far East or Africa who were incubating the disease when they arrived, but at least once (in 1973) by escape from a research laboratory! It is known to be transferred from patient to patient by physical contact – to be contagious – but it also spreads in an unpredictable way without contact. Diseases that do this are called infectious diseases, and with smallpox the manner in which it got around was random and haphazard. In the middle of the twentieth century, the World Health Organization initiated a worldwide vaccination programme for its elimination. By May 1980, they were able triumphantly to announce success: the last pocket of natural infection had gone. Stocks of the smallpox virus, called *Variola*, were held for research purposes in progressively fewer laboratories until, by the early 1990s, the WHO was satisfied that the world's last remaining supplies were being kept, under strict security, in only two: in specialized laboratories in Atlanta, USA, and in Siberian Russia. Most microbiologists felt, and the Executive Board of the WHO actually decided, that these final stocks should be destroyed in 1996. However, not all virologists agreed; some argued that further research on the live virus could provide new and important information on the ways in which virus diseases of several kinds progress. So the WHO postponed action on its decision until 30 June 1999. One hopes agreement will then be unanimous, because routine vaccination has been abandoned and humanity is now a sitting target for the disease. Even the strictest security cannot wholly exclude release due to human failure or natural disaster.*

Vaccination has eradicated smallpox as a natural hazard. Next on the WHO's hit-list is poliomyelitis, which it hopes a comparable worldwide programme of immunization will eradicate by the year 2000. Good progress can already be recorded: in 1996 the Organization was able to announce that polio had been eliminated from the Americas, because the last recorded case on that continent had occurred in Peru in 1991. Elsewhere local wars are making complete eradication difficult but the WHO's millennial target had not been abandoned at the time of writing. Poliomyelitis is rather like smallpox in its behaviour: there seems to be a tendency to contract the disease, at least in the temperate countries of the northern hemisphere, during late summer and autumn, and quite frequently only one person in a family or group will get it, though all have, as far as one knows, been equally exposed to infection. In the case of poliomyelitis the route of infection is still unknown, but it seems likely that the virus is airborne, floating around on dried droplets of breath, saliva or other body exudates. Colds and influenza are certainly spread in this manner and are highly infectious. One well-aimed sneeze from a snuffly baby, as is well known, can lay low a whole group of admiring

* Regrettably, in mid-1999, the WHO delayed destruction yet again, until at least 2002.

adults. During the early 1940s, colds and respiratory infections were causing serious losses of working time in Britain's hard-pressed wartime industries, so a nationwide campaign was launched in which hoardings, newspapers, trains, surgeries, even school notice boards, carried the slogan: 'Coughs and sneezes spread diseases – catch the germs in your handkerchiefs!' It was a wise move (even if those with childish minds, such as your present scribe, irresistably ended it as 'handkercheeses'), because a handkerchief does afford some protection to others, largely by trapping spray. However, it seems very likely that dried mucus from a cold, preserved in a pocket handkerchief, can remain infectious for a long time and can even re-infect the original sufferer. Even a handshake from a cold sufferer can be a risky thing. Many families know well that, in a bad winter, they can keep what seems to be the same infection travelling round from person to person from November to April, becoming, to their friends and relatives, a snivelling group of red-nosed horrors. Yet when the Common Cold Research Unit at Salisbury (which was closed in 1993) tried to reproduce this sort of dissemination in laboratory conditions, they found it very difficult. Does a family really keep the same cold running all winter? Or do they just become unusually susceptible to various colds that are doing the rounds? Does mood influence whether or not one catches cold? (I often think so.) The answer is probably 'yes' to all these questions, but the truth is that we do not yet know. Until we do, it is sensible to switch to disposable paper handkerchiefs as soon as a cold develops, and to BURN them when used, not to put them in litter bins, waste-paper baskets and so on.

Many bacterial and most virus diseases are infectious. The protein of mucus, present in cough or sneeze droplets, preserves the microbes from the lethal effects of drying. Most streptococcal throat infections are spread this way. But another important route of infection is through the intestinal tract, and an understanding of why this is so reveals some disconcerting truths about our social and domestic behaviour, even in this relatively hygienic age.

Most people in developed countries cover their mouths when they cough, or sneeze into a handkerchief, and they know why they do this: to protect others from the infectious aerosol of droplets that a cough or sneeze generates. Fewer people, though still a great number, understand that they must wash their hands when they have used the toilet. Toilet paper, after all, is permeable to bacteria, and excrement, not to put too fine a point on it, is a pullulating mass of bacteria and viruses, many of which are potentially pathogenic. Few people realize, however, that when a used toilet is flushed, a turbulence and spray of water and excrement is generated comparable to a sneeze: in any toilet one can isolate faecal clostridia and streptococci from the ceiling, walls and door handle as well as around and beneath the seat. British water closets certainly generate such infectious aerosols; it is probable that the vortex type favoured in the USA, depending on a swirl rather than a splash to flush the closet, is less generous in the matter of dispersing faecal microbes around the room. Moreover, many public and domestic lavatory suites are designed with the wash-hand basin in a separate room from the toilet, which is convenient if two people wish to use the two facilities at the same time, but which means that the occupant of the WC must use an unwashed hand to operate the flushing system and open the door. The idea that washing facilities should always be available in the same room as the WC is penetrating only slowly to the British, though it is realized fairly widely in the USA and Scandinavia. Incidentally, it is in Sweden that I have encountered the only sensibly-designed toilet paper: a two-ply roll of which one ply, the outside, was the old-fashioned British type, smooth and impermeable (but relatively useless for its main purpose), and the inner ply was the soft, absorbent type which is now increasingly popular because of its good wiping properties, though it is extremely permeable to microbes. This combination, used the right way round, is probably the most hygienic material available.

While I am on these important if unsavoury matters, let me consider the average gentleman's public urinal. Urine is, in fact, normally a sterile fluid. Unless one has a kidney or urethral infection, there are no bacteria in fresh urine. The great Lord Lister, the pioneer of elementary hygiene in surgery and hospital practice, used fresh urine as a readily available sterile fluid in some of his crucial experiments on the spread of airborne bacteria. A urinal, however, is far from sterile: it is a culture of bacteria especially rich in types capable of growing in urine and of releasing ammonia from the urea therein. The customary design of urinals is such as to ensure a generous splash-back of these bacteria on to the shoes and trouser-legs of anyone using them, again contributing to the spread of both pathogenic and harmless bacteria. The cup type of urinal, by reducing the splash, is to be preferred from the point of view of hygiene.

The British are a dirty nation, as anyone who has travelled northwards or westwards knows. (But those who travel south or east reach an opposite conclusion, and we should be grateful that, despite our national reputation for dirtiness, our disposition to litter public places and foul our public conveniences, we nevertheless reach a high level of hygeine by world standards.) Other nations survive in apparent good health; if we spray ourselves regularly with a fine mist of faeces, does it really matter? Are we not thereby building up an immunity to infections to which we would otherwise succumb? The answer, of course, is that one can certainly be too fussy about these things. Some exposure to infection is essential for the acquirement of immunity. But countries with lower hygienic standards than ours still have diseases such as typhoid, dysentery and cholera endemic among their populations, and it is mainly due to our fairly elementary standards of hygiene that we are now free of these scourges. But, with the increasing freedom of travel among nations, diseases of this kind are carried anew with greater and greater ease to parts of the world that had eliminated them. It would probably be impracticable, at least during this century, to eliminate cross-contamination by faecal microbes altogether, but some common sense in the design and use of lavatory facilities is essential if Britain is to preserve its freedom from the nastier infections that may spread from the intestinal tract.

We eat and drink microbes all the time. Sometimes in large quantities (for example, in fermented foods and beverages), more often in small, indeed tiny, quantities, as fortuitous contaminants of ordinary food and drink. Eating and drinking provide microbes with instant access to the alimentary canal and, though most are killed by the lysozyme of saliva or by the stomach's acidity and then digested, some may get through if ordinary food or drink is badly contaminated. If they are harmful, we may become ill. Typhoid, cholera and dysentery reach epidemic proportions when drinking water becomes contaminated with faecal organisms, and throughout history all three have been common (and usually fatal) hazards of life in cities and crowded communities. Advances in public health and hygiene improved the position during the nineteenth century, and Britain suffered its last epidemic of cholera in 1866. Serious outbreaks occurred elsewhere in Europe until the turn of the century and, in some tropical and subtropical countries, well into the twentieth century. These three diseases are in decline, but they re-appear in times of war, disaster or deprivation. Early in 1991 cholera struck in Peru. It was quickly detected but, because of the catastrophic economic situation in that area, measures to decontaminate the water supplies could not be taken and a serious epidemic developed, which spread to neighbouring countries. Modern treatments – use of antibiotics and taking measures to compensate for dehydration – have greatly lowered the death rate from cholera, but, as with most diarrhoeal diseases, the problem is essentially one of public hygiene.

In Britain, as in most of the Western world, water-borne diseases are rare. Public health authorities spend much effort assessing the levels of faecal pollution of potential water supplies and water companies have bacteriological control laboratories to give early warning of such problems. Modern water treatment, based on chlorination and filtration, can usually cope, but less tractable situations arise occasionally. In times of drought, special kinds of cyanobacteria may grow in the depleted waters of reservoirs and, though the microbes themselves filter out easily enough, they make a toxin which tastes, and is, nauseating. In Oxfordshire, in 1988–9, the water supply suffered a rather unusual kind of contamination, which was due to a protozoon called *Cryptosporidium*. It is common in cattle and causes diarrhoea in man. The microbe is usually resistant to chlorination (though ozone will tackle it) but it is killed by boiling – the locals had to boil their drinking water for some months until the problem was cleared up. In 1993 the water supply of Milwaukee, USA, became infected with *Cryptosporidium* and about 400,000 people became ill. Though no specific drug treatment is known for the disease, the diarrhoea lasts for only about a fortnight in healthy people; however, during that period they can pass the infection direct to others by careless hygiene. In Oxford, the infection probably originated from farm waste leaking into the water supply. *Cryptosporidium* presents two special problems: it is difficult to detect because it does not readily grow in laboratory culture media, and only very small numbers are sufficient to cause sickness, so quite a number of people can be in trouble before the agent is identified. In recent years techniques using antisera have made detection of *Cryptosporidium* somewhat easier.

Food-borne diseases are six or seven times as common as those carried by water, but their occurrence is nevertheless quite rare considering the hazards food encounters during its distribution in the modern world. I imagine that the disease in this class which is best-known among the general public is salmonellosis, the general name for intestinal infections related to typhoid, but milder. The reservoir of infection is cattle but, notoriously, salmonellae often colonize poultry; they are rapidly killed by heat, but imperfectly cooked chicken, or other food contaminated as a result of careless handling of raw chicken carcases, are the commonest causes of salmonellosis. In my boyhood I remember being warned never to eat duck's eggs ('They give you typhoid!'), though they were acceptable for baking, a myth which probably arose because some cases of diarrhoea had been traced to duck's eggs. Happily I took no notice, because I like them. During the Second World War, imported egg powder sometimes contained salmonellae, but in those protein-starved days the risk of occasional outbreaks of mild paratyphoid was considered worth taking. British hens' eggs had an excellent record until about 1988, when a new type of salmonella (a strain of *S. enteritidis*) became widespread among broiler birds and occasionally infected their eggs. Despite something of a panic in 1988–9, when a junior Minister in the government spoke carelessly of the matter and had to resign, the risk from eating an undercooked or raw hen's egg remains very small if common sense is used: avoid old eggs and do not leave fresh mayonnaise, mousse, beaten egg whites and so on for hours in warm places.

Nausea, diarrhoea and/or fever, usually mild but occasionally serious, can be caused by bacteria other than salmonellae in foods, for example by *Campylobacter, Listeria,* some clostridia and bacilli, as well as by viruses. Campylobacters were only discovered to be responsible for diarrhoea in the mid-1970s, yet *Campylobacter jejuni* is now known to be a more common cause of mild intestinal disorders in man than *Salmonella.* A campylobacter attack is usually brief; diarrhoea and sickness rarely persist for more than a week and most people recover completely, but it is disabling while it lasts. The natural reserve of *Campylobacter* is not clear, but it is common around farmyards, for example, and is usually present in the intestines of chickens, which it seems not to harm at all. Between farm and shop, chicken carcasses inevitably become contaminated with intestinal microbes, so, what with salmonellas and campylobacters, raw chicken can often be a fairly infectious cooking ingredient – and this is true of both broiler-house and free-range birds. Fortunately both types of bacteria are easily killed by disinfectants or by heat. There are two fundamental messages which cooks should take home from all this: (1) that, provided chicken is well cooked, it is a perfectly safe and wholesome food; and (2) if you have handled raw chicken (or any raw meat for that matter), wash your hands, your kitchen gloves, your utensils and surfaces well before handling anything else, especially if it is to be a raw food such as salad.

Changes in marketing practices over the last couple of decades have added new slants to microbial food-poisoning problems. *Listeria monocytogenes* can grow at low temperatures and has recently become troublesome because it can multiply in prepared food sold chilled, from refrigerated cabinets, for re-heating. *Listeria* has been in foods such as cheese and on salad for many years and most people are immune to it, or so trivially affected by it as not to notice. But those in poor health or with enfeebled immunity can become seriously ill and, though *Listeria* is now taken into account in setting the store lives of prepared foods, the wise consumer will ensure that commercial food for re-heating is heated thoroughly.

Campylobacter and *Listeria*, by the way, are new agents in the sense that they have only recently been recognized as pathogens, but, as with *Legionella, Helicobacter, Borrelia* and several others, this is because improved ways of detecting them have been developed; the chances are that both have been around for centuries, responsible for casual or mis-diagnosed intestinal upsets.

Inadequate hygiene at one or more stages from farm to kitchen is the major cause of food-borne illness in the developed countries as much as elsewhere. Some countries have greater problems than others: in Sweden a strict slaughter policy combined with import controls has apparently ensured that its relatively small poultry flock is free of salmonellae. But matters of hygiene are not the whole story. There are contributory factors, such as a growing public unease about food additives, which some say has obliged the food industry to cut down too far on antimicrobial preservatives. And, of course, people simply get slack; there is a continuing need to draw catering workers' attention to notices saying NOW WASH YOUR HANDS in their staff lavatories. But these matters reach too far into the mechanics of food technology to be discussed further here.

On the domestic front, times have certainly changed over the past couple of decades in the average British kitchen. Dishwashers, microwave ovens, refrigerators, deep freezes are common kitchen furnishings; synthetic detergents have replaced soap; working surfaces are plastic-coated rather than wooden. But the problems of hygiene in the modern kitchen differ little from those of our grandparents, who scalded their milk jugs and utensils, and scrubbed their wooden tables or draining boards, having first rubbed them over with yellow bricks of Sunlight soap. Most of today's precautions remain a matter of common sense. Cracked cups harbour mouth pathogens in the cracks; an infected cut finger can spread pathogenic staphylococci on prepared food and cause food poisoning; that old domestic stand-by, the wiping-up cloth can still spread more bacteria on a newly washed plate or glass than the detergent removed. Fortunately, the bacteria on a wiping-up cloth are mostly harmless. However, one hazard has grown dramatically over the last few decades, and it needs special emphasis. Convenience foods in all their varieties – chilled, frozen, pre- or part-cooked, some preserved in special atmospheres, in a vacuum, or with diverse additives – have become so popular that people must be reminded time and again not to handle them at the same time as they handle uncooked meat. Why? Because not only poultry but all kinds of uncooked meat are always contaminated by microbes, whether the meat is fresh or thawed from the freezer. The microbes are mostly bacteria and mostly harmless; they are inevitably picked up during food processing, at the farm, abattoir, distribution centre and shop. They multiply but slowly on the surfaces of fresh material, and they will soon be killed by cooking. However, they multiply like fun if they get on to cooked food, especially cooked meat. To them, a slice of cold lamb, some paté, some warm scrambled egg or whatever is a lovely habitat, as nourishing as a well-designed laboratory culture. And unfortunately among the kinds of bacteria present on fresh meats there are usually some human pathogens: salmonellae, campylobacters or unpleasant strains of *Escherichia coli* such as I have written about earlier. The wise cook should act as if all fresh meat, like every table bird, is contaminated with nasty microbes, however clean and wholesome it looks. I make no excuse for amplifying the advice I gave about raw chicken a couple of paragraphs ago: do not eat, drink, smoke or lick your fingers while handling raw meat; and before handling anything else, wash, scald or disinfect all utensils, surfaces, vessels and hands or kitchen gloves that have come into contact with it. Do not forget that the insides of the wrappings that the meat came in are infected too!

All this might seem something of an over-reaction; after all, you might well argue, have not most people thrived on carelessly handled meat for generations? Quite so; most of them have. But the trouble today is that techniques of intensive farming, mass packaging, distribution and handling, together with the widespread use of convenience foods, though they have collectively eased our daily lives, have also increased the ease with which food-borne illnesses can spread. A little common sense in the kitchen can do a lot to help our immune systems keep those illnesses at bay.

Food-borne and water-borne infections, when they do happen, are nasty and sometimes very serious, and lamentably they are on the increase. Yet on the positive side, as I said earlier, they are still remarkably rare considering the amazingly complex systems for supply and distribution which victual our crowded societies.

Certain diseases, notably skin diseases, are contagious: they only spread by contact between the infected part of one patient and the susceptible part of another. The disease called ringworm, due to various fungi collectively known as *Microsporon*, is an example of this kind, and so is its common manifestation, athlete's foot. Perhaps the most socially troublesome of the contagious diseases are the venereal diseases, which infect the genital organs and which are transmitted during sexual intercourse. Gonorrhoea ('clap' in vernacular) is a painful disease caused by a fragile coccus *Neisseria gonorrhoeae* and can be cured fairly readily by modern chemotherapy, but during certain campaigns towards the end of the Second World War so much do-it-yourself therapy was practised by the troops that drug-resistant strains of *Neisseria* emerged and, had not new drugs effective against the resistant microbes been developed, a critical situation could have arisen. Syphilis ('pox' in vernacular), caused by a spirochaete, *Treponema pallidum*, is a more drastic disease, because it is less easily detected and, if it proceeds unchecked, leads to physical, nervous and mental deterioration of a kind that cannot be cured. I wrote earlier of how it was introduced into the South Sea Islands by Europeans in the eighteenth century. It has been endemic there for generations, often retarding the mental and physical development of the population and providing a serious trap for visitors who, from Gauguin onwards, enter with too much enthusiasm into the free sexual *mores* of such societies. Venereal diseases can be cured if they are detected in time, but their involvement in the sexual conventions and taboos of Western societies makes them a particularly intractable problem of social hygiene. The type of person likely to contract venereal infection is, at least in Western society, not likely to be very responsible about noticing the disease in its early stages, about persisting with treatment and avoiding passing it on. Therefore foci of infection persist, notably in ports or areas with a considerable depressed or migrant population, and such foci can be the despair of social workers, doctors and medical officers of health. Moreover, the last few decades have been a period in which the sexual, religious and social tenets of Western society have been questioned rather as an earlier generation questioned its political assumptions in the 1920s and 1930s. An immediate consequence of the 'new morality' has been a marked increase in sexual permissiveness and hence in the spread of venereal diseases among adolescents. Patients now come from all sections of society in many parts of the world and mild venereal diseases, hitherto rare, have become much more common; examples are urethritis caused by a species of the bacterium *Chlamydia*, and genital herpes caused by a virus. Whether the 'new morality' is to be deplored or encouraged is outside the scope of this book; here one can only hope that whatever advances in enlightenment it brings, education in sexual hygiene will be among them.

The spread of infectious disease is normally an accidental process, though deliberate transfer of infection is sometimes encouraged. German measles (rubella) is a mild disease in childhood, but if it is contracted by an adult woman in the early stages of pregnancy it can have a teratogenic effect: which means that it can cause deformity in the foetus. Therefore the parents of young girls sometimes encourage them to play with infected playmates in the hope that they will get the disease over. I have done this myself. (My daughters remained doggedly healthy, developing the disease at an extremely inconvenient time many months later.) Others have encouraged mumps in this way, because the disease can have drastic effects on adults but is rarely serious in children. (I did not encourage mumps; they nevertheless caught it and duly infected me . . .)

It has not escaped military minds that the deliberate spread of disease among humans might be adapted as a weapon of war. Bubonic plague killed nine out of ten people in some communities during the Middle Ages, and the deliberate spread of comparable pestilence among an enemy could be an effective and demoralizing form of attack, one which would leave the enemy's wealth and industrial installations relatively undamaged. An example is the atrocity story told of early American pioneers, who are alleged to have sold blankets infected with smallpox to the Amerindians, knowing well that they had no immunity to the scourge. Biological warfare, as it is called, would strictly speaking encompass the use of swords or arrows tipped with biological poisons such as plant alkaloids or snake venom, a practice recorded among both ancient societies and modern though primitive tribes, but today it generally connotes the dissemination of pathogenic microbes or microbial toxins. It requires a highly virulent microbe, rapid and deadly in its action, against which the home troops and civilian population could be immunized. Among agents that have been considered are the plague bacterium itself, *Yersinia pestis*, which would be highly effective and lethal; the spores of the anthrax bacillus, which are tough enough to be used in shells or missiles and which are very persistent; various pathogenic viruses, which would be unaffected by drugs and antibiotics; and the purified toxin of *Clostridium botulinum*, among the most poisonous substances known. The agent of choice would be spread as an aerosol, because in that way it would reach more people and be less easily neutralized than if food and water were infected, or if infected insects or rats were distributed. It would be difficult to detect until it was too late. And it would be cheap and low-tech: a modest laboratory and the necessary know-how are all that would be needed to produce the weapon (to deliver it would be another problem altogether).

However, biological weapons present several compensating difficulties. The problems of preparing enough of the microbe, of keeping it in an active and virulent state, of immunizing one's own population, and then of spreading the agent in such a fashion that the aerosol travels in the right direction, are enormous, even given the resources that modern military organizations can command. A weapon that returns like a boomerang if the wind changes, and dies if it is not used quickly, does not readily commend itself to the military mind. Moreover, as I shall tell shortly, most airborne microbes are rapidly killed by sunlight, so effective biological warfare would rarely be practicable outside the hours of darkness. New pathogens might yet be discovered, or created by genetic manipulation, which would make even more dreadful weapons, even more difficult to detect, but it is difficult to see how they would overcome these operational problems.

In the days when nuclear holocaust loomed, biological warfare might have seemed a relatively humane form of warfare. For, however lethal the agent, there would still be a sporting chance that some people would survive, simply because there has not yet been a disease to which a few members of the population do not have a powerful resistance. This is not true of nuclear warfare, which could in principle make the whole of the planet uninhabitable by man and higher animals for decades. But in my opinion biological weapons, along with death rays, neutron bombs, phasers and so on, belong more to the realms of action novels and science fiction than to conventional warfare.

Terrorism, however, is quite another matter. As means of striking terror in a community, of killing randomly for political blackmail, biological weapons have become a real threat, and there is no doubt that, despite international agreements to ban such weapons dating from 1925, they have been stockpiled by a few countries, and by militant groups within countries, during the past few decades – and probably used, at least by Iraq. The fount of human malevolence seems inexhaustible, and civilized governments are wise to sponsor research on biological weapons – if only to seek remedies, and to work out means of monitoring and enforcing the ban.

I mentioned just now that aerosols of microbes are killed by sunlight, and this brings up the question of why diseases are seasonal. There are many answers to this question: one's natural resistance depends on one's nutritional status and on what other stresses one is putting up with (mental stress included), but such laboratory tests as have been conducted have provided very little evidence to support the popular view that damp and cold enhance one's susceptibility to disease. Instead it seems more likely that a damp atmosphere prolongs the life of microbes in an aerosol, and so does a low degree of illumination. Consequently, in winter, when the air is humid and the hours of daylight are short, persons living in communities get a heavier dose of live, infective organisms than they would in summer and thus stand a greater chance of catching disease. Sunlight kills most pathogenic microbes quite rapidly at ordinary temperatures when they are airborne in partly dried droplets, though spores are killed much more slowly; the main lethal effect is due to the ultraviolet component of solar radiation. Ultraviolet lamps can be used indoors to sterilize the air in operating theatres and pharmaceutical and microbiological laboratories. Even in diffuse daylight there is an appreciable amount of light of the effective wavelengths, though scarcely any penetrates glass in these conditions. Certain bacteria exist which are resistant to the sterilizing effect of daylight – they are, generally speaking, rich in the pigment carotene, which is also present in plants and protects the delicate chlorophyll pigment of leaves against damage by light – but happily they are not ordinarily pathogenic. As far as infective microbes are concerned, they mostly do not form spores, so the open air in daylight is a fairly safe place even in winter. The combination of snow and sunshine, with its high incidence of ultraviolet radiation, is most hygienic. No doubt this is the reason why a hard but sunny winter seems to entrain fewer respiratory infections than the typical British winter, cold, wet and grey.

Dryness, as I mentioned earlier, also hastens the death of airborne microbes. Though they survive for a while in dry conditions, they die off more rapidly than if the relative humidity is high. An interesting factor in the spread of infection in crowded communities is electric discharge. The London Underground, particularly in winter, might be expected to be a hotbed of all possible diseases, with literally millions of people crowded into it, twice daily, throughout the year. In fact it is not so. The air in the average tube system is remarkably free of live microbes, and the reason seems to be that the frequent electric discharges produced by the trains generate ozone and oxides of nitrogen, both of which are quite good aerial disinfectants. Machines to generate ozone deliberately were installed on the Central Line in 1908, but deliberate ozonization went out of use gradually and ceased altogether in 1956. The late Professor D. D. Woods, a most distinguished chemical microbiologist, used to relate how, in the early days of his career, he was astonished to find the air in his laboratory in a London hospital was virtually sterile *even with the window open*, when all sorts of spores and airborne microbes could be expected to drift in. The reason, he discovered, was that his window was close to the main outlet of the ventilation system of the London Underground. So many aspects of urbanization seem to enhance the risk of disease; it is refreshing to encounter one that operates the other way. The ozone content of the London Underground is today not much different from that of the outside air, so it is probably nitrogen oxides which protect travellers. It is interesting to speculate that, should the metropolitan underground transport system ever desert electricity for another power source, a consequence might be an epidemic of respiratory infections unparalleled even in these bronchitic islands.

I have discussed, so far, where the microbes of diseases come from, why, as far as scientists know, they cause diseases, how they are transmitted and how our natural defences act against them. In this chapter the case against the microbe has been very strong, so I ought perhaps to say a word about the harmless microbes that abound in civilized communities. The skin, for example, is populated by a menagerie of microbes including short, rod-shaped bacteria called pseudomonads, the harmless white staphylococcus mentioned earlier in this chapter, and various species of normally untroublesome yeasts; a brisk interchange of these microbes occurs among people all the time. But some of the species for which we provide a convenient surface can become a nuisance. The yeast *Pityrosporum ovale* is usually quite benign, but it can irritate the skin and cause dandruff if it grows too profusely on the scalp. Sweaty areas, such as under the arms or between the toes, tend to be richer in microbes and the characteristic smell of stale sweat is due to microbial action on sweat; some of the components of sweat have an anti-microbial action and help to keep the microbes down, but they are not wholly effective. Deodorants do not in fact deodorize: they contain disinfectants that prevent the development of microbes that would cause the odour.

Micrococci live in the normal nose and throat, and by serological methods it is possible to distinguish types and show that, on the whole, people retain their personal strains for many years. One is, as it were, adopted by a strain of micrococcus in early life which, in some completely mysterious way, repels other people's strains. The mouth has a flora including *Lactobacillus*, a milk bacterium which will appear again in Chapter 5, together with a fearsome-looking spirochaete (*Leptospira buccalis*) which is apparently quite harmless. On these feed a protozoon, *Entamoeba gingivalis*, which is probably beneficial in keeping the microbial population within bounds. The film that develops on the normal teeth and gums, and which is removed when one cleans one's teeth is called plaque, and consists of microbes embedded in a sort of glue. This is made by certain of the

BACTERIA ON THE SURFACE OF A TOOTH. An electron micrograph of the surface of a tooth to which numerous rod-shaped bacteria are adhering. In these numbers they are probably harmless, but aggregates which cover a lot of the surface form plaque and can promote decay. Magnification about 800-fold. (Science Photo Library)

bacteria, mainly streptococci, from carbohydrates which arrive in food. Sugar is a favourite diet of these bacteria and thus enhances plaque formation; that is why dentists are so much against sweet snacks and sugary drinks.

Under the microscope a specimen wiped from the surface of a tooth in a healthy mouth has a most alarming appearance to the uninitiated, but it is entirely fascinating to those who realize that this busy little microcosm of rods, short filaments, dots and occasional corkscrews is just as it ought to be. But sometimes all is not quite as it ought to be. Caries, the ordinary form of tooth decay, occurs because acids, formed mainly by a species of mouth bacteria called

Streptococcus mutans, attack the enamel of teeth. *S. mutans* is rare on healthy teeth but abundant round sites of decay. However, since most other mouth bacteria also form acid, yet tooth decay does not necessarily follow, it seems that caries is not quite as simple as that; it is probably caused primarily by the host's failure to cope with its normal population of microbes rather than by the appearance of new pathogenic types. Fluoride deficiency, particularly in early life, undermines the resistance of the teeth to the acids produced by mouth microbes; fluoride is mainly obtained from drinking water and most water supplies in Britain are now known to be deficient in fluoride. It is a tragedy that, in some localities, handfuls of faddists are ruining the next generation's teeth by opposing fluoridation of local water supplies.

I shall tell in Chapter 5 how the bacteria normally present in our intestines contribute in important ways to our nutrition; these are probably the most useful of the microbes which habitually live with us. It is probable, as I wrote earlier in this chapter, that there are numerous harmless viruses in the intestinal tract. The urethral tract should be sterile, but the vagina in females normally contains micrococci living harmlessly in its exudates. Babies develop nappy rash, not because urine is intrinsically harsh to their skin, but because bacteria grow in the wet nappy and form ammonia from the urea of urine. It is the ammonia that causes the rash, being a strong skin irritant. We live in fact with great numbers of personal microbes; they are ordinarily harmless and only become a nuisance if we behave in an unhygienic manner.

This is a book about microbes and ourselves, the species called man. But we are utterly dependent on the plants and animals that share the biosphere with us, and the interactions of microbes with plants and animals affect society in numerous ways, most obviously through agriculture and the environment. I cannot cover the diseases or microbial associations of animals and plants in even the limited way that I have treated man, and it must be sufficient to say that the same principles apply: microbes of all kinds impinge on the biology and ecology of animals and plants just as they do on ourselves. It is perhaps correct to say that fungi are more common agents of disease in plants than are bacteria, whereas the reverse is true of animals and people: since the spores of fungi are readily spread by wind and air currents, plant hygiene can present somewhat different problems from animal or human hygiene. Just to offer one example: stubble burning, which causes environmental problems in areas where cereals are cultivated intensively, is an excellent way of keeping fungal infections of cereals at bay year after year, but it also kills a variety of creatures, many quite harmless, that live among the plants – and it makes a nasty smoke. Wind is far from the only means of dispersing plant diseases. Sap-sucking and wood-boring insects are frequent carriers of both fungal and virus infections. A particularly sad example is Dutch Elm disease, which has transformed the British countryside during the last three decades by killing nearly all our elms. Even microbes themselves get diseases, from the bdellovibrios, which are tiny bacteria that parasitize and generally kill their more normal brethren, and the bacteriophages, which are simply viruses that attack bacteria.

With that brief acknowledgement of the fact that our fellow creatures share our involvement with microbes, I shall turn to the topic of chemotherapy, to the question of how the natural defences against microbial invasion can be aided.

The belief that specific substances exist that will cure disease has existed from time immemorial. The beneficial effects of herbal extracts, crushed oyster shells and alcoholic drinks on fevers and distempers, though often imaginary, form part of a strong tradition of folk medicine that persists to this day. In the seventeenth and eighteenth centuries, cookery books would contain as many recipes for dishes that would cure diseases as for dishes of gastronomic interest. I encountered a particularly unpleasing example many years ago in a seventeenth-century recipe book: it recommended snail water, the liquor obtained from prolonged seepage of live snails in water, as a certain cure for phthisis (tuberculosis). Some of these folk-medicine

THE DRASTIC EFFECT OF DUTCH ELM DISEASE. A mature elm tree has been killed by a virulent fungus which is carried from tree to tree by a beetle, which bores into the bark. Chopped-down fragments of such trees must be burned because they remain infectious for long periods. (Courtesy of the Forestry Commission)

cures undoubtedly had beneficial effects, but little logic underlay the discovery and prescription of such remedies until the end of the nineteenth century. At this time the germ theory of disease became widely accepted, structural organic chemistry was advancing with incredible speed and the stage was set for the emergence of chemotherapy, the science of controlling disease by specific chemicals. Paul Ehrlich, a German, was probably the father of chemotherapy: his most spectacular discovery, in 1910, was the drug salvarsan, or Ehrlich 606, which proved very active against syphilis. Previously, the only cure for this disease, and a most risky and uncertain one it was, was to feed the patient with poisonous derivatives of mercury. If the patient did not die, there was fair chance that the spirochaetes would, and that a cure would be effected. A similar situation arose in the treatment of trypanosomiasis (sleeping sickness, caused by a protozoon) with arsenical compounds, so Ehrlich set about preparing, quite deliberately, an organic material containing arsenic that would remain active against the trypanosomes yet be less lethal to humans. Salvarsan, which chemists represent by the formula:

$$(H_2N)(HO)C_6H_3{-}As{=}As{-}C_6H_3(OH)(NH_2)$$

was the 606th compound to be investigated, and although it was not very active in trypanosomiasis, it proved most effective against syphilis.

Ehrlich also noted that dyes, which bacteriologists were using to render these microbes visible under the microscope, were taken up very strongly by bacteria. If the dyes could be made poisonous, could they not be used to cure microbial diseases in the living patient? Acriflavine, a yellow dye that is still used for treating superficial wounds and skin conditions, was introduced by Ehrlich. It is a powerful bactericide, but is too poisonous for internal use. Other dyes such as methylene blue proved to have some microbicidal action (they are still used occasionally), but were still rather toxic. Domagk, in 1935, made the most spectacular advance in this direction by obtaining the first chemotherapeutic agent that was strongly active against bacteria, Prontosil:

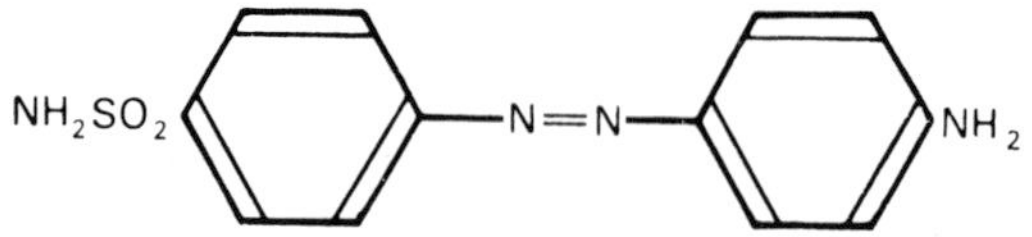

$$NH_2SO_2 \quad N{=}N \quad NH_2$$

A certain ingenuity went into the development of Prontosil, because, though it has the chemical structure of a dye, it is not in fact coloured. Domagk and his colleagues had realized that the property of being strongly absorbed by microbes was the important chemotherapeutic factor, not the property of being coloured. This ingenuity proved somewhat misplaced when it was found that Prontosil broke down in the patient's liver to sulphanilamide:

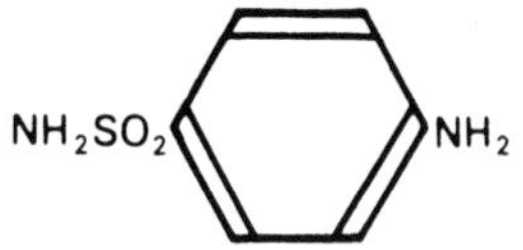

$$NH_2SO_2 \quad NH_2$$

a compound which is nothing like a dye, but which was just as active as Prontosil. This discovery released the floodgates, as it were, leading to the development of a variety of exceedingly potent anti-bacterial drugs called the sulphonamides in English (sulfa-drugs in American). These have the general formula:

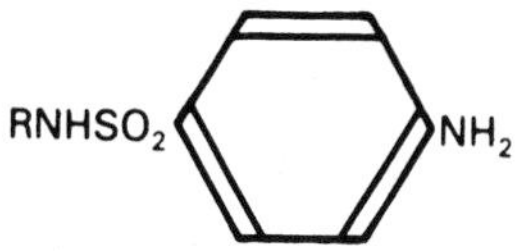

$$RNHSO_2 \quad NH_2$$

where R can be any of some two or three hundred atomic groupings, depending on the particular properties required. They could be tailor-made in chemical laboratories so as to stay in the gut or be absorbed in the blood stream; they were often more active against microbes than the original Prontosil; many of them were less poisonous to humans than the original materials.

Today few can recall the impact the sulphonamides made in 1935–7. Pneumonia, which had been the major killing disease in Britain during the twentieth century, abruptly became almost trivial. Puerperal fever, a systemic infection due to the bacterium *Streptococcus pyogenes*, which was often contracted in lying-in hospitals during childbirth, showed a dramatic drop in incidence and mortality. The sulphonamides were indeed a triumph for the chemotherapist.

Yet, for the scientist, there was this small, niggling query. They were nothing like dyes. Why, then, were they so marvellous? The answer, found by D. D. Woods working in Sir Paul Fildes's laboratories in the 1940s, was quite unexpected. Woods found that certain materials, such as serum, contained a substance that made bacteria immune to the sulphonamides, and eventually he isolated it. It proved to be a simple compound, called *para*-amino benzoic acid:

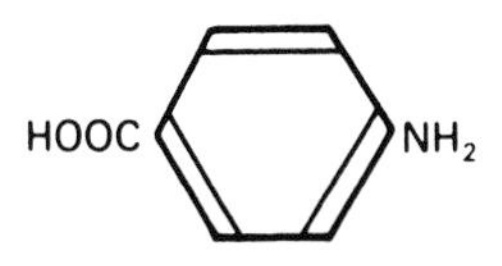

Now, a peculiarity of the effect of *p*-AB (which I shall call it for short) was that, if there was only a little sulphonamide present, only a little *p*-AB was needed to neutralize its effect on bacteria, but if a lot of the drug was present, a lot of *p*-AB was needed. A sort of competition seemed to exist, as far as the microbes were concerned, between *p*-AB and the drug. Woods also noticed that the formula of *p*-AB was rather like that general one I have written for the sulphonamides, and he proposed the following hypothesis to explain the situation. If all microbes are assumed to need *p*-AB in order to grow, perhaps sulphonamides seem so like *p*-AB to the microbes that they try to use them instead, and thus fail to grow. This theory has two obvious consequences. First, that sulphonamides would be found not to kill microbes but just to prevent them growing; secondly, that sooner or later some microbe might be found that would be unable to make the *p*-AB it needed and would require to be provided with it for growth. In the second instance, *p*-AB would turn out to be a vitamin for some microbe.

Both of these deductions proved brilliantly correct. Sulphonamides do not kill bacteria: in the infected patient they stop their multiplication and give the body's defence mechanisms time to deal with them. And several microbes are now known that require *p*-AB as a vitamin. The discovery of *p*-AB as a vitamin led to enormous advances in both microbial and general biochemistry, stretching into realms of biosynthetic chemistry that I cannot possibly deal with here. From the medical point of view it opened a new, rational approach to chemotherapy: if one knew the sort of vitamins and growth factors needed by microbes, one could make chemicals in the laboratory that were rather like them (called in laboratory jargon structural analogues) and hope they would inhibit microbial growth and thus be valuable chemotherapeutic agents.

This hypothesis proved abundantly true in most respects. The 1940s and early 1950s were a period of intensive research into microbial nutrition: vitamins and vitamin-like compounds were discovered and isolated, structural analogues prepared and, in test-tube experiments, these frequently proved to inhibit microbial growth in the competitive manner that the sulphonamides showed. It is ironical to have to record that not one of the drugs made was of practical chemotherapeutic value. They were too toxic to humans, or the kidneys eliminated them too well, or the blood and tissues contained too much of the vitamin they were antagonizing, or the infective bacteria did not need the vitamin. One of the few successful drugs made according to the rational approach proved not to obey the rule at all. It arose from studying analogues of vitamin B_2 (riboflavin), which is required by many microbes. By altering analogues in various ways a group of research workers at Imperial Chemical Industries' laboratories ultimately developed a drug, paludrine, which was highly effective against malaria. But by then it had been so much altered that it had no competitive action against B_2 at all.

The next major advance in chemotherapy, really a step back to the early 1930s, occurred in quite a different way. Most people are familiar with the story of penicillin: how Fleming recognized it when a stray mould grew in a culture of micrococci and started to dissolve the colonies; how he attempted to isolate the active material, failed and gave up; how Chain, a refugee working at Oxford, took up the problem and succeeded in extracting the material; how it proved fantastically active, more so than any drug known hitherto, and was prepared in milk churns at Oxford; how, because the Second World War was on, development was transferred to the USA, with the ludicrous result that, after the war, the British had to pay patent royalties to use methods of making it developed there; how it became universally available after the war, but its use led to the appearance of penicillin-resistant strains of microbe and pencillin-sensitive patients. These stories, with their ramifications into politics, personality, vested interest and carelessness cannot detain us here. Penicillin is one of a class of products made by moulds called antibiotics. They have strong antibacterial actions and this property is probably of value to moulds in nature, since moulds and bacteria tend to compete for the same types of nutrient.

The discovery, development and success of penicillin led to a burst of research activity on the part of the pharmaceutical industry during which tens of thousands of moulds, actinomycetes, even bacteria and algae, were screened for anti-microbial activity. Over the last fifty years, well over a hundred have come into general medical use. I shall write more about them in Chapter 6; for the purposes of this chapter I shall note that, like sulphonamides, they tend not to kill bacteria but only to stop them growing. It seems clear that they act in a variety of ways, some interfering with the microbes' genetic apparatus. Penicillin prevents bacteria from making their cell envelopes properly.

Here I must say a little more about drug resistance in microbes, something I first mentioned in Chapter 2. Microbes can acclimatize themselves to such substances as sulphonamides and antibiotics if they encounter them in small doses: variants, called mutants, arise spontaneously (see Chapter 6) and are unaffected by the agent. Therefore, when using these drugs in practice, it is important to give as massive a dose as the patient will tolerate right at the start and to sustain a high level throughout the treatment. There is a second way in which microbes can become drug-resistant: they can sometimes acquire resistance by gene transfer from other, naturally resistant, organisms. The way this happens cannot detain us here – it would lead me into premature discussion of the biology of genetic elements called plasmids, which I shall deal with in Chapter 6. The present message is that, by gene transfer, microbes can sometimes acquire resistance to more than one antibacterial substance at a time. If a patient relapses, or develops a new infection after treatment with a chemotherapeutic drug, an entirely different drug should be used lest the infective microbes be resistant to the earlier one. If possible, the new drug should be one to which resistance is not known to be conferred along with resistance to the first by gene transfer. Where possible, physicians today try to use a 'cocktail' of drugs because the chance of resistance to all of them being acquired is generally much less than the chances of resistance to one of them appearing. I mentioned earlier the disastrous effect of ill-considered use of sulphonamides to treat gonorrhoea during the Second World War. This is the reason why antibiotics and sulphonamides are not, and should not be, made available except on prescription. Unfortunately, there are countries in the world where this is not the rule, and some nasty antibiotic-resistant diseases are around. The sort of penicillin-resistant bacteria that appear in practice, as distinct from in the laboratory, seem to be those that are able to destroy penicillin, and in recent years partly synthetic penicillins have been made industrially that are insusceptible to penicillinase, the enzyme that destroys ordinary penicillin. These new products therefore work on the naturally resistant strains so, with this drug, one problem of resistance has receded – for the time being. But among the anti-malarial drugs resistance has again become a serious problem. Quinine, the traditional remedy for malaria, had been replaced in general use, since the Second World War, by two synthetic drugs, chloroquine and amodiaquine. These, unlike quinine, are prophylactics (they protect against the disease) as well as remedies and, for two decades, they proved remarkably successful. Unfortunately, in the mid-1960s, reports appeared from places as distant from each other as Brazil, Colombia, Malaya, Cambodia and Vietnam of infections resistant not only to these drugs but to others like them. The basic reason for this seems to be careless use of the drugs in therapy. Now, in many localities, it has become necessary to return to the traditional quinine, to which resistance is very rare. Drug resistance is today one of the focal points of research in chemotherapy.

Nevertheless, chemotherapy, at least in regard to bacterial and protozoal infections, made dramatic strides in the mid-twentieth century. In retrospect it seems something of a comedy of errors – Ehrlich missed the importance of salvarsan for two years because he was concerned with trypanosomiasis and it was his colleague Hata who realized its value in syphilis; the sulphonamides were developed as dyes and were successful for the wrong reasons; the structural analogue theory proved perfectly correct but only really useful retrospectively; the best-ever antibiotic, penicillin, was the first one to be discovered. Yet it was one of the most productive comedies of errors in the history of mankind: not only were the resulting advances in medical, biochemical and chemical knowledge quite spectacular, but the bacterial and protozoal diseases have, as a result of chemotherapy, largely come under control. Tuberculosis, bacterial pneumonia, typhoid, plague, anthrax, cholera and so on are all curable, given accurate diagnosis and facilities; even the dreaded leprosy can be controlled, in fact it is high on the WHO's 'hit list' of diseases that it hopes to eliminate from the planet – as with smallpox, but using chemotherapy rather than vaccination.

Chemotherapy and immunotherapy, together with improvements in all aspects of hygiene – all of which, remember, stemmed from advances in basic understanding of how microbes work and behave – have brought about dramatic changes in the social patterns of disease. In the early 1900s the major killer diseases in the developed world were bacterial: pneumonia, tuberculosis and diarrhoea; today they are heart disease, cancer and strokes. However, though bacterial diseases are no longer the scourges they once were, there is no justification for complacency.

The recent history of tuberculosis conveys a chastening message. In outline, the bacterium responsible for what was once the most dreaded of mankind's microbial scourges, *Mycobacterium tuberculosis,* has long been known to be selective in its pathogenic effect. During the first half of the twentieth century, when 'TB' was still widespread, *M. tuberculosis* could often be isolated from the respiratory tracts of perfectly healthy people. These people were carriers, in whom an infection was present but completely contained by their own immune systems. In 1944 the discovery that the antibiotic streptomycin, taken over many months, could cure TB changed the medical scene; supplementary drugs which enhanced and hastened the cure were discovered during the next decade or so, and by the mid-1960s TB came to be regarded as a rare and not especially serious disease. The number of carriers, too, probably decreased as casual contact with TB sufferers became rarer; it really seemed as if the 'white plague' of many centuries was at last dying out. But there was a catch. Treatment was not cheap, and reservoirs of TB persisted in impoverished and under-developed areas, such as inner cities and remote parts of Africa and Asia. Moreover, treatment still entailed taking a cocktail of drugs for a year, for most of which time the patient would feel perfectly well; so patients tended to abandon treatment, especially patients from impoverished, shifting or drug-abusing backgrounds. Abandoning antibiotic treatment too soon is a surefire way of generating an antibiotic-resistant pathogen; within a decade or so strains of *M. tuberculosis* resistant to the drug cocktail were being reported. Then, within a few more years, came the HIV pandemic. The combination of an immune system damaged by HIV and infection by *M. tuberculosis* proved disastrous: carriers lost their immunity and developed the disease, new infections spread, numerous new reservoirs of the disease were created. As the 1980s progressed, drug resistance and HIV independently caused a worldwide up-surge of cases of TB, especially in Africa but significantly among deprived and alienated groups in cities such as New York and Edinburgh as well. In 1993 the WHO declared the disease a 'global emergency' – this despite the fact that the disease is scientifically well understood and, in proper conditions, curable.

The message of tuberculosis applies to other fundamentally manageable diseases, too: defective immunity among patients and drug resistance among pathogens are increasing. The crucial problems are too often social rather than medical: a need for facilities, which include transport, hygiene, adequate education, decent food and water and, of course, competent medical help. In some parts of the world, especially those afflicted by ethnic and religious strife, these facilities are very poor indeed, with increased scope for interrupted, incomplete or careless chemotherapy.

On a global scale, in 1996 diarrhoeal diseases such as cholera, typhoid and dysentry killed some 3.1 million people, mostly in developing countries, and tuberculosis killed about the same number, figures exceeded only by pneumonia, which can be viral or bacterial and killed about 4.4 million. To complete the catalogue of major global diseases, the gross world mortality assignable to microbial infection in 1996, according to WHO data, was 17 million, the protozoal disease malaria contributing some 2.1 million and the remaining 6.5 million largely viral, with hepatitis B in the lead at about 1 million deaths. In global terms AIDS is rapidly catching up to join the major killers – and the impaired immunity of HIV-positive patients causes them to contribute to the maintenance of other scourges by becoming reservoirs of diseases such as tuberculosis and varieties of pneumonia.

Poverty, war, deprivation and HIV are contributing to a resurgence of at least some bacterial diseases, but it remains true that our *bêtes noires* are the virus diseases. Sulphonamides and antibiotics act on bacteria by interfering with their multiplication: they stop the bacterial cells from growing and dividing. In an infection, the invading microbes are multiplying rapidly whereas the cells of their victims, if they are growing at all, are doing so extremely slowly in comparison. Hence these drugs, though they may act on both host and microbe, influence only the microbe's growth seriously and thus they enable the host to recover. Viruses, however, grow in quite a different way from other microbes. They actually get inside the cells of their hosts and pervert the metabolism of those cells. They cause a defect in the mechanism controlling the cell's own machinery, so that it uses its metabolism to make the wrong thing. Thus, instead of keeping themselves in good repair, the cells make lots more virus. One can put this point another way: the way a cell functions is controlled by its genetic structure, which means that the precise chemical composition of its genes programmes it for the period of its existence. Genes consist of substances called nucleic acids and so, mainly, do viruses. A virus infection programmes cells to make more virus and thus the chances of an effective chemotherapy of virus diseases are limited, because any effective agent would be equally damaging to the healthy cell. But all hope is not lost: nucleic-acid analogues have been made that were active in practice against herpes and types of Asian 'flu and there are encouraging possibilities for the development of similar agents effective against HIV. Mainly, however, it is the natural defences, interferons and immunity in particular, which provide our bastion against virus infection. A pretty soggy sort of bastion they can become during a cold, damp British winter.

4 Interlude: how to handle microbes

The time has come, I think, to say something of how all these things are known. So often, in books of this kind, the author tells readers the results of scientific progress, paints a sort of panorama of present-day knowledge, without giving them any idea of how this progress came about, of how the knowledge was obtained. This, you may say, is just fine: as a lay reader you are prepared to take my authoritative word that everything written in this book is based on sound, well-conceived experiments.

Would that it were! The trouble with science is that its day-to-day aspects are extremely uncertain. Occasionally great and obvious advances in knowledge take place, but generally research and its applications progress by repetitious and tedious experiments, almost always giving negative or useless results, from which, slowly and over a long period, a picture of the behaviour of whatever-it-is the scientist is studying emerges. No scientific finding is 100 per cent certain; the majority of observations that find application are more than 90 per cent certain. Today, living in a society which depends for its existence on a highly developed technology, it is most important that laymen should be able to view critically the results of scientific research – or at least the claims made for them by journalists, scientists and scientific administrators. It would be very surprising, for example, if some of the information I have written in this book is not falsified by recent research before it is published. How, then, is one to judge what is likely to be well established and what is a little dicey?

The answer, which applies to scientists as much as to laymen, is to develop a feeling for the subject. That may seem a highly unscientific statement, but I make no apology for it, for it is not as unscientific as it sounds. If one knows something of the way in which experiments are done, one can distinguish those that are rigidly precise from those that are merely suggestive. Since scientists make use of both, a knowledge of the sorts of experiments on which a subject is based gives one, in due course, a sort of instinctive understanding of what beliefs are reliable and what should be accepted with reserve, to be modified or abandoned in the light of future experiments.

The whole of microbiology is based on the belief that living matter does not generate itself from non-living matter: that a truly sterile broth, for example, will never go bad if it remains uncontaminated by a microbe. This belief, which was not widely accepted before the late nineteenth century, rests on a number of very simple experiments in which broths were sterilized and left, exposed to air and warmth, in vessels designed so that airborne microbes could not enter them. Some of John Tyndall's original broths, set up in the late nineteenth century, could still be seen in the 1960s at the Royal Institution off Piccadilly in London. Yet the chances are that life originated spontaneously on this planet at some time, as I shall tell in Chapter 10, so the view that spontaneous generation does not now occur does not imply that it never could occur. It implies merely the acceptance of a belief that it is an event of such extreme improbability, in this day and age, that it may be disregarded for the purposes of ordinary scientific research.

The principle that sterilized material will, if suitably protected, remain sterile unless one infects it is basic to microbiology. Microbiologists prepare sterilized broths, sometimes jellified, in which microbes can grow, and infect these with particular strains and species so as to keep them 'pure', which means uncontaminated by other microbes. These are called cultures of microbes, and every now and again a small portion of the population is transferred, or subcultured, into a new lot of broth or jelly to keep the strain live and multiplying. The compositions of these broths range from a simple solution of a few chemicals, through soups and milk preparations, to most complex brews of blood, meat and vitamin supplements.

Whole textbooks have been devoted to their preparation and I shall not discuss details of them here, but there are certain basic principles used to devise such broths that are important. First, however, let me introduce the technical word medium, which is used a lot by microbiologists. A medium (plural: media) is an environment, usually a broth or jelly, in (or on) which microbes are allowed to grow. Most microbes require, in order to grow, a solution containing traces of elements such as iron, magnesium, phosphorus, sodium, potassium, calcium, a source of nitrogen such as an ammonium salt and some kind of carbohydrate food, sugar, for example. A balanced chemical fertilizer mixture of the kind used in gardening will, for example, make a fine medium for many bacteria if a little sugar is added, and if such a solution were to be infected with a grain of soil and put in a warm place, a positive menagerie of soil bacteria, mostly little rods called *Pseudomonas*, would quickly grow. Since they would be using dissolved oxygen to oxidize the sugar, they would rapidly exhaust their supplies of dissolved air, except at the surface of the liquid, so that deep in the culture medium anaerobic bacteria such as *Clostridium* would start growing. One would soon have a horrible, evil-smelling mess, probably bubbling as the anaerobes generated carbon dioxide from the sugar; the microbial population would be very mixed and of little use to anyone who wished to learn something about the types of organism present.

Microbiologists use two general techniques to obtain pure cultures of a single type of microbe. The first is called *enrichment culture*, a procedure that makes use of a *selective medium*. Supposing, for example, one wants some nitrogen-fixing bacteria, one could make up the sugar medium as I have described but leave out the ammonium salt. In these circumstances, if it were infected with a little soil, only those microbes that could use atmospheric dinitrogen could grow, and thus the culture would become rich in nitrogen-fixing bacteria. Once they started growing, of course, some of the nitrogen fixed would become available to other microbes in the soil inoculum, these would start to grow and the population would become pretty mixed. But it would be enriched in nitrogen-fixing bacteria, at least to start with. If one wanted sulphur bacteria, one could leave the ammonium salt in but add sulphur instead of sugar; some ferrous sulphate in place of the sugar would encourage iron bacteria. Instead of changing the composition of the medium one could change its acidity: a weakly acid sugar medium favours the growth of yeasts and moulds rather than bacteria. Or one could exclude air, simply by using a bottle filled to the brim and stoppered, and so enrich the medium in anaerobes. (With a sugar medium, however, this is not a very good idea, as clostridia often generate gas and blow the stopper out.) A little sulphate in such a medium enriches the population in sulphate-reducing bacteria; nitrate selects for denitrifying bacteria. Use of a high temperature selects for thermophiles: heat your soil before inoculating the medium and only microbes that form heat-resistant spores will grow.

Obviously the possibilities of enrichment culture are almost limitless; readers may devise for themselves media suitable for enrichment culture of microbes utilizing alcohol, disinfectants, rubber, shoe leather, plastics and so on. Not all the obvious media work and, correspondingly, one can observe microbes in natural environments that respond to no simple enrichment technique. But, generally speaking, enrichment culture is the primary step used by scientists who are seeking pure cultures of microbes.

Medical microbiologists, however, do not make much use of enrichment culture methods, for the simple reason that their enrichments have been done for them. An infected patient is already an enrichment culture, so that the medical scientist jumps at once, as it were, to the second step, that of isolating a pure strain of the enriched microbes.

Once one has an enriched population or, in other words, has obtained a culture of microbes containing a majority of the organisms one is interested in, how does one get the population pure? The easiest way to do this is to use a jellified medium, and to spread a tiny drop of the enrichment culture over it in such a way that each individ-

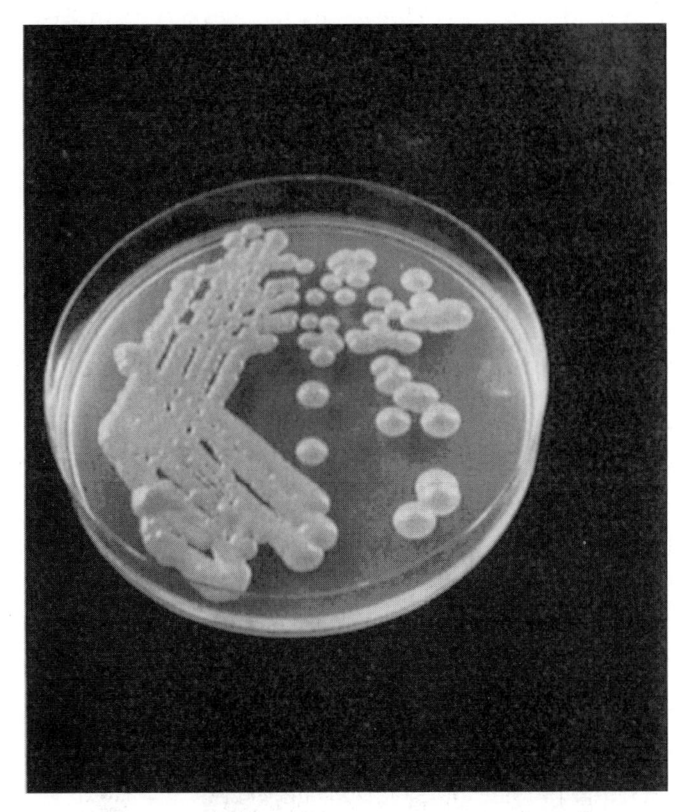

COLONIES OF BACTERIA IN A LABORATORY. A tiny drop of a culture of bacteria was spread, with a sterile wire, on the surface of a suitable jellified medium. After incubation, separated colonies appeared where the population had become sparse. The organism is *Derxia gummosa*, from tropical soil. (Courtesy of Dr Susan Hill)

ual microbe is well separated from its neighbour. When the jellified culture is allowed to grow, each separate microbe will multiply to form a colony of similar microbes, and those colonies that are widely separated will come mainly from the predominant organisms in the enrichment culture. It is then a simple matter to infect a new culture with a fragment of one of those pure colonies and thus obtain a pure culture of microbes.

This is the principle of a process called plating. Microbiologists generally use covered dishes called Petri dishes (after their inventor) containing media set with a gelatinous seaweed extract called agar for plating cultures; for anaerobes they use tubes of media set with agar. Other procedures, such as micro-manipulation, can be used to obtain pure cultures. They all depend on causing single microbes from large populations to form colonies of progeny well separated from their neighbours.

For plating to be successful, not only must the medium be suitable to the microbes in composition, but the medium, glassware and instruments must be sterilized. In addition, the operations must be carried out so as to reduce the possibility of contamination by airborne microbes to a minimum. Such aseptic technique, in microbiologists' jargon, requires working conditions to be free of draughts and both media and glassware must be pressure-cooked or sterilized in ovens for some time. Not all media can be pressure-cooked, because cooking may decompose their constituents and make them unsuitable for fastidious microbes. In such cases a very fine filter may be used to remove extraneous microbes, or irradiation with γ-rays or ultraviolet light may be used. Sterilization by γ-radiation is used to provide sterile plastic ware for microbiologists, because plastics, though cheap and disposable, rarely stand the temperatures needed for heat sterilization. Pressure-cooking, in live steam above the boiling point of water, or baking at high temperatures, may seem rather drastic procedures to eliminate microbes that are mostly killed at temperatures above 50 °C, but it is necessary because the spores of certain microbes can be very heat-resistant. It so happens that the common airborne bacteria include some of the toughest spore-forming types.

It is usually important to know not only what microbes are present in a sample of material, but how many. In principle, culture in jellified enrichment media, either on Petri dishes or in air-free tubes, is adapted for this purpose: one counts the numbers of microbial colonies which appear when a representative sample of the material, if necessary homogenized and appropriately diluted, is dispersed in or on an appropriate medium and incubated. It is a procedure fundamental to microbiology, with many subtleties and variations, but this is not a practical handbook and I shall not go into detail; for present purposes the important point is that detection and estimation of microbes normally takes as long as the colonies take to grow, which might be anything from overnight for *Escherichia coli* to a few weeks for *Mycobacterium tuberculosis*. Microbiological information can sometimes be frustratingly slow in coming.

Enrichment culture and single colony isolation will yield pure cultures of most microbes capable of growing on laboratory media, but some microbes are not domesticated so easily. *Mycobacterium leprae*, the bacterium that causes leprosy, has never been grown away from living tissue, and pure cultures of protozoa have often been obtained only in association with live bacteria on which they feed. Research on the pathogen *Legionella* was held back for several years because it did not grow well in the laboratory except in complicated environments such as the yolk sac of a developing hen's egg. The discovery of a laboratory medium, albeit a complicated one, sped things up and, in particular, led to the discovery that *Legionella* is sensitive to antibiotics such as erythromycin or rifampicin. Particularly intransigent in these respects are the viruses, which just have to be grown on living tissue. They can be separated from bacteria and other microbes quite easily: a filter of suitable fineness will let viruses through but hold up all larger microbes. But once filtered, viruses must be provided with living hosts in which to multiply. Tissue cultures, fertile chicken's eggs or cultures of bacteria are most often used; animal hosts are sometimes necessary and in at least one case – the common cold viruses – human volunteers are the best means of culturing them. The isolation of pure lines of such viruses depends basically on diluting enriched populations such that only a few infective units remain which, since the population was enriched initially, are presumed to be similar and representative of the predominant type. With exigent viruses, such isolations can be very difficult, and sometimes impossible. In fact, microbiologists have to live with the sad truth that many more bacteria and viruses exist than have ever been cultivated in the laboratory. These days scientists can extract and examine the genetic material, the DNA in particular, from all kinds of creature, and recognize within that DNA chemical structures which are on the one hand uniquely characteristic of the species from which they originated, or on the other hand generally characteristic of its class of organisms. I shall tell more about the nature of those structures when I get to biotechnology a couple of chapters further on; the message here is that DNA from viruses or bacteria, for example, is easily distinguished from that from higher organisms by various features of its structure. By such means seemingly dormant virus-like structures have been found in the genomes of higher organisms such as pigs and humans and, correspondingly, DNA extracted from natural sites rich in bacteria, such as watery sediments or soil, has indicated that an abundance of bacteria exist in nature that have not been grown in a laboratory and are not assignable to any known species.

There seems to be a substantial population of microbes out and about which remain beyond reach of conventional microbiological study and manipulation. Sometimes one gets hints of what the missing microbes are capable of from the effects they have on the environment, and this can stimulate a search, sometimes successful, for ways of isolating and handling them. For example, when I was a younger researcher in the 1950s, conventional methods of enrichment and isolation yielded only two or three species of sulphate-reducing bacteria, though there were hints that in some special localities strains were present which doggedly eluded detection. Then new media and procedures for seeking such bacteria were developed and by 1995 more than 50 species were recognized. Microbes can be very elusive at times; this being so, how far can microbiologists in general be sure that the behaviour of microbes that they actually can grow in the laboratory parallels their behaviour in nature? The answer is that they cannot be sure. The late Professor Kluyver, a distinguished Dutch microbiologist, used to argue that all cultures of bacteria are laboratory artefacts, strains whose characters had become altered as they adjusted themselves to grow in laboratory media. He was, of course, quite right. Microbes have remarkable properties of adaptability, and microbiologists must always keep in mind the reservation that the behaviour of their material in the laboratory might be quite misleading as regards its behaviour in its natural habitat. A simple example is the typhoid bacterium, *Salmonella typhi*, which almost always needs to be provided with the amino-acid tryptophan for growth when it is freshly isolated from a patient with typhoid. It readily loses this character in the laboratory – apparently it learns to make its own tryptophan very easily – and almost the only way to make it regain a need for tryptophan is to infect an experimental animal with the strain and re-isolate it.

Since microbes cause disease, this sort of question becomes particularly important in medical microbiology. Is a microbe obtained from a diseased patient really the cause of the disease? In some cases there is little doubt: bacteria obtained from blood, which is normally sterile, may justly be regarded as the cause of a septicaemia. But in mouth disorders, for example, there are so many microbes around already that, unless a quite unusual type is present, it is difficult to be certain which is causing damage. Indeed, in this particular instance there is still no general agreement about which of the flora of the mouth are responsible for tooth decay. One of the earliest bacteriologists, Dr Robert Koch of Berlin, crystallized this dilemma in a set of conditions known as Koch's postulates: a microbe may be accepted as the cause of a disease if (1) it is present in unusual numbers when and where the disease is active, (2) it can be isolated from the diseased patient, and (3) it causes disease when inoculated into a healthy subject. These conditions, for obvious reasons, are not easy to apply in practice, but if they are not adhered to, the scientist can be badly misled. The common cold, now known to be a virus infection, causes secretions which encourage the growth of all sorts of bacteria, some harmless, some irritant. In the early years of this century these bacteria were thought to be the cause of colds, and all sorts of antibacterial preparations were offered as remedies, but they are now known to be secondary bacterial infections developing as a result of the primary viral infection. Intestinal disorders cause gross changes in the intestinal flora which are often a consequence rather than a cause of disease. Koch's postulates apply throughout microbiology and are by no means restricted to its medical aspects: I shall discuss in Chapter 7 the corrosion of stone, a phenomenon that certainly can take place through the agency of sulphur bacteria but which often has nothing to do with them. Koch's postulates, regrettably, are sometimes forgotten even by those in the best position to make use of them.

Having obtained a pure culture of microbes, the first thing one generally does is to look at it. Are the microbes rod-shaped, spherical, twisted or comma-shaped? Do they swim about? Lie in chains or clusters? Form filaments? Form spores? Do they show an internal structure: granules and nucleus? One can conduct cultural tests: do they grow in milk, broth, a simple sugar and salts mixture? Do they form gas and/or acid? What do their colonies look like on jellified media? An important test for bacteria was devised in 1884 by Christian Gram, based on whether the organisms, killed, dyed and treated with iodine, did or did not retain the dye after washing with alcohol or acetone. This test, the Gram reaction, is still very valuable for dividing bacteria into two great groups which, by some coincidence that is still not clearly understood, correspond in several other important properties: Gram-positive bacteria (those which retain the dye) tend to be particularly sensitive to drugs such as penicillin and sulphonamides and to have other physiological properties in common. Tests based on appearance, culture and staining reactions give, with the aid of published keys and guides, some idea of the nature of the microbe and, if that is their aim, microbiologists can follow up these clues with more tests, including those with antisera, and identify the microbes completely. But again I must emphasize the element of uncertainty mentioned earlier: if the microbe is a well-known and important one, such as a *Salmonella*, it may be possible to obtain a very precise identification, but if it is one of the myriad of rod-like bacteria that inhabit soil, for example, it is likely that only a rather vague classification will be reached.

Sometimes a microbe does something useful, makes an antibiotic, a vitamin or other chemical, in which case the scientist may wish to grow large amounts of it. Mass culture of microbes on a production scale is called fermentation. This is an incorrect name, actually, because fermentation strictly refers to the transformations of substances brought about by microbes growing in the absence of air – the classical example is the fermentation of sugar to alcohol by yeasts – but today industrial microbiologists refer to any large-scale cultural process, with or without air, as a fermentation. In principle all fermentations are laboratory culture procedures scaled up, but this can be easier said than done. The engineering problems of handling, containing and sterilizing large volumes of culture fluid, and of incubating, harvesting and extracting the products, are so great that a whole technology called biochemical engineering has grown up around them. Even the procedure of supplying air to several thousand gallons of microbial culture is more of an engineering feat than it sounds: on that scale the microbes tend to consume oxygen faster than the engineer can persuade it to dissolve. Biochemical engineering has become a distinct discipline, largely as a result of the expansion of industrial microbiology consequent on the development of antibiotics. Here I can do little more than note its existence, but I should mention one important concept which is now being accepted among biochemical engineers: continuous culture.

Suppose you are an industrialist producing yeast for the baking industry. By traditional methods you have to keep a stock culture of yeasts, grow a large seed culture from it, prepare however many thousand gallons of medium your fermenter holds, sterilize it, inoculate it, wait for it to grow and harvest it. Then you must clean up and start again. How much better if the culture grew continuously! This is what happens in continuous culture: a fermenter is designed with an overflow, so that sterile medium is pumped in at a rate that is rather slower than the fastest at which the microbes can grow. Once established, the culture continuously overflows into a collecting vessel and can be harvested continuously; the microbes, so to speak, multiply as fast as they are fed. The process has the advantage that the production process can be automated, the plant works night and day and, once established, it is less prone to contamination than traditional procedures. But its adoption by industry on a large scale has been slow for rather a mundane reason: the fermentation industries have invested a lot of capital in batch fermenters and are loath to write off such expensive equipment while it can still be used.

Continuous culture is of great value in research too. If the microbes are growing as fast as they are fed, one can choose which one of the nutrients fed to them shall be the one that limits their growth. If a bacterial culture is grown in a simple medium of sugar and salts one can, by keeping the concentration of sugar low and ensuring that the salts are plentiful, arrange things so that the bacteria use up all the sugar provided. The concentration of bacteria in the culture is determined by the concentration of sugar in the medium provided; in microbiologists' jargon the sugar is said to limit their growth. If, now, one grows similar bacteria in a medium with plentiful sugar but limited by the supply of ammonium, one obtains ammonium-limited organisms. And one finds they are different in several ways. They are rich in carbohydrate and are tough – they do not die easily. Their enzyme balance and chemical composition have changed. By choosing diverse limiting nutrients, one can alter the biochemistry of microbes to remarkable extents. This gives microbiologists an experimental control over the physiology of their material which is unique in biology and which is still producing valuable basic knowledge.

Continuous culture makes use of the fact that microbes need certain nutrients to multiply and that their growth can be controlled by adjusting the supply of these nutrients. Many microbes need quite complicated nutrients, such as amino-acids or vitamins, and it is sometimes difficult for analytical chemists to analyse foodstuffs and other materials for such compounds. Microbiologists have used microbes for this sort of analysis: if one has a material containing vitamin B_{12}, for example, and wishes to know how much it contains, one of the least complicated ways is to add a little to a culture of microbes that require B_{12} and see how well they grow. This process is called microbiological assay and it is the only method of measuring amounts of some of the vitamins. Microbiological assay was responsible for the unexpected discovery that sewage is one of the richest sources available of vitamin B_{12}; protozoa are particularly useful for B_{12} assay. In the 1940s most of the amino-acids were assayed with microbes, but today chemical methods for their analysis have been developed to such an extent that microbiological methods for amino-acids are obsolete.

Microbes, like all living things, die. Cultures have, therefore, to be subcultured to keep the strain alive, and this can be tedious if one has a large collection of microbes. There are, however, two ways in which they can be preserved without subculture: deep-freezing or freeze-drying. Both of these operations must be carried out in rather special ways, but if they are done properly the microbes go into a state of suspended animation and can be stored for very long periods. To deep-freeze a live bacterial culture, for example, the organisms are suspended in quite strong (20 to 50 per cent) solutions of glycerol (but a number of other compounds of the class chemists call non-polar are also effective). When such a suspension is frozen, nearly all the bacteria remain alive, whereas most would have died in an ordinary medium. If they are stored at a really low temperature, at −70 °C or, better, at −200 °C, they die only very slowly. In this respect they are like the tissue and blood-cells which can be cold-stored in glycerol in banks for surgery or blood-transfusion, but the whole living microbe can be so stored. Protozoa, perhaps because of their greater internal complexity, do not respond well to such storage.

Protein, such as white of egg or blood serum, also protects against freezing damage, and so do sugars. If one suspends bacteria in a mixture of serum and the sugar glucose, it is not only possible to freeze them without damage but also to dry them. The ice in the frozen mixture must be sublimed off under a high vacuum. This process, known as freeze-drying, is very useful because, once dry, the cultures need not be refrigerated. It is used by culture collections such as the National Collection of Industrial and Marine Bacteria referred to at the beginning of Chapter 2. If one orders a culture of bacteria from such a collection, one receives a little ampoule containing a speck of a dried serum–sugar mixture with the dormant microbes in it; on transfer to a suitable sterile liquid medium they will revive and multiply.

Freeze-drying has only been widely adopted for about fifty years and, though some microbes are known to die out over several years even when freeze-dried, others that were dried in 1950 are still alive. (Or perhaps I should say capable of being revived, so that I do not beg the question whether a freeze-dried population is truly alive!)

Microbiologists, industrialists and research workers may wish to keep their microbes alive, but in many day-to-day circumstances the problem is the converse: how to kill them. I have referred to sterilization throughout this chapter. How, in fact, does one sterilize something? I mentioned pressure-cooking and baking, and touched upon filtration and γ-irradiation, but these are of rather limited general application. Disinfection is quite a serious problem in general hygiene and is often carried out rather inefficiently, so I shall survey the matter briefly here. For completeness I shall repeat the processes mentioned earlier.

Pressure-cooking: In hospitals and laboratories, instruments, culture media and infected material are sterilized in large pressure cookers filled with steam – called autoclaves – so that everything reaches a temperature of 120 °C for at least fifteen minutes. This is long enough and hot enough to kill even the most heat-resistant spores, but one must remember that, even in steam under pressure, the middle of a heap of blankets, or of a large bulk of liquid, takes a long time to reach the temperature of the steam.

Steaming: All vegetative microbes, that is microbes that have not formed spores, are killed by steam, so if one steams a material, waits for the spores to germinate and steams it again, the chances are that few if any spores will remain. This process is sometimes used for delicate media that would not stand pressure-cooking. Usually one steams a third time for luck.

Boiling: Boiling in water is a rough and ready way of sterilizing utensils, sometimes used with dental and surgical instruments. It is perhaps the most practical procedure for domestic emergencies.

Pasteurization: Milk and some other foods can be spoiled by steaming in the sense that their flavour is impaired. If they are heated to about 70 °C for a short time, nearly all vegetative microbes are killed and only spores remain. Thus, though they are not sterile, they last longer than they otherwise would have done without going bad. Beer, cheeses and milk are often protected, though hardly sterilized, in this way.

UHT: Milk is particularly susceptible to changes in flavour and quality on boiling or pasteurization. This problem has been overcome in an ingenious way by ultra-high temperature (UHT) treatment. The milk (or cream) is heated to a high temperature (well above boiling point) for a very short time (about four seconds) and cooled rapidly. The heat lasts long enough to kill virtually all the bacteria but is too short to affect the flavour and quality perceptibly. Packaged in good sterile containers, such milk lasts several months in near pristine condition.

Ultraviolet irradiation: I mentioned in Chapter 3 the lethal effect of sunlight on airborne microbes. The most active wavelengths are in the short ultraviolet range, around 260 nm, and by irradiating transparent objects with a UV-lamp one can sterilize them. The air in operating theatres and in bottling plants in the pharmaceutical industry can be sterilized in this way; the radiation is rather damaging to human skin and particularly to the eyes.

γ-irradiation: this is as lethal to microbes as to any other living things, though highly resistant species such as *Deinococcus* (earlier called *Micrococcus*) *radiodurans* exist. Such radiation can be used to sterilize opaque but heat-sensitive materials: electronic equipment, for example. The process is used to sterilize certain research materials such as the plastic Petri dishes mentioned earlier and has been used to treat components of space vehicles, to avoid interplanetary contamination. It can be used to sterilize food, thus increasing the shelf-life of fresh vegetables, for example, and is very effective because it rarely affects flavour. The objection to it is a lamentably human one: that it might be misused to disguise food which was already sub-standard, or even bad.

Filtration: Very fine filters are available that will sterilize liquids by filtering out microbes. They are rarely used outside the laboratory except in preparing sera for injection.

Chemical sterilization: Disinfectants, such as phenol (carbolic acid) and its numerous proprietary variants, are simply poisons that are more lethal to microbes than to people and animals.

As I indicated in my examples, these various procedures provide a range of choices for use in different contexts. A century or so ago microbiologists were rather casual in the way they handled microbes – an in-joke among microbiologists today is that the beards of the great figures of the nineteenth century were unrecognized culture collections. In those days, dealing with a pathogen could be dangerous, and episodes in which microbiologists became infected by their research material, and sometimes died, were not unusual; escapes of pathogens from microbiology laboratories were also known. Such incidents are fewer today because a rational technology directed towards microbiological safety has grown up.

Consider, for example, how many of the techniques I mentioned contribute to a well-run operating theatre. Here the primary objective is to protect the patient from pathogens, which are inevitably abundant in hospitals and which may be in the air, on the walls and floor, on the instruments, on and in the medical staff. So the incoming air will be filtered, with the ventilation arranged so that the air is at a somewhat higher pressure than outside, which prevents outside air, with its quota of dust and microbes, from being sucked in under doors or through leaks (there will be no windows). Walls, floors, installations and so on will have been swabbed with a chemical disinfectant to remove microbes left over from last time, and ultraviolet lamps will cope with any microbes which are missed and stirred up by human activity. The operating staff will wear clothing that has been autoclaved, including masks and hair-covers to prevent their own germs from being spread around; they will wear plastic gloves which have been γ-irradiated and use instruments which have been steamed. The patient will wear autoclaved garments and the skin around the site of the operation will be swabbed with a chemical disinfectant, to kill the microbes inhabiting that skin. When it is all over, left-overs will be sterilized or burned, and the theatre disinfected for the next patient.

Asepsis in all aspects of hospital work is extremely important. One of the few disadvantages of the era of antibiotics has been that they are so good that medical staff have come to depend on them to protect against 'casual' infections. Medical microbiologists are today concerned because microbiological standards in general hospital practice have become less rigid now than they were in mid-century, and in-house infections, sometimes by antibiotic-resistant strains, occur much more often than they ought; I shall return to this matter in Chapter 9. An especially troublesome example is MRSA, which stands for methicillin- (or multiple-) resistant *Staphylococcus aureus*. It is a special strain of the familiar yellow staphylococcus which originated in Australia in the late 1970s and which has spread to hospitals in many parts of the world, including Britain. The bacteria have acquired resistance not only to an otherwise useful type of penicillin called methecillin, but also to half a dozen or so unrelated antibacterial substances. The staphylococci do little harm to healthy people, but they can cause very serious and sometimes lethal blood-poisoning infections in patients with injuries or weakened immunity. Every surgical operation, even a minor one, causes injury. Therefore, if MRSA is detected in a hospital, the wards are usually closed and thoroughly disinfected. Until 1997 at least one further line of defence against MRSA was available: the bacteria were sensitive to an antibiotic called vancomycin. Sadly, in May of that year, a sub-strain resistant to that antibiotic, too, was reported in Japan. As was widely foreseen

– even in the last edition of this book – its appearance had been only a matter of time; it will become a serious hazard if it is not prevented from spreading.

Research microbiologists who handle pathogens have somewhat different concerns: their objective is to protect themselves, and to prevent their research microbes from escaping from the laboratory. Many of the precautions I described for an operating theatre apply, with the important difference that the air pressure in the laboratory will be *below* that outside. In extreme cases protective gear, total changes of clothes, disinfectant showers and so on are needed.

Fortunately, most microbes are harmless, and the concern of the researcher is to protect the culture from contamination by air- or breath-borne 'little strangers'. Precautions are needed, but they are less drastic; the Health and Safety Authorities of most countries now specify precautions – called containment levels – for various classes of microbiological research.

In the home, of course, all this palaver would be intolerable. But asepsis is sometimes called for and chemical disinfectants are most appropriate. Chlorine is a good disinfectant; as I mentioned before, it is used in domestic water supplies and, while it cannot be used in concentrations that would completely sterilize drinking water, it keeps gross microbial contamination under control. In swimming pools it is used to prevent cross-infection in crowded conditions; it is useful in preventing transmission of infection between adults and bottle-fed babies. Yet it must be used sensibly: I have seen a mother religiously sterilize bottles and teats with Milton, pasteurize the feed and then, at the last moment, touch the teat to her hands or lips to see if the temperature is correct! Thus she nearly subcultures her skin or mouth flora into her baby. Splash, by all means, but do not touch, should be the rule.

The phenols are good general microbicides and can usually be used to swab floors and walls safely, but they are quite powerful poisons and should be kept away from skin and food. They do not, as many people believe, kill smells; that belief has arisen because they have a strong smell of their own. Of course, by killing the bacteria responsible, they may stop the smell of putrefaction. Ordinary soaps and detergents are only moderate disinfectants, but there is a class of detergents, the cationic or quaternary detergents, which are excellent disinfectants for use on the skin. They form the basis of many proprietary creams. Disinfectant powders, as I wrote in Chapter 3, are the basis of deodorant creams and powders; they kill the microbes that ferment sweat and cause it to smell.

Certain simple chemicals such as copper salts have some disinfectant action and are used in horticulture (as Bordeaux mixture). They are rather poisonous.

Disinfectants need time to act. It is no good, for instance, pouring carbolic acid down a smelly sink and washing it away at once; likewise it is a waste of money to chlorinate a WC and immediately flush it. Because disinfectants are selective poisons they rarely act instantly. The quaternary detergents seem to be something of an exception to this rule, in that they act at once if at all, but generally it is wise, when using a chemical disinfectant, to expose the material being treated to the disinfectant for as long as is reasonable. They function, moreover, by reacting in a chemical fashion with the living microbes, so if there are a lot of other materials present for them to react with there will be less disinfectant available to kill the microbes. One would need far more phenol to kill 10 million bacteria in soil than one would need to kill the same number of bacteria in water, simply because much of the phenol would react with the soil particles and be neutralized as far as the microbes were concerned. Or, to return to my earlier domestic example, a baby's bottle with encrustations of milk will require much more Milton to sterilize it than a clean bottle, because the chlorine reacts with the milk solids as readily as with the microbes. That is why clinics are so insistent that the Milton method be used only with clean bottles.

Disinfectants are rather different in principle from antibiotics and chemotherapeutic drugs. They are general biological poisons which, as I said, kill microbes more effectively than they kill higher organisms. They are, to use microbiologists' jargon again, microbicides (a word analogous to insecticides) whereas many drugs and antibiotics do not kill microbes at all, but merely prevent them from multiplying. Though disinfection is an important and necessary part of the day-to-day hygiene of civilized communities, it is important not to become obsessed by it. As I have already pointed out, we need some exposure to infection to develop any resistance to disease at all. Perhaps the mother who touched the baby's teat to her lips was effectively wiser than we who raise our hands in horror at the thought? The answer, of course, is that she was not, because she did it out of ignorance, and she might, for example, have had gingivitis. The wise thing is to know what one is doing and why one is doing it. One can take liberties with microbes if one knows one is doing so; to take them in ignorance is to court disaster. Where have you heard that before, you ask? It is as true of atom bombs as of microbes, so remember that scientists, while they may know more of their subjects than politicians or housewives, have gained with their knowledge a realization of how profoundly ignorant they really are. Science has made fabulous advances during the present century, yet each fragment of knowledge teaches us how much more we have yet to learn.

5 Microbes in nutrition

I imagine that few people are unaware that beers, wines, cheeses and so on are prepared by allowing microbes to act on foodstuffs; even fewer can have failed to recognize that food goes bad through the action of microbes. But these two kinds of microbial activity are relatively minor aspects of the importance of microbes in the whole field of human and animal nutrition. In this chapter I shall naturally deal with food preparation; its spoilage will crop up in Chapter 7. But I shall start with a topic which is quite different, yet which is perhaps the most important stage in nutrition: the assimilation of food.

Assimilation, technically speaking, is the process that follows digestion. Once food is eaten, digestion starts, and enzymes of the mouth, stomach and intestines break the food down into chemical fragments which the organism can absorb into its blood stream and use for its biochemical purposes. Carbohydrates are broken down to sugars, proteins to amino-acids, fats are partly broken down, partly emulsified. Some components of food – woody matter, for instance – are not readily broken down by the digestive enzymes, and it is here that microbes come in. Ruminant mammals, such as sheep or cattle, have a primary stomach (called the rumen) in which grass, which is almost the only food they eat, quietly ferments. The rumen is a sort of continuous culture of anaerobic microbes, including protozoa and bacteria, which collectively ferment the starch and cellulose of grass to yield fatty acids, methane and CO_2. Rumen juice is extremely rich in microbes – up to 10 billion organisms/ml is commonplace – and they are very active: an ordinary cow produces 150 to 200 litres of gas a day and a large, well-fed, lactating cow is almost a walking gasworks at 500 litres a day. (The gas, by the way, emerges from the mouth, as a belch, not from the rear end.) Some of the microbes are quite difficult to culture in the laboratory because they are so sensitive to air: with all this gas production the fluid in the rumen is completely anaerobic. This culture is diluted steadily by the animal's saliva and by water from the grass eaten; thus the contents of a typical sheep's rumen are replaced once every day. Therefore the rumen discharges into the further intestine a mixture of mainly bacteria, fatty acids, gas and a few unfermentable fragments of the food. The animal assimilates almost entirely fatty acids and fragments of dead microbes. Since fatty acids are equivalent to carbohydrate, it is the microbes that provide the animal with the vitamins and amino-acids necessary for its growth. Sulphate-reducing bacteria, which are also present in the rumen, assist by generating sulphide from any sulphates ingested with the grass, and a sheep can apparently use this sulphide to form part of its protein.

Kangaroos, tree sloths and several other animals also conduct foregut fermentations. One of the most strange, to my mind, is the hoatzin, a pheasant-sized Venezuelan bird (*Opisthocomus hoatzin*). It is vegetarian, and its crop and oesophagus together act as a rumen, housing a fermentative microbial population that is especially good at detoxifying plant alkaloids. This enables the bird to enjoy a wide-ranging diet including the leaves of plants which are poisonous to most other creatures. The plumage of the hoatzin is exotically beautiful and it is said to be unusually tame for a wild creature, but apparently it has, for humans at least, a personal problem: its ruminant habit causes it to smell unpleasantly of cow, which has earned it the colloquial names 'stinky cowbird' or 'stinky pheasant'. (I confess to never having smelled one.)

The hoatzin illustrates a subtle advantage of having a rumen or comparable anatomical fermenter: the friendly microbes therein destroy substances which would otherwise poison their host. Plants rely on poisons – including alkaloids, phenolic substances and proteins called lectins – to protect themselves from large animal predators as well as from microbial pathogens and insect parasites; in fact,

A HOATZIN OR STINKY COWBIRD ON ITS NEST. (Newcastle University Expedition 1992/Bird Life International)

it is fair to say that the majority of plants are in some degree poisonous to most animals. The leguminous tree *Leucaena*, for example, is very useful in tropical forestry and agriculture, not least because it grows very rapidly and is a renewable fuel resource which, since it harbours symbiotic nitrogen-fixing bacteria in its roots, requires no added nitrogenous fertilizer. Not surprisingly, its leaves are naturally rich in nitrogen and decay to a good compost where they fall. The leaves ought also to be a very nutritious food for ruminants, too, but in Australia they proved to be poisonous to goats. Apparently they contain a substance which, in the stomach, becomes converted to a powerful poison called dihydroxypyridine. However, they do not kill Hawaiian goats. It transpired on investigation in the late 1980s that the rumen flora of Hawaiian goats includes bacteria which break down dihydroxypyridine and render it harmless. The species of bacteria responsible was duly isolated, cultivated in the laboratory and, when injected into the rumina of Australian goats, established itself there and enabled them to eat *Leucaena* leaves safely. Moreover, Australian sheep, when provided with the necessary microbe, could also feed on hitherto poisonous *Leucaena* leaves. It is a success story: the microbe, *Synergistes jonesii*, is now used deliberately by Australian sheep farmers. The exploitation of intestinal microbes, natural or bred by genetic manipulation, to protect vegetarian animals from plant toxins, and to aid their nutrition in other ways, is a promising new research direction for agricultural and veterinary science.

Wood-eating insects such as termites also rely largely on populations of cellulose-decomposing bacteria in their guts to decompose wood into materials they can assimilate. Shipworms, which are marine molluscs that bore holes in wooden ships, have cellulose-decomposing bacteria in a gland off the intestinal tract. An interesting example of this sort of nutritional interdependence involving two microbes occurs in the protozoon *Crithidia oncopelti*. This unicellular microbe has symbiotic bacteria inside its cell, actually within its protoplasm, which apparently supply their host with an amino-acid, lysine, which it needs for growth. The bacteria are sensitive to penicillin but the protozoon is not. If *C. oncopelti* is freed of its bacteria with penicillin – cured of its infection, so to speak – it dies, unless it is provided with lysine.

Carnivores, and omnivores such as man, are obviously less dependent on microbes for their nutrition. For one thing, they tend to eat sheep and cattle, thus by-passing the problem of converting cellulose and starch to protein. Though they eat vegetable matter, the cellulose, which constitutes the major part of it, is almost entirely excreted. Nevertheless, the mouths and lower guts of humans and other animals are also microcosms of microbes. Man, for example, lives with two continuous cultures: the mouth and the colon. I discussed the flora of the normal mouth in Chapter 3. Many of its inhabitants survive the acidity of the stomach and are to be found in the lower intestines: the lactobacilli and streptococci are usually there. But a new flora is also present: a rod called *Escherichia coli*, methane-forming bacteria, gas-producing bacteria of the group *Clostridium*, the ubiquitous rod-shaped anaerobes called *Bacteroides* and, usually, yeasts. New types of lactic organisms and streptococci are also found and, sometimes, non-pathogenic protozoa. The combined activities of these microbes can, after a starchy meal for instance, cause discomfort, because the gas they produce from incompletely digested food is the main source of flatulence or wind. Such gas – called flatus – is mainly nitrogen from swallowed air with about 25 per cent of methane and hydrogen and a little carbon dioxide. Though they are not consistent inhabitants of the human colon, sulphate-reducing bacteria are sometimes present and the hydrogen sulphide which they produce decreases flatulence by inhibiting methane production. Regrettably, the smell of the hydrogen sulphide makes the flatus especially offensive. The regular intestinal bacteria are beneficial, however, because they synthesize several substances, while growing and fermenting, that are invaluable to our nutrition. These are all members of the B group of vitamins and it is, indeed, quite difficult to render normal, healthy individuals deficient in B vitamins: during the Second World War, volunteers remained perfectly healthy for weeks on diets of polished rice which ought to have given them beri-beri in a matter of days. Given a brief course of a sulphonamide drug, which killed off much of their intestinal flora, they rapidly succumbed to deficiency diseases. This is the reason why doctors, if they know their job, look out for vitamin deficiencies in patients who have been treated with antibiotics: though the important site of action of the antibiotic may have been elsewhere, the drug usually has a fairly drastic effect on floras of the mouth and intestine. Mysterious gut disorders and irritations which sometimes occur after a course of antibiotics often have a similar origin: the intestinal flora becomes unbalanced and produces troublesome physical reactions as the population returns to normal.

One of the important vitamins synthesized by the intestinal flora of both humans and animals is vitamin B_{12}. This is a complex chemical containing the metal cobalt which, among other functions, is concerned in blood formation; its discovery revolutionized the treatment of pernicious anaemia. The intestine contains organisms which synthesize B_{12} and also organisms that break it down, and in young animals the amount of B_{12} actually assimilated depends on the balance between these two types. In the early days of antibiotics, the left-over mould, being a perfectly wholesome form of vegetable matter as far as anyone could see, was tested as feed for pigs and chickens. It seemed a near-miraculous discovery, like having one's cake and eating it, when such animals were found to grow and put on weight dramatically. They did not grow into giants, but they reached adult weight uncommonly rapidly and economically. Antibiotic wastes appeared to be a sort of 'Food of the Gods'. The precise mechanism of this action is still uncertain, but the major factor is simple: the bacteria that destroy B_{12} are more sensitive to antibiotics than those that make it. The wastes used as feeds contain traces of the original antibiotic and the upshot was that the B_{12} balance in the animal's intestine was shifted in favour of assimilation by the animal. This is not the whole story: antibiotic residues usually contain B_{12} themselves, which, though it is of a slightly different chemical composition from the usual bacterial vitamin, assists the animal's nutrition. Antibiotic residues containing traces of antibiotics are now routinely used in intensive animal husbandry and have, in fact, proved to be a mixed blessing. There is no doubt that their widespread use has had two undesirable consequences. The first is that antibiotic-resistant pathogens have been selected for and have caused antibiotic-resistant epidemics among farm animals and (allegedly) man. The second is that traces of antibiotics (e.g. penicillin) have come through into animal products such as milk. The proportion of patients showing penicillin sensitivity steadily increased during the first two decades for which it was in general use. (In 1970 some 7 per cent of patients in the USA reacted allergically to this antibiotic.) The probable origin of such sensitivity is the continuous consumption of small amounts of penicillin in, for example, milk, and this can lead to an allergic response when the person receives the large dose needed to treat disease. The widespread use of antibiotics in food production is risky, not only because it can cause selection of antibiotic-resistant bacteria, but also because it decreases the effectiveness of our medical resources against other microbes. In the light of their effect in cheapening food and making a decent standard of protein food available to more and more people, one can argue that the risk is worth taking; the best compromise, which is now being widely adopted, is to stick to antibiotics which have no medical use for meat production and to avoid those which persist in the product.

An understanding of the function of microbes is extremely important in agriculture. Obvious examples occur in the diseases of farm animals. Enteric diseases, usually mild, are endemic in poultry and can cause salmonelloses in man. Although bovine tuberculosis was all but eliminated from cattle by tuberculin testing in the 1930s, it keeps reappearing; there is quite strong evidence that badgers are the natural reservoir of bovine tuberculosis. Conservationists resist plans to poison the badgers, creatures which have gained a cuddly public image despite some curiously unprepossessing social habits. I sympathize, but the moral position is a difficult one. Drastic measures are certainly necessary when virus diseases such as foot and mouth disease or fowl pest get established: with such scourges there is nothing to be done but to massacre and destroy the infected cattle or poultry, as the case may be.

One of the most unexpected, and unfortunate, developments in animal health occurred in Britain during the 1980s, when, it seems, the scrapie agent of sheep acquired the ability to infect cattle (I write 'it seems' because a possibility remains that it was a rare, unrecognized, kind of scrapie endemic in British cattle that became widespread). It was unexpected because cattle had co-existed with scrapie-infected sheep for over two centuries and no transfer of infection had occurred. The new disease, called bovine spongiform

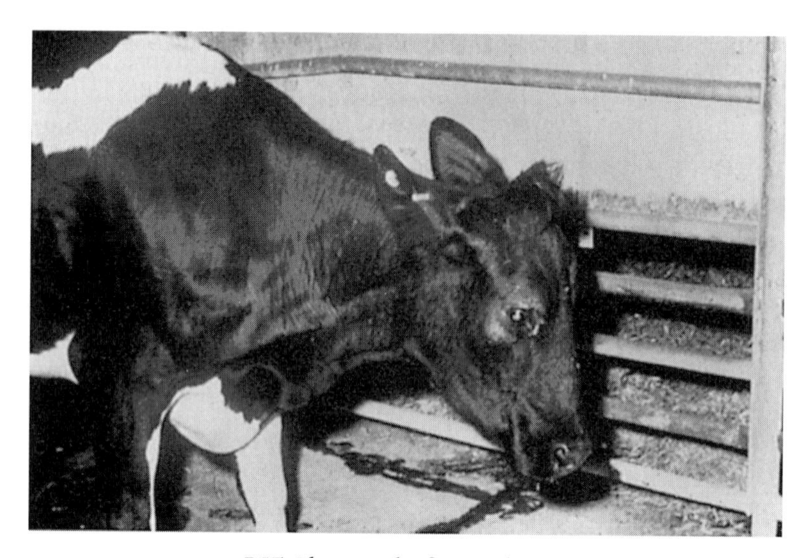

A CASE OF BSE. Photograph of a cow showing symptoms of bovine spongiform encephalopathy ('mad cow disease'). Characteristically, the head is carried low and the ears are held back. (Courtesy of G. A. H. Wells, Central Veterinary Laboratory)

encephalopathy, BSE for short, was first detected in Sussex late in 1984. It afflicted the bovine's brain and nerve tissue, which then became sponge-like; it was dubbed 'mad cow disease' by the press because it caused the animal to lose coordination and stagger about before it eventually died.

How did it happen? The consensus is that a sadly misguided agricultural practice led to it. Cattle food supplements which contain protein augment growth and meat production, and after the 1950s farmers in Britain, and some other countries, fed concentrates containing cooked offal, including sheep and cattle meat, to their stock. Although this was an unnatural food for a vegetarian ruminant, the practice seemed logical because the meat protein was well cooked and largely denatured, and, anyway, meat protein is not wholly alien to herbivores: they normally eat the placentas after birth of their offspring. However, probably because of a change in the technology of preparing such concentrates, undamaged scrapie agent survived the treatment and proved able to infect the hitherto immune cattle by way of the gut. This probability was only slowly understood, but in 1988 offal was banned from cattle feed and infected cattle had to be destroyed. These measures were not always pursued with adequate efficiency at first, and numerous consequent problems emerged. For one thing, the disease takes at least five years to incubate, so vast herds of seemingly healthy cattle were at risk; for another, the possibility existed that a cow incubating BSE could infect her calf – the possibility is still not firmly excluded, though if it happens at all it is very rare. Again, some people feared that the infective agent (a prion; see p. 26), having 'jumped' from sheep to cattle, could also jump to other species, and indeed something very like BSE was early detected in antelopes which had been fed protein concentrates in zoos. Might it then spread to humans?

That fear proved to be devastatingly probable when in 1994 a new kind of spongiform encephalopathy in humans was recognized. It took the form of Creutzfeldt–Jakob disease (CJD) but attacked younger people than did regular CJD; it also showed a rather different pattern of symptoms and killed more quickly. Prions are all so similar biochemically that the only way in which they can be distinguished is by the responses they elicit when injected into the brains of sensitive experimental animals, a slow and unsatisfactory procedure at best. However, preparations of the new form of CJD (shorthand name: nvCJD), tested in laboratory mice, caused a disease pattern indistinguishable from that caused by preparations of the BSE agent, while preparations of regular CJD caused a different disease pattern. The conclusion seems inescapable that nvCJD is caused by the same agent as BSE, and that it arose from the unwitting consumption of BSE-infected beef.

The number of cases of nvCJD was only about forty-one by July 1999, but it is lethal and currently incurable. Its incubation period is not known; it could be long, because in mid-1997 a case of nvCJD was diagnosed in a patient who had become vegetarian over a decade before the disease struck her. The new fear at the time of writing is that there may be many more cases of nvCJD incubating among Britain's population.

The BSE epidemic originated in Britain with over 150,000 cases recorded by January 1996, distributed among over 33,000 farms; at its peak over 200 new cases a month were appearing. In countries such as France, Switzerland and Spain confirmed cases were fewer, numbering only tens or hundreds by mid-1996. Nevertheless, as most readers will know, BSE has undermined beef consumption in Western Europe, and had catastrophic effects on the British beef industry, effects that have spilled over to beef from verifiably BSE-free herds. Many people no longer eat beef at all. Arguments have raged about how well the government actually enforced its seemingly stringent regulations about handling brain and nerve tissues during the butchering of meat, and the European Community quite reasonably followed the lead of the USA and banned the import of British beef and cattle. On the domestic scene the extent to which the nation's cattle must be culled, how, and who is to pay, have all generated political turmoil.

The good news is that feeding, culling and butchering regulations seem to be taking effect. Vast numbers of cattle, many barely at risk at all, have been or are being culled, and by 1998 the rate at which new cases of BSE were turning up was declining impressively. The chances are that BSE will be eliminated from the British cattle herd by around 2005.

But what of nvCJD? And are there other, less dramatic, new prion diseases around? Could BSE have 'jumped' back into sheep, rendering them a more dangerous source of an encephalopathy transmissible to other creatures than once they were? Prospects for both animals and people are far from clear, because we still know too little about the behaviour and action of prions. Not for the first time, a social situation has out-run the underlying science, but common sense suggests that steps should be taken to eliminate both BSE and scrapie from all agricultural herds – a formidable but not impossible task.

There are about 200 recognized microbial infections of farm animals, including such drastic diseases as tuberculosis, brucellosis, trypanosomiasis and rinderpest; in 1956 the US Department of Agriculture estimated that livestock production could be doubled in the developing countries if control of communicable microbial infections could be made adequate. Within limits, quarantine regulations and chemotherapy keep these diseases under control, but occasional epidemics can occur and the situation will not be satisfactory until they can no longer happen. In the 1960s there was an outbreak of African swine fever, a virus infection of pigs, which obtained a foothold in Spain and Portugal and seriously threatened the European bacon industry. In 1991 'blue ear disease', which makes sows abort and piglets die, appeared in Germany, Holland and Belgium; it resembles a plague, probably caused by a virus, which had already afflicted pigs in the USA for several years. African horse sickness has appeared briefly in the Middle East, and also a South African strain of the foot and mouth virus. The rapidity and ease with which people and animals can travel about the world today make these persistent foci of infection a source of danger not only to the developing countries but also to more developed communities. The question of the control of livestock diseases is one of the urgent topics being faced by the Food and Agriculture Organization of the United Nations.

These matters are probably familiar, in principle if not in detail, to any intelligent reader of the newspapers. What is less well known, perhaps, is the importance of plant pathogens in crop production. Plant diseases, spread by insects, dispersed by wind or transmitted from root to root in soil, can cause enormous losses in agriculture: a figure of £1,900,000,000 was quoted in 1965 as the annual loss to the USA due to plant pathogens. Rusts, primitive types of fungus that damage cereal crops, destroyed enough cereals in New South Wales in 1947–8 to feed three million people; in 1935 a third of the banana crop of Jamaica was destroyed by Panama disease, caused by a pathogenic variety of the fungus *Fusarium oxysporum*. In 1956 nearly 40 per cent of the rice crop in a part of Venezula was damaged by a disease called *Hoja blanca*, due to a virus. Bacteria belonging to the genus *Erwinia* will rot plants and cause the dreaded 'fire blight' of fruit trees. Citrus stubborn disease, which leads to stunted growth and mis-shapen fruit, is attributed to a cousin of the mollicutes called *Spiroplasma citri*; in 1969 it afflicted over a million trees in California.

Catastrophes of these kinds look remote as paper statistics, but in practice, particularly in the less developed countries, they mean that great numbers of people will go hungry, starve and possibly die. Generally speaking, crop plants have a fairly high natural resistance to pathogens or they would not be in use; disastrous infection results from bad or unfortunate husbandry. Proper attention to the organic content and alkalinity of soil can often prevent infections spreading drastically. In recent years the possibility of using antibiotics on crops has been considered seriously and they can be effective. However, they are prohibitively expensive for all except the richest countries, which normally have least need for them. The use of other dressings, sulphur dressings, Bordeaux mixture and so on, to prevent fungoid blights in horticulture and viticulture is, of course, traditional, and it is interesting that sulphur dressings are effective because of a microbe. *Thiobacillus*, the sulphur bacterium which I first mentioned in Chapter 2, slowly oxidizes the sulphur to sulphuric acid on the surface of the plant, gently producing an environment that is too acid for the development of pests such as the fungus *Oidium*, yet which is not so strong that it damages the grapes.

Plant pathogens can thus cause serious social and economic disasters, as well as being a constant irritant to farmers. But everything has its obverse side: in the USA and Australia carefully chosen pathogenic fungi have been used successfully to keep weeds in check. For example, there is a weed evocatively named 'stranglervine' which damages citrus groves, and preparations of a fungus called *Phytophthora palmivora*, which can be used to control it, have been available commercially in Florida since the early 1980s. Conversely, certain strains of *Bacillus* make antibiotics which act against fungal pathogens of plants, and have been used successfully to control 'damping-off' disease of seeds and seedlings.

Biological control of agricultural pests, by deliberately encouraging the spread of microbes antagonistic to the pest, is now a serious prospect. A striking example of so-called microbial biocontrol is the impressive biological warfare conducted, unofficially in fact, against rabbits after the Second World War. In May 1952 a few rabbits in Eure-et-Loire, France, were infected with the virus disease myxomatosis and released. The disease spread rapidly and in October 1953, it reached Britain, spreading from near Edenbridge in Kent. By the end of that year the disease had spread through twenty-six departments of France and had reached Belgium, Holland, Switzerland and Germany, killing 60 to 90 per cent of the rabbit population. It is now endemic throughout Europe. Resistance to the disease has been acquired by rabbits only slowly, and though resistant populations are on the increase, local epidemics still occur in the countryside. It is a disgusting disease in its symptoms, but there is little doubt that the post-war revival of European agriculture would have been much slower without it. In some areas productivity increased three-fold. In the well-fed communities of Europe today, we can afford to share the doubts of animal lovers about destroying the rabbit, once an engaging feature of the country scene, and officially the deliberate spreading of myxomatosis is frowned upon in Britain. But the farmer has no such doubts: in 1964 I was told that the black-market value of a well-myxomatosed rabbit was £50, a lot of money in those days.

Rabbits were introduced into Australia from Europe in the mid-nineteenth century, and, with no natural predators, rapidly became a plague. In the 1950s Australians deliberately introduced myxomatosis, but with only short-lived success: within little more than a decade the population had acquired resistance. However, a wholly different rabbit virus, called a calcivirus, endemic in Central Europe, proved in the 1990s to be very effective. In 1995, while it was being field-tested on Wardang Island off the South Australian coast, it escaped to the mainland and spread locally. It killed millions of rabbits and so impressed farmers that infected rabbits were rumoured to change hands at A$400 each. In the summer of 1996 it was released intentionally at numerous sites, despite some misgivings about possible environmental side-effects. Those misgivings caused the New Zealand authorities to reject its use but, predictably, it was smuggled in from Australia in 1997 and has now taken hold there. Though warm weather seemed to slow its spread in Australia, possibly by influencing some insect which assists its transmission, it promises to be a valuable form of biological control – until calcivirus-resistant rabbits appear!

Smaller pests can also be controlled with microbes. There exist predacious fungi that trap and digest potato eel-worms, and there are several preparations of bacteria available which can be used against insects. *Bacillus thuringiensis* has proved to be especially valuable in the control of agricultural pests. When it forms spores it also generates proteins which are extremely toxic to the larvae of numerous species of insect. Caterpillars, for example, will inadvertently eat spore-forming *B. thuringiensis* residing on leaf surfaces and be rapidly killed. Many different strains of *B. thuringiensis* are now known, and a corresponding variety of toxins are produced; one can match the toxin to the particular pest causing trouble. Preparations of spores of *B. thuringiensis* for crop-spraying have been marketed since the 1970s and are admirable for organic farming, being harmless except to certain insects. Despite worldwide use, cases of resistance among the target insects have been remarkably few. Care must be taken to use pure strains: *B. cereus* closely resembles *B. thuringiensis* but can, albeit rarely, be pathogenic to man.

A success story of the 1980s was the use of a variety of *B. thuringiensis* called *B. israeliensis* to control black fly in the African tropics. These flies are the vectors of river blindness, a particularly crippling disease (caused by a nematode worm, not by a microbe), and they had become resistant to chemical insecticides at just about the time that *B. israeliensis* was discovered in Israel. Within six years it was in successful practical use and by 1994 some 30 million people had been protected from the disease. Moreover, 15 million hectares of cultivatable land, hitherto too dangerous to use, had been created. When they work, insecticidal bacteria can be excellent biological control agents. A very promising development of the last decade has been the successful use of genetic engineering (of which more in Chapter 6) to create plants able to make *B. thuringiensis* toxin for themselves, so that their leaves become intrinsically toxic to caterpillars or other plant pests. However, there is, as so often, a catch: a spray of toxin has an immediate effect on the pest and the toxin itself soon disappears, but a plant which makes toxin continuously would provide ideal conditions for the selection of toxin-resistant insects.

Insect viruses, too, have been used successfully as agents for biological pest control, for example, on sawflies and gypsy moths in forestry. They are not difficult to apply, they can safely be sprayed, and in an ingenious development bees have been recruited to spread them around crops, but viruses have to be prepared in, and be extracted from, larvae of the target insects, which is an expensive process. Fungi pathogenic to insects have also been used: *Metarhizium flavoviride* shows promising activity against locusts, plagues of which cost over US$400 million in organophosphorus insecticides alone between 1985 and 1989. Research in these directions is very timely, as people begin to realize that some chemical pesticides – not all, but some – can be persistent and dangerous, both to man and to the natural ecology.

I have already mentioned the importance of the nitrogen-fixing bacteria to agriculture. Generally speaking, the important process is symbiotic nitrogen fixation: the process whereby microbes infect a plant and settle in a nodule, and the combination of plant and microbes fixes nitrogen. The best-known combinations are the leguminous plants, clover, beans, lucerne and so on, with their associated bacteria *Rhizobium*. Since some strains of *Rhizobium* form more effective nodules than others, inoculation of leguminous crops with good strains of rhizobium is sensible agricultural practice. But though the legume + rhizobium pair is the most useful agricultural

ROOT NODULES CONTAINING NITROGEN-FIXING BACTERIA. The photo shows lobed nodules on the roots of a pea plant. Within the nodules, symbiotic bacteria, *Rhizobium leguminosarum*, become dormant and convert atmospheric dinitrogen to a form which the plant can use. (Courtesy of Professor John Beringer)

combination, other symbiotic systems exist and can be more important in nature. The alder has a symbiotic nitrogen-fixing microbe called *Frankia* (an actinomycete) and this enables it to colonize arid and mountainous areas. *Shepherdia* and bog myrtle are hardy plants with a similar symbiont that colonize poor soils – heath or bog lands – and when such plants become established they create more fertile conditions and enable other plants to establish themselves. Nearly 140 non-leguminous plants and shrubs are known that fix nitrogen with the aid of nodules. Certain lichens, those combinations of an alga and a fungus of which I wrote in Chapter 2, can, if the algal partner is a cyanobacterium able to fix nitrogen, render a tiled roof so fertile that ordinary flowering plants will grow on it – thus gracing the rural scene. Cyanobacteria are very widespread and many types can fix nitrogen. In 1883 the volcanic island of Krakatoa in the Malayan archipelago more or less exploded, killing many thousands and causing, incidentally, splendid sunsets (due to atmospheric dust) for many years after. The eruption virtually sterilized the island, and the first living things to return were the nitrogen-fixing cyanobacteria. As these renewed the fertility of the soil, other plants, birds, insects and animals slowly returned and the island is now fully recolonized.

The nitrogen-fixing bacteria are still fundamental to the world's food production, and in all but the most highly developed agricultural communities their activities still determine the amount of food produced. Many nitrogen-fixing bacteria are not symbiotic: *Azotobacter*, *Clostridium pasteurianum*, *Klebsiella* and about 80 other types of free-living microbe fix nitrogen in the absence of a plant host and have no need of a nodule. One of these, *Beijerinckia*, was once considered important in the fertility of tropical soils, but the truth of the matter seems to be that they consume so much carbohydrate or similar carbon source to fix a little nitrogen that, from an agricultural point of view, they are not much use. There is just not enough carbonaceous matter available in ordinary soil to enable bacteria of this kind to fix useful amounts of nitrogen; if there were, other bacteria, non-nitrogen-fixing, would consume it more rapidly. Why they are such inefficient nitrogen-fixers is a complicated problem – indeed, they may be more effective in nature than in the laboratory, and their known inefficiency has not prevented agronomists, particularly in the former USSR, from trying to improve farming yields by deliberately infecting the land with such bacteria. Great things were once claimed for soil dressings of azobacterin, a preparation of *Azotobacter* used in the former USSR, but now both Russian and Western scientists doubt its value. The soils treated were often so poor that the peat on which the *Azotobacter* was usually added would probably have done quite as much good on its own, and the situation is additionally complicated by the fact that *Azotobacter* produces auxin-like substances – materials that stimulate plant growth – without in fact augmenting their nitrogen content at all, and thus without improving their value as protein sources for food.

In the present state of knowledge it is safest to regard most free-living nitrogen-fixing bacteria as relatively unimportant to man's economy. The cyanobacteria, however, are very important in arctic regions, where they seem to be the primary source of soil fertility. In waterlogged rice fields of the Far East they provide the main source of nitrogen for the crop. Their importance has been well understood in India and Japan, where methods of farming cyanobacteria to produce a green manure for rice production have been developed. Particularly valuable in rice culture is a symbiosis between a tiny water fern called *Azolla* and a cyanobacterium (*Anabaena azollae*) which forms a marketable green manure rich in nitrogen: it is one of the most efficient nitrogen-fixing systems known. Cyanobacterial nitrogen fixation is both a practical and an intellectually satisfying process: with the aid of sunlight the microbes convert atmospheric CO_2 and nitrogen to the raw material of the basic food of the Eastern hemisphere.

Whole books have been written about the importance of microbes in agriculture, their effect on soil structure and fertility and their role in the decomposition and recycling of vegetable matter. As the late

A MARKET FOR GREEN MANURE. The photograph shows farmers at a market in Vietnam where baskets of damp-dried *Azolla*, with its symbiotic nitrogen-fixing bacteria, are on sale for use as fertilizer for rice. (Courtesy of Dr I. Watanabe)

Professor Hugh Nicol pointed out years ago, the major raw materials of agriculture (soils and manures) are either microbial products or substitutes for microbial products. In a survey such as this I can only be selective and mention what seem to me to be the highlights. After all, this chapter is concerned with nutrition and, though there would be virtually no science of nutrition without agriculture, there are things other than bread and meat.

Beer, for instance. It is curious that, when I mention to laymen the industrial importance of microbes, the first thing they think of is beer. The importance of yeasts in the production of fermented alcoholic drinks has impressed itself on successive generations and if, in a serious book such as this, one cannot truly maintain that liquor is essential to nutrition, one must admit that it certainly enhances the pleasure of nourishing oneself.

Beer is produced by such a complex process that one wonders, when one considers it dispassionately, how anyone ever thought of it.

Yet some kind of beer was made by the ancient Babylonians around 6000 BC, according to a tablet found in 1981, and beer was apparently a unit of currency in cash-less Pharaonic Egypt. The Ancient Britons made a beer from malted wheat before the Romans introduced barley. Essentially the process of making beer is this: barley is caused to germinate by steeping it in water for a day or two and leaving it in a warm, damp place for between two and six days. This process is called malting; gibberellins, mentioned later in this chapter, are used to control it. The grain sprouts and develops enzymes that hydrolyse the starch stored in the seed to sugars; the malt is then killed by gentle heating but, since the enzymes are not wholly destroyed, breakdown of the starch to sugar continues to take place. It is then steeped in water once more, so that the sugars soak out, together with amino-acids and minerals needed by yeasts for growth. This extract, the malt wort, is in due course boiled, to inactivate the residual enzymes, and hops are added to impart a bitter flavour. Though this was only realized in the 1950s, hops also introduce materials that hinder the growth of bacteria in the extract. When the wort is cool, yeasts are introduced and the whole is allowed to ferment for a week or so. The material is not stirred and so, though the yeasts start off by growing aerobically, they rapidly exhaust the oxygen and the bulk of the population grows without air. In these circumstances they convert the sugar of the wort to alcohol and to the gas carbon dioxide; they stop growing when sufficient alcohol has accumulated to be inimical to further growth. After a period of storage, to settle out the yeast, the fermented liquid may be drunk.

Many refinements are employed, according to the type of beer being brewed, but essentially only two types of yeast are used, *Saccharomyces cerevisiae* and its close relative *S. carlsbergensis*, selected for their tolerance of alcohol, their flavour and settling properties. A variety of treatments is used to stabilize the beer (so that it will last), to retain its gaseousness and to prevent precipitation on storage; in addition, extraneous bacteria, lactobacilli and acetobacters, have to be kept under control, because they can spoil the beer by forming lactic or acetic acids. These matters form the bulk of the craft of brewing, for despite advances in our understanding of the process, brewing is far from being a science.

Wine, the fermented juice of grapes, is produced by a similar fermentation process, but nothing analogous to malting is required. Grapes are crushed, traditionally by treading with bare feet but nowadays mechanically, and their juice is collected and fermented with wild yeasts, yeasts which appear naturally on the fruit or which contaminate the wine vats from year to year. These yeasts are usually close relatives of *S. cerevisiae*, though wine specialists often call them *S. ellipsoideus*, and sometimes (mainly in California), pure strains are used. Extraneous bacteria are kept down by sulphuring: treating the grape juice (the must) with sulphur dioxide (or with sodium thiosulphate, which forms sulphur dioxide in contact with the acids of grape juice), which is more toxic to bacteria than to yeasts. Part of the craft of wine-making lies in adding enough sulphur to yield a good fermentation and yet not sufficient to spoil the taste. White wines, which are made from grape juice separated from the skin, pips and stalks at an early stage, often suffer from over-sulphuring, recognizable as a flat after-taste (as of a London fog, if you are old enough to remember). Red wines, which are red because the fermentation is conducted in the presence of skin and pips, so that colouring matters are extracted, seem less prone to troubles from over-sulphuring, possibly because they have a higher content of tannins than white wines: tannins are slightly antibacterial, so less sulphuring is necessary. *Rosé* wines are made by exposing the must briefly to the solid grape debris, and therefore they too are sometimes prone to over-sulphuring. The thought that *rosé* wine can be made by mixing red and white wine is, naturally, unthinkable to any self-respecting wine manufacturer. One must conclude that cheap *rosés* are sometimes made unthinkingly.

The basic process in wine-making is, just as in brewing, a yeast fermentation of a sugar solution. With certain wines, notably the Burgundies, the must is unduly acid, because it has a high content of malic acid. In these circumstances lactobacilli, present on the fruit or in the vat, convert malic acid to the weaker lactic acid and thus lower the total acidity. Oenologists call this process the malo-lactic fermentation. Part of the quality of the sweet white wines called Sauternes is due to the use of grapes that have been partly dehydrated as a result of attack by the mould *Botrytis cinerea*. As wine-making depends on a microbial fermentation, it should be possible to produce wine continuously by continuous culture techniques and, indeed, in the Bodega Cyana region of Argentina this has been done. But the great wines, those with noble names such as *Château Latour, Château Lafite, Château Mouton-Rothschild*, are artistic triumphs rather than works of craftsmanship, and their quality depends almost entirely on details of the viticulture, the clarifying procedures after fermentation, the storage and maturation procedures. These details require a degree of experienced human intervention that is still far beyond the capacity of an automated, continuous process. In maturation the microbes play little or no part, though the contaminant mould *Oidium* usually found on grapes is said to add important flavours to a good wine. During maturation certain interactions between fruit acids and alcohol take place, and a certain amount of precipitation of insoluble matter occurs: the crust of a good, matured wine is a precipitate of tartrates and tannins. Red wines can improve steadily over up to fifteen years, though immediately after bottling they may deteriorate briefly (become bottle sick). White wines show little improvement once bottled.

Reinforced wines, such as madeira, sherries, port and vermouths, are basically wines to which sugar, extra alcohol and sometimes herbs have been added. Microbes play no part in these treatments. In the case of sherries an interesting sort of continuous culture process is used. The wine, duly fermented, is admitted to a sequence of casks in tiers. One cask leads to the next and as many as ten casks may be arranged in sequence. Wine is drawn off from the bottom cask, called the solera, and the process of travelling from first to last cask may take several years. Each cask develops a floating scum or *flor* of yeast, related to *S. cerevisiae* but called *S. beticus*, and though this microbe has little effect on the alcohol content of the wine, it adds certain flavours and aromas that give sherry its famous character. After withdrawal from the solera, sherry is reinforced with brandy and may be sweetened with sweet, fresh wine.

Brandy, and all other spirits, are distilled from wine or malted cereal fermentations. Microbes play no part in their preparation after the fermentation stage and I shall not discuss them further.

Champagne and other sparkling wines are of some special interest, because they make use of a double fermentation process. Fermentation, as I explained, yields not only alcohol but the gas carbon dioxide and, indeed, the gas produced by a bubbly, fermenting vat of malt wort or must can be lethal if a worker accidentally gets trapped in a vat full of it. In the manufacture of champagne some of this gas is deliberately trapped in the bottle. White wine, suitably blended, is mixed with a little syrup and bottled in specially strong bottles with bolted-on corks. It is left in racks to ferment slowly (a special strain of *S. cerevisiae*, a champagne yeast, grows) and sufficient gas is formed to carbonate the wine while yet, if the process is performed properly, avoiding explosion. Over several months the bottles are slowly turned over in racks (called pulpits) until they are upside down and the yeast and sediment have settled on the inside of the cork. At this stage the cork is removed and replaced rapidly, so that a plug of yeast and debris is expelled (sometimes the necks are frozen to facilitate the process) and an equivalent volume of syrup and brandy, in variable proportions, is introduced. This process produces a stable wine which retains its sparkle for a long time after opening. Cheap sparkling wines are produced merely by high-pressure carbonization of still wines, as in the preparation of soda water. They tend to go flat rapidly.

Pétillance, a slight prickliness found in some young, local wines of Portugal, France and Italy, can arise from a related cause: the wine has been bottled before fermentation was complete and a slow residual fermentation has mildly carbonated it.

Great wines are made from fermented grape juice, preferably by the French, though it must be admitted that creditable products are made by certain Rhinelanders, Australians and Californians. Generally speaking, it is the character of the grape and the ability of the wine-maker, rather than the microbe, that determine the quality of a wine.

Cider (from apple juice), perry (from pears) and a variety of fruit wines, root wines or beers and even flower wines can be made. All depend essentially on the fermentation of fruit sugars by yeasts and, except for commercial ciders and perry, the yeast used is usually wild, that is, one that is introduced naturally with the fruit. Pulque, a Mexican beer produced by fermenting the juice of the succulent plant *Agave* (the spirit distilled from it, tequila, is perhaps more familiar to Europeans), is viscous because it contains bacteria of the *Lactobacillus* species as well as yeasts. Saki, the Japanese rice wine, is made by a yeast fermentation, after the mould *Aspergillus oryzae* has broken down the starch of steamed rice to fermentable sugars. Home-made wines can have somewhat knock-out effects, because wild yeasts may produce mildly toxic by-products of fermentation such as acetaldehyde; the results may be a source of transient pride to those who have so painstakingly manufactured them, but it must be admitted that the pleasure of home-made wines usually lies more in the sense of accomplishment than in their gastronomic qualities. Yet accomplishment is not to be sneered at and, if only to prove that even a cursory survey such as this can yield information of practical value, I include the following recipe for a flower wine which, at the appropriate season, can be prepared in the home and can be drunk within ten to fourteen days (it does not keep). It illustrates the principles of champagnization without any of its hazards and, most important, is a pleasant, low-alcohol drink which, because of its short fermentation period, has none of the side-effects of more elaborate brews.

*Elderflower champagne**

Collect, or cause your children to collect, about nine healthy blossoming elderflower heads. They should not come from a site close to traffic. Steep them in one gallon of cold tap water containing one lemon, cut up, two tablespoons of white vinegar and one and a half pounds of sugar. After twenty-four hours strain and bottle. Drink after it has become active – about ten days. Activity is signified by slight effervescence when the stopper is removed and a turbidity in the wine due to the growth of wild yeasts.

This wine illustrates three of the basic principles discussed. Yeasts are present in the nectar of the elderflower (provided exhaust pollutants have not eliminated them); acidity, provided here by the vinegar but normally present naturally in fruit juices, favours their multiplication and prevents growth of bacteria which would otherwise cause distasteful flavours. Secondly, the raw materials must be removed after brief steeping or they too introduce unpleasing flavours. Finally, confining the fermentation in stoppered bottles prevents the escape of carbon dioxide and makes the brew slightly gaseous (the principle underlying champagne fermentation). It is advisable not to forget about the bottles or they may burst.

Fermented milk products have as long a history as wines and beers. Cheeses, for example, were offered to the gods by the Ancient Greeks, possibly as a substitute for ambrosia. Milk is an ideal material for the growth of many kinds of microbes, and some, such as *Streptococcus agalactiae*, a causative organism of bovine mastitis, or *Brucella*, responsible for contagious abortion in cattle, can cause troublesome infections in man. Bovine tuberculosis used also to be a hazard of milk consumption, but modern dairy hygiene has virtually eliminated such risks. Careless handling in the home can, however, re-introduce infectious organisms, and the economical practice of returning tasted but unfinished milk to the jug, particularly chil-

* A tested recipe, for which I am indebted to Mrs Beryl Kelly.

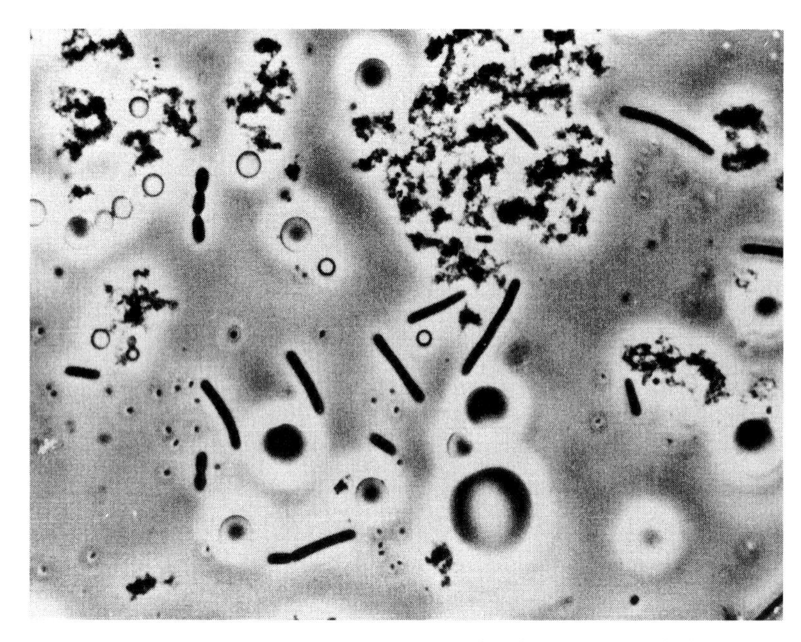

UNSEEN FRIENDS IN YOGHURT. The photomicrograph shows lactobacilli and short chains of streptococci in a wholesome specimen of yoghurt. The spheres are fat globules and the debris is clotted milk protein. The magnification is about 800-fold. (Courtesy of Dr Crawford Dow)

dren's milk, is a cause of many minor domestic catastrophes. Fortunately for most of us, the commonest microbe to sour natural milk is the harmless *Streptococcus lactis.*

Lactobacilli are normal inhabitants of unpasteurized milk. Yoghurt is traditionally made by allowing the subspecies *Lactobacillus bulgaricus* to ferment the milk sugar (lactose) to lactic acid; this clots the milk as well as making the environment unsuitable for the growth of many pathogenic bacteria; another lactic bacterium, *Streptococcus thermophilus,* adds a characteristic creamy flavour; sometimes yeasts are also present. The product is a wholesome food, widely eaten in the Middle East and Balkans and now very popular in Western Europe and the USA. Many supermarket yoghurts are pasteurized to increase their shelf-life, and the acidity of fruit juices is entirely compatible with yoghurt, a fact which enables manufacturers to add a variety of fruit flavours to their product as well as helping to stabilize it. 'Live' yoghurts, those which contain living lactic bacteria, can be beneficial during recovery from gastro-intestinal illness, largely because of their digestibility and vitamin content. Lactic acid bacteria are major inhabitants of our lower intestines, but most of those in yoghurt are killed by the high acidity of the normal stomach. Nevertheless, a few survive, partly because the milk proteins in yoghurt seem to provide a degree of protection from the stomach acid. A recent variant of the traditional yoghurt called bio-yoghurt has become popular because the main fermenter is not *L. bulgaricus* but a lactic acid bacterium called *Bifidobacterium*, a Y-shaped anaerobe which yields a milder flavoured yoghurt believed to be especially health-giving. The idea is that bifidobacteria are also plentiful in healthy intestines, so, provided those in the yoghurt survive passage through the stomach, they displace nastier bacteria such as salmonellae from the gut lining. Not all authorities are convinced; it is indeed likely that large doses of bifidobacteria could have this so-called 'pro-biotic' effect on a bowel disturbed by a much-altered microflora, but one would not expect an occasional pot of bifidobacterial yoghurt to have much effect on a normally fit individual.

Leben is incompletely fermented sheep's or goat's milk; buttermilk is partially fermented skimmed milk that has become viscous because *Leuconostoc*, a filamentous relative of the lactobacillus, has grown in it. Butter owes its flavour to a slight growth of streptococci during its preparation (their growth leads to formation of a flavoursome chemical called acetoin), and most dairies keep starter cultures that are good at forming this material.

Acidity causes milk to clot, and the clot, or curd, arising from ordinary milk fermentation is the basic material of cheese manufacture. Though originally prepared by a microbial fermentation, the curd has for centuries been made by treating milk with the enzyme rennin familiar to many as the rennet used in making junkets. Rennet was traditionally made from the stomach juices of calves, but in recent years, to appease vegetarians, rennet is often obtained from bacteria which have been modified genetically (see Chapter 6) to make calves' rennin. Essentially the curd is a mass of casein, the protein of milk and, after removal of the whey, cheeses are formed simply by allowing further microbial action to take place on the curd. As with alcoholic drinks, whole books have been written on cheese manufacture. Here I can only take a brief look at the subject.

Cream cheeses or *cottage cheeses* are simply the fresh curd, or one which has been allowed to age slightly so that the lactobacilli cause some decomposition of the protein. They do not last long. As such cheeses age, protein breakdown continues further, traces of ammonia are formed, more whey is released and the curd becomes more dense. True *curd cheeses*, familiar as Cheddar or Cheshire cheeses, are made from compressed curds. In those delicious, unsavoury-looking cheeses such as Camembert, Carré de l'Est and so on, decomposition proceeds nearly to the stage of putrefaction, aided by fungi such as *Geotrichum* which grow on the surface of the curd. Considerable amounts of ammonia and amines derived from amino-acids appear. Species of *Penicillium* also grow in the veined cheeses, such as Stilton or Gorgonzola, and in these cases they spread throughout the curd. The veins are due to the spores of the moulds, which are coloured. In *Swiss cheeses*, such as Gruyère and Emmenthaler, *Propionibacterium* grows in the curd, forming propionic acid, responsible for their characteristic flavour, and carbon dioxide, responsible for the holes. *Processed cheeses* can be made from any of the above. They are normally homogenized with fresh curd, plus preservatives, and then pasteurized and packaged to prevent further microbial action. They are good, perfectly nourishing forms of food, but are generally of minimal gastronomic interest.

Cheese-making, like wine-making, is a craft of great subtlety, even in these days of enlightenment about food processing; real Stilton, for example, is only produced near (but, curiously, not at) the village of Stilton, Cambridgeshire. The formative organisms of Camembert are available all over the world, from Australia to the USA, yet somehow the perfection of the authentic Norman product is rarely reached; it is a happy circumstance for the British that the economical Normans choose to export the best of their products and that good, ripe Camembert is found far more often in Britain than in France. It is, on the other hand, a regrettable circumstance that the economical Normans have chosen to stabilize some of their great cheeses by pasteurization, thus stopping the maturation which was one part of their greatness.

Meat or fish that has gone bad through microbial action is unsavoury and dangerous, yet processes in which microbes flavour and help to preserve meat or fish protein have been in use for many centuries. Continental hard sausages, for example, develop their special flavours because they are allowed to age, during which time a population of non-pathogenic staphylococci within the meat ferments some of the protein to form lactic acid, which acts as a preservative, and to yield various flavoursome by-products. Sauces and pastes made of fermented, salted fish and/or shrimp have been part of Mediterranean cuisine since Roman times, and the production of comparable fish sauces constitutes a substantial industry in South East Asia (one of the species of bacteria among the fermentative microflora rejoices in the name *Staphylococcus piscifermentans*).

A major use of microbes in the food industry arises in the baking of bread. Conventional dough is a thick paste of wheat flour in water to which live yeast has been added. During the few hours in which the dough is left to rise, the yeasts ferment sugars in the dough, and the carbon dioxide so produced forms tiny bubbles which lighten (leaven) the bread as it is baked. The process can be imitated by skipping the fermentation step but adding a touch of baking soda instead of yeast (thus making soda bread), but the nutritional value of the yeast – more of that later – is then lost. Making bread with added yeast is historically rather a recent innovation, about five centuries old. A more traditional method, which is essential even today when rye flour is used, is the sour-dough process, which ancient tomb paintings indicate was used in Egypt in the 13th century BC. Rye flour needs more acidic conditions than wheat flour to bake nicely, so its dough is fermented for several hours without added yeast, whereat lactic acid bacteria grow and generate lactic acid; wild yeasts grow alongside and leaven the dough with carbon dioxide. Modern sourdough bakers conserve special 'starter' cultures, populations of lactobacilli plus yeasts, to set the process off. Most modern sourdoughs, such as that yielding the celebrated bread of San Francisco, are made with mixtures of wheat and rye flour. I am told that a pale imitation of the real thing can be made by omitting the fermentation steps and acidifying the dough with citric acid.

Sauerkraut, the standby of many a *Brasserie d'Alsace* in France, is made by allowing shredded fresh cabbage to ferment by the agency of lactobacilli, of a kind closely related to those used to make traditional yoghurt. The lactic acid they form gives the vegetable its characteristic pickled taste and protects it against further microbial decomposition. Actually, the process used to make silage in agriculture is essentially similar: grass is treated so that lactobacilli multiply and the acid they form preserves the material from complete putrefaction. In this case other microbes including clostridia grow alongside the lactic bacteria and contribute to its distinctive smell.

Vinegar, widely used in pickling as well as in everyday food preparation, is a dilute solution of another important food acid, acetic acid. Much commercial vinegar was once produced synthetically by appropriately diluting industrial acetic acid, but this is now illegal in Britain; the name vinegar now means the traditional *vin aigre* (sour wine), made by the action of acetic acid bacteria (*Acetobacter* and *Acetomonas*) on wine. These bacteria usually grow spontaneously when wine is exposed to air and oxidize its alcohol to acetic acid. Traditionally, vinegar is made by allowing rough wine to trickle down towers of birch twigs, or other woody materials, on which a film of acetobacters grow. Thus a sort of continuous culture is formed: the twigs allow access to air and from the bottom vinegar may be tapped off. Any alcoholic beverage can serve for vinegar manufacture: wine (or Orleans) vinegar should properly be made from wine, malt vinegar from beer and cider vinegar from cider, and these beverages are the commonest sources of vinegar in France, Britain and the USA respectively. Many bacteria other than acetobacters are to be found in vinegar towers and the microbiology of the process is not understood in detail.

Microbial fermentations are involved in some other aspects of food production. Cocoa and chocolate are made from the fruit of the cacao tree by a process in which the cocoa bean is first fermented (yeasts starting things off, lactic and acetic bacteria joining in later) and later dried, heated and processed. The fermentation is necessary to develop the chocolate flavour. Tempe is an Indonesian food which is made by part-cooking beans (usually soya beans) that have previously been soaked and damp-dried, inoculating the mash with the mould *Rhizopus* and letting cakes of the mash ferment for some days in packs. The mature tempe is fried in slices, roasted or cooked in other ways; it is nicer than the original beans but no more nutritious. But *Rhizopus* can up-grade the nutritional value of some foods. Cassava, being almost pure carbohydrate, is of relatively low nutritional value; fermentation with *Rhizopus* plus a little ammonium salt, so that the mould uses the ammonium to make protein, has been used to up-grade it. Probably the best-known fermented bean product, however, is that standby of Chinese–American cuisine, soy sauce, the product of the fermentation of soya beans, rice and cereal by (principally) the mould *Aspergillus oryzae*.

Modern methods of food processing and treatment sometimes, though far less often than food-faddists would have us believe, lead to products that are less wholesome than they might be. White bread, for example, is well known to lack several vitamins (E and many of the B group) present in wholemeal bread. So the practice has grown up of manufacturing nutrients of this kind in order to replace those lost in food processing, as well as to enrich foods of limited nutritional value, and for use in medicine. Lysine, which human beings cannot synthesize, is an amino-acid derived from protein: a certain amount must be provided in the diet, and it has been made industrially for the enrichment of bread. One process is interesting because it made use of two microbes in succession: one, a special strain of *Escherichia coli*, cannot make lysine because, as a result of a mutation, it lacks a certain enzyme. It can only make an immediate precursor of lysine called diaminopimelic acid (DAP for short). If, then, this mutant is grown with only a little lysine (for it has to have some or it will not grow at all), its whole lysine-synthesizing system works normally up to the DAP stage, but stops there, with the result that relatively large amounts of DAP collect in the culture. Another organism, *Klebsiella aerogenes*, contains plenty of the enzyme necessary to convert the DAP to lysine, so, in the industrial process, this organism was grown, killed with toluene and the extracted enzyme used to convert the DAP, accumulated by the *E. coli*, to lysine.

This process is gratifying to microbiologists, because, instead of arising from some traditional procedure which microbiologists later came to understand, it developed as a direct result of an increasing understanding of the biochemistry of bacteria. It arose particularly from fundamental work conducted by Dr Elizabeth Work at University College Hospital, London, who originally discovered DAP as a curious amino-acid encountered only in bacteria. It is rare for fundamental research to pay off in so clear-cut a fashion. It has, however, been abandoned in recent years in favour of simpler processes using lysine-producing mutants of *Corynebacterium glutamicum*.

Vitamin C, or ascorbic acid, is one of the few vitamins that can be consumed fairly safely without medical guidance, and it is widely used to accelerate recovery from illness as well as for the treatment of genuine cases of vitamin C deficiency (scurvy). Industrially it is made from a plant product, sorbitol, by a series of chemical transformations, and for one of these steps the material is exposed to a bacterium called *Acetobacter suboxydans*, which oxidizes it more gently than is possible by conventional chemistry. Vitamin B_2, or riboflavin, is produced microbiologically using the yeasts *Eremothecium ashbyii* or *Ashbya gossypii*: in 1957 the latter microbe provided the USA with

400,000 pounds of riboflavin. Vitamin B_{12}, or cobalamide, which I discussed at the beginning of the chapter, is essential in the treatment of pernicious anaemia and, though it was originally isolated from fresh animal liver, it is now produced exclusively using microbes. The demand for it in medicine is small, but its use as a supplement to animal feed makes its commercial production profitable. The streptomycete *S. olivaceus* and the bacterium *Bacillus megaterium* have both been used to produce it industrially; the bacteria involved in sewage fermentation form considerable amounts of B_{12} and extraction from sewage may ultimately prove the cheapest source of this vitamin. Carotene, a precursor of vitamin A, is present in certain coloured yeasts and bacteria as well as in algae; in Brazil it has been produced industrially with the aid of the fungus *Blakeslea trispora*; in Israel and in the former USSR it is extracted from natural blooms of the halophilic alga *Dunaliella* and marketed; such 'natural carotene' is rather costly. Ergosterol, a relative of vitamin D, is present in yeasts, but this source has not been exploited industrially so far.

Though they are not strictly nutrients, I should mention here the gibberellins, which have proved valuable in agriculture and brewing. Originally discovered as the causative agents of a fungus disease of rice, they are substances that are excreted by a certain fungus, pathogenic to plants, called *Gibberella fujikuroi*. They have a hormone-like action on plants, accelerating growth and cell division, and, when uncontrolled, they cause rapid death of seedlings. Controlled application of gibberellins can, however, accelerate growth of crops, decrease the dormancy period of potatoes and so on; in brewing, gibberellin treatment to accelerate malting is now almost universal.

There are other microbial products that are important in food production. Citric acid is used in enormous quantities in the soft drinks industry. The current annual production in the USA is believed to exceed 100 million pounds, all of which is made by the action of the mould *Aspergillus niger* on sugar. The citric acid industry is one of the most secretive, and details of the process are difficult to come by, but essentially a mat of the mould is allowed to grow on a sugar solution, containing certain salts and at a controlled acidity, and after a few days nearly all the sucrose becomes citric acid. Lactic acid, which I mentioned earlier in connection with the milk fermentations, is also used in the soft drink industry. It can be manufactured using whey as the raw material and *Lactobacillus casei* as the microbe, and other strains of *Lactobacillus* adapted to raw materials such as maize sugar or potato sugars are also used industrially. Another substance which can be prepared from microbes is glutamic acid, used as an additive to enhance the flavour of canned and packaged foods (as the monosodium glutamate of dried soups, for example). Many bacteria and some moulds can be used to produce this material, but most of the 15 million pounds consumed annually in the USA is still of plant origin (from sugar beet).

The examples I have presented so far in this chapter show the importance of microbes in the assimilation of food, in food preparation and processing and, most important, in the basic processes of agriculture. If, then, we are so dependent on microbes for almost every aspect of our nutrition, why do we not dispense with agriculture, give up eating plants and animals, and live on microbes? The question, absurd as it may sound, is perfectly reasonable and has, in various forms, been asked many times. Yeast is one of the most nutritious of foods, being rich in protein and vitamins of the B group and having a reasonable quota of fats; waste brewers' yeast is marketed and used in most Western countries as a food supplement under various trade names. In some parts of the world, such as East Africa, parts of Malaysia, of Indonesia and of China, there is not so much a food shortage as a protein shortage: most of the population gets something approaching a reasonable minimum of carbohydrate, but less than the minimum of 16 per cent protein in the diet required by a healthy human (children need rather more, mature adults rather less). This protein shortage could be alleviated by converting spare carbohydrate to protein, and the obvious way to do this is to allow a micro-organism to grow on it, such as a yeast.

During the Second World War a process was worked out in Britain for producing food yeast (*Candida utilis*) from molasses by a sort of primitive continuous culture procedure. The product had a pleasant taste, a sort of toasted, meaty flavour. Experiments performed by Britain's Medical Research Council, with the cooperation of the armed forces, showed that yeast could indeed provide much of the dietary protein needed by an ordinary individual. A problem was that the yeast was so rich in B vitamins that, if it formed too great a part of the diet, a risk of hypervitaminosis arose. The tests were so successful that, after the war, a plant was set up in Trinidad to prepare food yeasts from wastes of the sugar industry. The material was used in East Africa, India and Malaya but – and here is the human side of the problem – enormous resistance to its widespread use was encountered, even among starving populations. Unfamiliarity, as most parents learn from their children, is an extraordinarily powerful deterrent to eating what is good for one. The food yeast production plan foundered, partly because of conservatism on the part of the consumers, partly for simple economic reasons: the demand for the protein, such as it was, was half a world away from the supply, and the consumers were far too poor to pay for a procedure requiring a moderately advanced technology. Molasses could be put to more uses, or even thrown away, more cheaply. Petroleum fractions have been used for the culture of yeasts, grown in water in which crude petroleum is emulsified. The yeasts utilize the waxy components of the petroleum and actually improve the fuel quality of the petroleum that remains after growth. Moreover, since the waxes are pure hydrocarbons, unlike the sugars of molasses, they are much richer in terms of carbon, so that one gets nearly twice as much yeast per pound of wax as one would per pound of sugar. The product is said to have a nasty taste of petrol, but this can be removed, and its economic basis is the sounder because it assists in the refinement of petrol as well as forming protein. Dr Champagnat, a leading protagonist of the process, estimated that diverting 3 per cent of the world's oil production to preparation of food yeast could double the world's protein supply. Yeasts are not the only microfungi that have been considered as proteinaceous foodstuffs. A balanced protein food can be made from the filamentous *Fusarium graminarum*, grown on syrup prepared from cereal starch. After many years of evaluation, its developers initiated a company called Marlow Foods to market it in Britain under the name of Quorn mycoprotein, emphasizing its qualities as a health food: it is not only a good protein source, it is fashionably high in fibre and low in calories and saturated fats. It has been a commercial success with sales approaching £150 million per annum.

Mass-cultured bacteria have also been taken seriously as potential food. The Shell Petroleum Company once developed, but later abandoned, a project for obtaining bacterial protein from natural gas. This gas is methane, so one can use it for the mass culture of methane-oxidizing bacteria, which could then be useful as an animal feed, a fertilizer or even a food supplement for humans. Imperial Chemical Industries have actually marketed a bacterial protein as a cattle feed, derived from methane which has been converted chemically to methyl alcohol before being fed to the bacteria. Dr Seymour Hutner once suggested that such bacteria might be used as food for the mass culture of protozoa, which in turn should be offered as fodder to fish and thus yield increased amounts of a protein food that is intrinsically agreeable to humans.

Such projects for growing food yeasts or bacterial protein are short-term measures in the sense that they make use of plant material such as molasses, or of fossil materials such as petroleum or methane. On the one hand their yield is limited by the world's productivity of sugar, on the other we know that the oil and natural gas resources of this planet will last for only a few more generations. More satisfactory for the long-term interests of the Earth's population are projects for the mass cultivation of algae, studied by the Carnegie Institute of Washington and the Tokagawa Institute of Japan in the 1950s. Algae, such as *Chlorella* and *Scendesmus*, use sunlight and carbon dioxide as their main nutrients and thus take the place of plants. They produce greatly increased yields per acre, properly handled, and they are in many ways as valuable as yeasts: they provide wholesome food supplements, though they would never, one hopes, be offered or accepted as a complete diet. The problem, which applies to all microbial foodstuffs which are required on a large scale, is that a fairly advanced technology is required to produce and harvest them, and communities enjoying such technologies have, so far, been able to corner enough of the world's more conventional food supplies to keep themselves reasonably well fed. The essential problem is simple. Agricultural products, grain, meat or vegetables, are reasonably concentrated foods when produced: the best, richest cultures of yeast, *Chlorella* and so on, contain less than 1 per cent of the harvest. The rest is water. Removing that water, by sedimentation, centrifugation or filtration is what requires the technology: it is an expensive and power-consuming process. Nevertheless, the late Professor H. Tamiya calculated that *Chlorella* protein could be produced in Japan at less than one-third of the cost of milk protein. Though, to the writer's taste, *Chlorella* recalls slightly fishy spinach, Professor Tamiya claimed it can be made delicious and has given recipes for *Chlorella* cakes, biscuits and even ice-cream. But Japan is technologically the most advanced of the Far Eastern nations and, over-populated though it is, it has at the present time insufficient demand to make a *Chlorella* industry viable.

Cyanobacteria are microbes which sometimes form long filaments. The species *Spirulina* is one such. It grows in salty lakes (e.g. Lake Chad) and has been eaten by the natives of Chad and Mexico, because it is not difficult to harvest: it can be collected as a tangled mat, dried in the sun and eaten as a kind of biscuit. It is nourishing food, particularly when it comes from salty water, because it is then quite rich in carbohydrate and protein. Since the 1980s its consumption has become quite a cult among whole-food and health-food fanciers; in the sunshine of Southern California *Spirulina* is now cultivated in huge artificial ponds, which yield harvests of hundreds of tonnes per year.

However, generally speaking the world shortages of protein foods which were foreseen in the 1950s and 1960s, and which provided the impetus for much of the research which I have described in the last few pages, did not materialize, at least on the scale expected. There have indeed been shortages, some shocking and prolonged, but civil strife and politics rather than agricultural productivity have been their root cause. Though microbial foods have a niche market among health-food enthusiasts, they have not proved to be promising basic foodstuffs to the food industry. Yet people have got to be fed, or they fight. Microbes are already established constituents of animal feeds; it is probably just a matter of time before microbes become an accepted part of the diets of ordinary people.

In the last three chapters I have been rather subjective in my consideration of the microbes. I have looked at their role in our sickness and our health. I have considered how scientists handle them and I have been concerned with their importance in what we eat and drink. These are certainly serious matters which concern every one of us directly. But the importance of microbes to mankind stretches far more deeply into our social structure and economy than such day-to-day considerations would suggest. In the next three chapters I shall look at microbes from the point of view of society rather than of the individuals who compose it; I shall consider their involvement in industrial production, in the manufacture, storage, distribution and disposal of products. Health and food will, of course, crop up again and again, but my attention will be mainly focused on the impact of microbes on the economic machinery that keeps society going.

6 Microbes in production

MICROBES AND RAW MATERIALS

Despite today's massive production, through the agency of microbes, of fermented foods and drinks, of antibiotics, vitamins and chemicals, it remains true that the most important microbial products, as far as mankind is concerned, were laid down millions of years ago.

Before I justify that assertion, I have a digression to make. Industry consumes energy, by using manpower, animal power, wind power or machinery driven by fuel such as petrol or electricity. The more sophisticated an industry becomes, the more it is likely to use mechanical power and therefore fuels. I am sure I need not offer the reader examples of this principle: choose almost any industry and reflect on its history. Indeed, it has become a commonplace of sociology to express the level of industrial development of a country or community in terms of the energy (excluding manpower and animal power) consumed per head of the population. Today, energy is needed for every facet of industrialized civilization, from boiling potatoes to pressing steel. This is why developing countries become obsessed with hydroelectric schemes, power stations and so on. Even food production, the basic process of society, becomes increasingly energy-consuming as it becomes more mechanized. Manpower, helped by domesticated animals, can support small agrarian communities, but as soon as the population increases and social organization becomes at all complex, communities come to depend on machines, fertilizers, chemicals and, therefore, on the energy that drives and produces these things.

The major energy sources of the world today fall into two classes.

These are the renewable resources, ones which renew themselves as they are consumed, including hydroelectricity, solar energy, wood, tidal and wind power. With the demise of the windmill, only hydroelectricity is important in this part of the twentieth century (although wood, for cooking, is important in parts of the Third World). There are also the non-renewable fuels: the fossil fuels (coal, natural gas and oil), which accumulated in earlier geological eras, and certain radioactive elements (uranium-235 and plutonium), which provide nuclear energy. Industry takes these sources of energy, either directly or through the electricity-generating industry, and uses the power to transform natural materials such as wood, coal-tar products, metal ores and so on into economically useful products.

A valid and very instructive account of the world's economic situation can be produced in terms of the availability of energy, as anyone knows who remembers the dramatic changes in the economies of the world precipitated in 1972 by the oil-producing nations (OPEC). Their decision to increase the price of crude oil and to regulate its production caused the price of petroleum, in particular, to rise suddenly, and the knock-on effect was a prolonged period of economic stagnation. The essential equivalence of energy and money has rarely been illustrated so clearly. The world's supplies of fossil fuels are limited, although they are good for quite a few decades yet, but they will inevitably become more expensive to win as the easier reserves become exhausted. Again, oil provides an instructive example: the days of the backyard gusher in Texas or California are long past and costly devices to seek and extract oil beneath the ocean bed are now needed. Beware of pundits who say 'oil supplies will run out' and 'copper reserves will become exhausted' and so on. They will not. But they will become so expensive to win that it will not be worth trying, and the effect on society will be the same. Fossil energy cannot but cost more in future, and it will not simply be an increased cash cost; as environmentalists have repeatedly made clear, there is an additional environmental cost which will also increase, arising from the carbon dioxide and other end-products their combustion produces. Looked at in that light, the development of nuclear energy, despite its own waste disposal problems, becomes something to be regarded with hope rather than with the dismay with which contemporary argument so often surrounds it.

Though I obviously cannot go into greater detail about the world's power resources here, there are two more basic principles that should be recognized. The first is that concentrating a material is energy-consuming and therefore expensive. To take a simple example, if one wishes to obtain common salt from the sea, one has to evaporate away or otherwise remove about 32 grams of water for every gram of salt recovered. Now, no matter how one does this – by boiling, by using sunshine or by some sophisticated process such as electro-dialysis – a lot of energy has to be used in getting rid of that water. If the energy derives from sunshine, or a dry wind, then it is cheap – but one waits a long time for one's product and one gets it in small amounts. If one wants a lot and in a hurry – and it is almost axiomatic that highly developed societies want a lot of almost everything in a frantic hurry – then it is cheapest to find a natural salt deposit and expend energy in digging it out and carting it to where it is needed. To generalize, then, it is always more economical in terms of energy to use as concentrated a raw material as possible. Given unlimited power we could extract all the raw materials of industry – iron, copper, nickel, sulphur, uranium and so on – from dilute sources such as the sea or ordinary rock and soil. But we do not have unlimited energy, nor shall we have it until well into the 2000s, if at all. Therefore, to express the principle differently, any concentrated ore, such as a sulphur deposit, soda deposit or bed of iron ore, represents a saving of energy.

Industry makes things from concentrated raw materials, and this brings me to the second principle. The effect of using materials such as iron, copper, sulphur and so on is to disperse them about the world and so to dilute them: the whole trend of industry is to take concentrated materials and, as a result of using them, to make them become diluted. I shall discuss several examples of these principles in this and the next chapter.

To return to the opening sentence of this chapter, then, a major importance of microbes in industry is that, over geological eras of time, they have provided mankind with several concentrated reserves of industrially important materials. Microbes were concerned in important respects with the genesis of the two fossil fuels and were responsible for the deposition of several important minerals. A whole subject, called geomicrobiology, has grown up around the study of microbes in the formation and treatment of fuel and mineral resources. For the first part of this chapter I shall look at this subject and observe how microbes, millions of years ago, contributed to our basic industrial needs today.

Sulphur is one of the best established of the microbiologically produced minerals. Nearly every major industry that exists consumes sulphuric acid for one reason or another – it is used for pickling metals, electroplating, treating artificial fibres, preparing fertilizers, manufacturing all kinds of chemicals and pharmaceutical products, extracting ores and so on. It has been said that the national demand for sulphuric acid in a country is a measure of its degree of industrialization. Much the easiest way of making sulphuric acid is to burn sulphur to form sulphur oxides and then to react these with water. One can make sulphur oxides by other means, such as burning iron pyrites (FeS_2) or heating calcium sulphate (the mineral gypsum) with coke and sand, but native sulphur is the most concentrated source of sulphur possible and, consistent with my first principle enunciated above, it is the most economical raw material from the point of view of the power expended in obtaining it. A little elemental sulphur is needed industrially as a vulcanizing agent in rubber production, for making matches or in certain chemicals; a little is used in medicine and horticulture too, but by far the greater part is needed to make sulphuric acid. In this respect sulphur provides a very good example of my second principle: though used on an enormous scale, very little sulphuric acid appears in the final products of industry. One can think of electrical accumulators, which contain free sulphuric acid, or certain detergents which are organic derivatives of sulphuric acid, but by and large sulphuric acid is used during the process of production and does not form part of the finished product. Hence, when it is used, it goes, figuratively (and sometimes actually) down the drain. It is disposed of by some means or another and eventually finds its way into the sea, usually as sodium or calcium sulphate. A lot of sulphur escapes into the atmosphere. Nearly all fuels – coal, oil, wood, petroleum – contain sulphur compounds which, when they burn, pollute the atmosphere as sulphur oxides. (This is why curtain fabrics, stone and metalwork corrode so rapidly in towns; it is also one of the reasons why town dwellers are so prone to bronchitis, because sulphur oxides damage the lung membranes.) Over Britain, five million tonnes of sulphur pollute the air each year, ultimately being washed into soil, rivers and seas as an important component of the environmentally damaging 'acid rain'. It is possible that the sulphur cycle, which I discussed in Chapter 1, results in a net movement of sulphur from the land to the sea. In general, the pattern of sulphur transfer today provides a very clear example of industrial civilization taking a concentrated natural resource and diluting it.

The demands of industrial countries for sulphur, which is mainly converted to sulphuric acid, are enormous. In 1951 the USA was consuming nearly five million tonnes a year and Britain needed nearly half a million tonnes – though, because of a world sulphur shortage, it was not getting it. The world's deposits of native sulphur are mainly located around the Gulf of Mexico, in Texas, Louisiana and in parts of Mexico itself. Other deposits exist, in Sicily, Ireland, North Africa and the Carpathians, but something like 95 per cent of the world's supplies come from the Gulf area. The mineral is located in rather confined deposits, called domes, and is always found associated with calcium sulphate and, usually, oil is not far away. The question arises, how did it get where it is, and why is it always associated with a particular geological pattern? The answer, which is reasonably well established, is that it was formed as a result of intense microbial activity during a geological era of warmth and sunshine, probably while a sea was drying up. The Caribbean is known to have stretched far into the Southern States of the USA and into Mexico some 200 million years ago (whether it was the Permian or the Jurassic period is still uncertain), and it was probably about that time that the world's major deposits were laid down. Sulphate-reducing bacteria living in the drying, concentrating sea used organic matter for the reduction of calcium sulphate in the water to calcium sulphide. This in its turn became oxidized to calcium carbonate and free sulphur, probably through the agency of the photosynthetic sulphur bacteria. Hence the sulphur cycle progressed as far as sulphur but no further, so that sulphate was reduced and sulphur accumulated. The reason why it did not progress further is almost certainly that there was no air available: the drying up of the sea caused organic matter to become concentrated, whereupon microbes grew and used up all the dissolved oxygen. In addition, the coloured bacteria generated more organic matter photosynthetically from CO_2, using sunlight, so a huge, anaerobic salt pan developed, with calcium sulphate crystallizing out and sulphur sedimenting; deposits of microbial organic matter also formed which, later, may well have contributed to oil formation.

How do scientists know this? There are two lines of evidence, one of which is that, in certain parts of the world, one can see such a process happening even today. In Libya there are a number of lakes (near the hamlet of El Agheila) where warm artesian water, rich in calcium sulphate and containing hydrogen sulphide, comes to the surface through springs. One of these, called Ain-ez-Zauia, is about the size of a swimming pool and is slightly warm (30 °C). It is saturated with calcium sulphate and contains about 2.5 per cent sodium chloride – a reasonable approximation to a warm, drying-up sea, if a little weak in salt. Under the Libyan sun, this lake produces about 100 tonnes of crude sulphur a year, formed as a fine, yellow-grey mud which is, in fact, harvested by the local Bedouin. (They export some to Egypt – or did when I was there in 1950 – and use it as medi-

A SULPHUR-FORMING LAKE IN LIBYA. A view across Ain-ez-Zauia, near El Agheila, with a Jeep for scale. In the warm, saline water, a combination of several kinds of sulphur bacteria make sulphur from sulphate, using solar energy. The salty encrustations around the edge consist largely of calcium sulphate and carbonate.

cine themselves.) The way in which the sulphur is formed is this: sulphate-reducing bacteria reduce the dissolved sulphate to sulphide at the expense of organic matter formed by coloured sulphur bacteria, which in their turn have made the organic matter from carbon dioxide using sunlight and sulphide, some of it from the spring waters, some formed by the sulphate reducers. Thus we have sulphur formed from sulphate by two interdependent types of bacteria, the whole process being propelled by solar energy. The bed of the lake consists of red, gelatinous mud made up almost entirely of coloured sulphur bacteria; the bulk of the lake is a colloidal suspension of sulphur rich in sulphate-reducing bacteria; the whole system smells strongly of hydrogen sulphide.

My colleague the late K. R. Butlin and I took samples of this lake back to our British laboratory in 1950. Artificial lake water (corres-

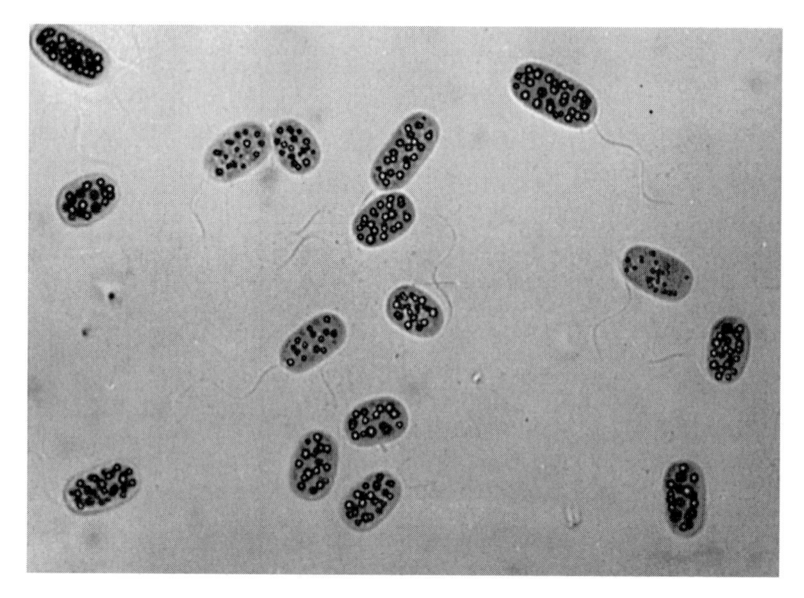

A RED, PHOTOSYNTHETIC SULPHUR BACTERIUM. A photomicrograph of *Chromatium okenii* showing its tail (flagellum) and many shiny granules of sulphur within the cells. Bacteria like these were abundant in Ain-ez-Zauia, oxidising sulphide to sulphur and using sunlight to make carbohydrate from carbonates. Magnification about 500-fold. (Courtesy of Professor N. Pfennig)

ponding to the analysis of the real thing) was prepared, and a small mock-sulphur lake of about ten gallons was set up in which, when illuminated, the red gelatinous mud grew and sulphur was formed. By altering the conditions somewhat it was possible to accelerate sulphur formation quite considerably.

There were other lakes of this kind in the neighbourhood and similar lakes and sulphur springs exist in various parts of the world. The fact that one can isolate the appropriate bacteria from them, and even duplicate biological sulphur formation in the laboratory, is circumstantial evidence in favour of the belief that this is how the majority of sulphur deposits arose. But there is stronger evidence. Professor H. Thode of Canada showed, in about 1950, that during biological sulphide formation, some separation of the natural sulphur isotopes took place, and that this did not occur during chemical sulphate reduction.

Perhaps, for non-chemists, I should digress for a moment and explain what an isotope is. Nearly all elements, such as hydrogen, oxygen, nitrogen and sulphur, exist in nature as mixtures of atoms, the majority of which have a certain mass but a few of which have a different mass. Sulphur, for example, consists mainly of atoms that are thirty-two times as heavy as a hydrogen atom, but about 2 per cent of its atoms are heavier, thirty-four times as heavy as a hydrogen atom. These isotopes can be detected and measured readily in a device called a mass spectrometer and, no matter what form of chemical combination the sulphur occurs in – as sulphide, sulphate, thiosulphate, organic sulphur compounds, for example – the ratio of the isotopes will be similar. Similar, but not identical. Because what Thode observed was that sulphides and sulphates found in meteors or volcanoes, where no possibility of biological action existed, had identical isotope ratios among their sulphur atoms. So did sulphur-bearing minerals taken from geological strata laid down before life originated on this planet. But sulphide formed in cultures of sulphate-reducing bacteria, or in natural environments where sulphate-reducing bacteria were active, was richer in the lightweight isotope, and the residual sulphate was richer in the heavier isotope. For some reason, it seemed, bacterial sulphate reduction separated the natural isotopes of sulphur appreciably. Volcanic sulphur had the natural or meteoric isotope ratio, but the Texas and Louisiana sulphur deposits, as well as those in Sicily and samples sent from Ain-ez-Zauia, had the biological ratio.

Hence isotope experiments provide very good evidence that bacteria were responsible for the conversion of sulphate to sulphide during the formation of the world's major sulphur resources. But they provide no evidence that bacteria were involved in the next step, the oxidation of sulphide to sulphur. Some authorities believe that bacteria had nothing to do with this, that oxidation by air, or a slow chemical reaction between sulphide and sulphate, provide a sufficient explanation of sulphur formation. Russian workers have provided good evidence that 80 per cent of the sulphur found in Carpathian deposits arises through the action of thiobacilli (which I introduced in Chapter 2: colourless bacteria that can oxidize sulphide in air to sulphur and usually further, to form sulphuric acid). As far as what happened geological eras ago, the point will probably never be settled, but the beauty of conceiving the second step as biological is that it explains how the sulphate-reducing bacteria obtained energy for sulphate reduction. They obtained it from the carbon compounds manufactured from carbon dioxide by the coloured sulphur bacteria or the thiobacilli as the case may be.

I have described sulphur formation as taking place while the sea was receding from what is now Texas, Louisiana and parts of Mexico. I should add that some geologists believe that the evaporation occurred first and that, long after the salt beds were buried in later deposits, bacterial sulphate reduction took place as petroleum, which some authorities believe these bacteria can use as an energy source, seeped into the beds of calcium sulphate. I cannot discuss the pros and cons of these views here, and merely record that no one disputes that sulphate-reducing bacteria were involved in the primary step of forming sulphide from sulphate.

The biological nature of sulphur formation has given rise in proposals for manufacturing sulphur industrially using bacteria. I shall discuss these possibilities later in this chapter.

A second important mineral deposit which is formed through bacteria action is soda, which is mined in various parts of the world. The sulphate-reducing bacteria are also involved in this process, which takes place when sulphur formation fails to occur on any large scale. If, for some reason, massive bacterial sulphate reduction takes place in nature, it is usually calcium sulphate that is reduced, because this salt, responsible for permanent hardness in water, is one of the commonest mineral sulphates. Though I have, for brevity, talked so far of the reduction of sulphate to sulphide, it is usually calcium sulphate that is reduced to calcium sulphide. In chemical symbolism one can write:

$$CaSO_4 \rightarrow CaS$$

If carbon dioxide is present, as it always is as a result of the respiration of the microbes, some of this calcium sulphide reacts with it, giving hydrogen sulphide:

$$CaS + CO_2 \xrightarrow{\text{in } H_2O} CaCO_3 + H_2S$$

The hydrogen sulphide has the characteristic bad-egg smell of badly polluted environments. The other product is calcium carbonate or chalk. In certain environments the main sulphate mineral is sodium sulphate – the Wadi Natrun in Egypt is such a place – and in this case the end product is sodium carbonate, which is soda. The late Professor Abd-el-Malek in Egypt studied the Wadi Natrun and produced circumstantial evidence that this view of the formation of soda is correct: the numbers of sulphate-reducing bacteria in the environs of the soda deposits increase as the deposits get stronger.

Sulphides are formed wherever sulphate-reducing bacteria become active, but, because these bacteria are strict anaerobes and do not function in air, their activity tends to be rather localized. They need a good supply of organic matter and sulphate to become established, though once established they tend to keep themselves going, because sulphide is rather poisonous to other living things, which therefore die and tend, by decomposing, to augment the organic matter available to the sulphate-reducers. Other sulphur bacteria may develop alongside, the limited ecological system based on the sulphur cycle called the sulfuretum (p. 15) arises. The world's sulphur deposits were probably formed as parts of gigantic sulfureta, and as I have just described, soda can be formed if a sulfuretum is established in certain environments. Now, most terrestrial waters contain dissolved iron, and some have copper and lead in solution as well. When such waters encounter a sulfuretum, an immediate chemical reaction occurs in which the dissolved metal reacts with the H_2S to form a metal sulphide. This material is deposited as a precipitate. In this manner, some believe, have many of the world's resources of sulphide minerals been formed. Uranium ores may have become concentrated in this manner; copper and lead occur mainly as sulphide ores and it has been possible to mimic their formation in the laboratory. But laboratory experiments aimed at imitating nature do not necessarily prove that nature actually ever behaved that way: isotope distribution experiments of the kind that established the biological origin of native sulphur have not given unequivocal results with copper, lead and other metal sulphide ores, so the theory that they were formed biologically is not well supported.

Except in the case of iron. Iron sulphide is found in many marine sediments and areas where sulfureta have been functioning, and usually the sulphur in such iron sulphide deposits has the biological isotope distribution. An important mineral containing iron is iron pyrites, and this is known to be formed geologically from sedimentary iron sulphide by way of a partly hydrated mineral called hydrotroilite. The precise chemistry of the process need not concern us here; the upshot of the matter is that iron pyrites (which has the chemical formula FeS_2 in contrast to the FeS of iron sulphide) has a biological origin, again largely owing to the sulphate-reducing bacteria. Pyritized fossils – fossils that have been transformed into pyrites while retaining their original form – probably arose because the decaying organism allowed a little sulfuretum to become established and thus a replica of the more rigid parts of the dead creature built up as, atom by atom, dissolved iron seeped into the sulfuretum. But the major importance of iron pyrites is as an alternative to sulphur for manufacturing sulphuric acid. Pyrites can be burned to give iron oxides and gaseous oxides of sulphur, the latter being easy to convert to sulphuric acid on an industrial scale. In 1950 about a sixth of Britain's million-and-a-half tons of sulphuric acid were made from pyrites. The process is not as economical as that using native sulphur, but as such sulphur becomes scarcer and more expensive it is being used increasingly.

The sulphate-reducing bacteria, then, are extremely important in the genesis of two, three and possibly more of the world's mineral resources, but they are not the only microbes so involved. There is a specially pure kind of iron ore, known as bog iron and found on the edges of marshy areas, which is formed through the action of iron bacteria, of which I wrote briefly in Chapter 2. These bacteria have the property of oxidizing dissolved ferrous iron to ferric iron which, being less soluble in water, precipitates out as a deposit resembling rust. In chemical terms this can be formalized as:

$$FeX_2 + O_2 \xrightarrow{\text{in } H_2O} Fe(OH)_3 + 2HX$$

where X is a monovalent anion such as an organic derivative. (Non-chemists may need reminding that in Chapter 1 I explained that, when dissolved in water as a chemical derivative, iron exists in two forms, one of which is formed by the action of oxygen on the other and is less soluble.) The seepage waters from a peat bog, for example, are relatively rich in dissolved ferrous iron and are rather acid. Where such waters flow, say, into a chalky area and become neutralized, iron bacteria grow in great numbers and, in due time, can form massive deposits of ore. It is not clear why the bacteria do this, and the idea that the oxidation of ferrous iron enables them to grow autotrophically seems to be mistaken. However, the product is a very pure ore and was, because of its ready availability and purity, probably the first metal ore to be used by man. *Sphaerotilus, Leptothrix* and other iron bacteria provided mankind with the means of transition from the Stone Age to the Iron Age.

Nowadays, of course, other types of iron ore are mainly used industrially, because there is just not enough bog iron left. But the process of bog iron formation can often be seen on a small scale where peaty and iron-rich waters flow out of springs and bogs, forming a brown rusty deposit on stones and rocks. It is probable that some deposits of manganese oxides were formed in a similar way.

Lest I give the impression that bacteria are the only microbes involved in mineral formation, let me remind readers that the chalk, which is almost pure calcium carbonate, and of which a considerable proportion of the British Isles is constituted, was formed over geological time from the compressed shells of types of spiny protozoa collectively known as Foraminifera, together with the shells and skeletons of a variety of tiny multicellular animals. And micro-algae along with cyanobacteria apparently play a major part in the formation of another calcareous mineral called travertine, which forms in lime-rich streams and springs.

A complicated process due to bacteria occurs in the natural leaching of pyrites. Later I shall tell how coal and gold mines contain strata of iron pyrites, and how certain bacteria (the best-known being a sulphur bacterium, *Thiobacillus ferro-oxidans*) oxidize this to form, among other things, sulphuric acid, which corrodes piping and damages mining machinery. In dumps of pyrites outside mines these organisms grow and turn the environment acid. More of the pyrites dissolves – because acid helps to decompose pyrites – and one of the products of this reaction is free sulphur. This, too, is oxidized, by bacteria such as *Thiobacillus thio-oxidans*, forming yet more sulphuric acid. Thus one gets an interesting set-up in which rainwater permeates the dump and, with the aid of bacteria, washes out dissolved iron and sulphuric acid. The effluent waters are brown and rusty-looking. Now, all pyrites deposits contain copper in small amounts, which is valuable, and this washes out as copper sulphate. A minor industry has developed for extracting the copper by running the leached water over scrap iron, when the iron dissolves and copper is precipitated. Expressed chemically:

$$Fe + CuSO_4 \rightarrow FeSO_4 + Cu$$

Iron bacteria later convert the ferrous sulphate to ferric oxides, which settle as the deposit called ochre, which is used in the paint industry. Though the production of ochre by this process far exceeds demand, the copper found is sufficiently valuable to make the process worthwhile. Reports in American literature indicate that molybdenum, titanium, chromium and zinc may be concentrated from pyritic strata by the action of *T. ferro-oxidans*. Of particular importance for the future of atomic energy is the fact that uranium may be leached out of low grade sulphide ores in a similar manner. Gold, too, is often associated with pyritic minerals. In 1993 there were press reports that a population of bacteria, including presumably *T. ferro-oxidans*, had been developed for use in the Ashanti gold mine of Ghana to release gold particles entrapped in pyrites and thus augment the mine's output.

A rather remarkable earlier claim involving gold had been made in France in 1964, to the effect that an aerobic sporulating bacterium – clearly not a thiobacillus – found in tropical soil had proved capable of releasing gold from combination in the mineral laterite, a major component of such soil. The quantities released were very small and no microbiological gold rush ensued. Some strains of the fungi *Aspergillus* and *Penicillium* form chemicals such as citric and oxalic acids, which can dissolve minerals such as nickel, cobalt or aluminium from non-sulphide ores, but whether they are of value in practice is not clear.

To return to my opening theme – that the most important economic activities of microbes took place geological eras ago – consider the case of coal, the basic fuel of the industrial revolution. The genesis of coal is fairly well understood nowadays: huge forests of plants, mainly related to present-day mosses and ferns but gigantic in size, flourished about 300 million years ago in a geological period called the Carboniferous era. The environment was warm and humid, with swamps and bogs abounding, and as the vegetation died and decayed it formed a sort of vast compost heap in which any oxygen that penetrated was immediately consumed by putrefactive bacteria. Thus an anaerobic fermentation took place and methane – marsh gas – was formed while the plant debris became converted to materials of rather indeterminate chemical composition called humic acids. This process still occurs today: the product of such decay in bogs, when it dries out, is peat, itself a valuable fuel. 'Will-o'-the-wisp', a flame of burning methane dancing over a peat bog, is an important element of Irish folklore which has the unusual status of being a genuine natural phenomenon, if a rare one. Humic acids are distinctly related, in a chemical sense, to phenols, the disinfectants I mentioned in Chapter 3, and they have preservative properties in that, though formed by microbial action on plant material, they tend to prevent further bacterial action. This is why metals, wooden objects and even corpses, when recovered from peat bogs, often show remarkably little decay.

Peat, then, is an early stage in the formation of coal. From the chemical viewpoint it is plant material which consists mainly of carbon, hydrogen and oxygen, though depleted in oxygen and enriched in carbon and hydrogen, so that, when dry, it burns readily in air. During the Carboniferous era, as millennium succeeded millennium, the peat deposits became overlaid by sand and rocks and were thus compressed. As the pressure increased, the peat turned into coal, first becoming brown coal or lignite, which is structurally rather like peat, later forming the familiar bituminous coal widely, and extravagantly, used by the domestic British (to pollute, if readers will forgive a petulant aside, their atmosphere by sucking warmed air up chimneys, in the belief that they are warming their houses). Under very high pressures the very pure coal called anthracite was formed. During the compression process a one-foot stratum of peat yielded about an inch of coal, and the mineral underwent further chemical change, becoming enriched in carbon and depleted in hydrogen, so much so that anthracite is almost pure carbon. Quite why pressure should have this effect on peat is not at all clear, but it is fairly certain that, in the early stages, residual action by bacteria resistant to the disinfectant action of peat assisted the removal of hydrogen. Anyway, the important point from the economic point of view is that the primary process leading to coal formation was the putrefaction of plant material by methane bacteria which, as I trust the reader will recall, are strongly anaerobic bacteria: they do not grow in air.

Methane is marsh gas. If you find a pond that has had regular seasonal deposits of leaves and other vegetation in it and poke a stick into the bottom mud, bubbles of marsh gas will emerge. This is formed by methane bacteria and can be caught in a jam-jar and burnt. If it ignites spontaneously, it forms Will-o'-the-wisp mentioned earlier. Methane must have been formed in enormous quantities during coal formation and it is the main component of natural gas, which became an increasingly important energy source in the latter part of the twentieth century. Beneath the North Sea are reserves of subterranean methane that may well exceed in energy value the whole coal reserves of Britain. (A most satisfactory prospect, if I may digress once more, since both coal mining and coal burning are hazardous to health, and the burning of coal acidifies rain and damages the countryside as well as wasting valuable coal-tar products.) Natural gas is a relatively clean fuel and an increasingly useful one: the USA consumed over 2.5 *billion* cubic feet in 1996. Methane is normally found in coal mines – it is the hazardous fire damp, the cause of many tragic explosions in mines. About a quarter of the vast reserves of subterranean gas that are now being tapped resulted from the action of the methane bacteria over geological eras, and it seems likely that the remainder has been there since the Earth originated, because methane is one of the few interplanetary gases (Jupiter consists largely of methane together with ammonia). The primitive atmosphere of the Earth, before life originated, very likely contained methane, some of which could have become entrapped as the earth cooled and settled down. A joint origin of this kind could account for the presence of such gases as ethane and propane, which are found in small amounts in natural gas but which are not formed naturally by any known microbes.

In the 1990s considerable interest centred on a newly found mineral which may well be a potential resource of fossil fuel. At high pressures and near-freezing temperatures, methane combines with water to form a so-called clathrate compound known as methane hydrate, in which methane molecules become trapped among water molecules (one methane to between 5 and 6 waters) and together they freeze to form crystals. This item of chemistry has been known for decades; what is new is the discovery that, at depths of 500 metres or more, the world's oceans have vast deposits of methane hydrate holding, according to one estimate, an amount of methane equivalent to twice the world's other reserves of fossil fuels. There is little doubt that the bulk of that methane is biogenic: it has been generated from organic detritus over geological time by methane-producing bacteria, which are always abundant in marine sediments. Instead of escaping as marsh gas does, the gas has become entrapped by the icy water as the hydrate. The 'methane ice' crystals so formed are rich in cold-tolerant bacteria, which are preyed upon by marine worms ('ice worms') that bore and inhabit holes and crevices in the mineral, and higher organisms such as mussels and starfish feed on the worms: the mineral supports a dark and chilly world of living creatures down there! But apart from intriguing biologists, this seemingly massive reserve of natural gas is of immense economic interest to many countries, especially those, such as India and Japan, which have very little indigenous fuel reserves of their own. Methods of exploiting it have yet to be devised, but it can only be a matter of time: one has but to warm up or decompress methane hydrate to release the gas.

Man's third main fossil fuel is oil together with the distillation products of oil collectively called petroleum. The question of whether oil originated as a product of microbial action is still not settled, largely because no one has successfully caused bacteria to form oil in laboratory conditions – at least in important amounts. Oil hydrocarbons could have been formed chemically by the action of water on metal carbides during the infancy of this planet, but oil deposits have the characteristics that make a biological origin highly probable. First, they are rich in anaerobic bacteria, particularly the by now familiar sulphate-reducing bacteria, and they are associated with sulphur deposits which are known to have a biological origin. Moreover, when scientists have succeeded in detecting oil-like compounds in microbial cultures, they have been formed in mixed populations that included sulphate-reducing bacteria. Secondly, in crude oil one can detect compounds called porphyrins which are chemicals derived from the respiratory enzymes of living organisms and which are not known to occur away from living things. Thirdly, certain of the hydrocarbons of petroleum are optically active, which means, in non-chemical terms, that their molecules have a special kind of configuration that is only known to result from the action of biological systems. (I shall explain the type of configuration more precisely later in this chapter.) None of these points is conclusive: all could have resulted, for example, from the action of microbes on oil *after* it was formed, an action which is quite familiar to microbiologists. But the chances are that oil was formed by microbial processes analogous to those which led to the sulphur, coal and natural gas reserves of this planet.

If the responsibility of bacteria for oil formation is not proved, there is little doubt about their role in the coalescence of oil deposits. Much oil in deposits is absorbed on rock – called oil shale – and this usually consists largely of calcium sulphate. Professor ZoBell of California has shown very clearly that our old friends the sulphate-reducing bacteria, when grown in the presence of oil shale, cause the absorbed oil to be released from the rock and to coalesce as droplets. The bacteria do this by a variety of mechanisms, one of which is to reduce the rock chemically to sulphide, thus changing its configuration and releasing the absorbed material, another of which is to form a detergent-like substance which cleans off the oil. Other anaerobic bacteria contribute to the effect, and the great oil deposits of Texas and California, which are enormous subterranean pools of oil released from shale, are believed to have formed as a result of bacterial action on the shale. Spent oil wells, which are wells that have ceased gushing because the pressure under which they existed before they were tapped has been released, still have much useful oil in them, and some of this can be displaced by injecting brine or sea water under the oil stratum and floating the oil out. (This process is called secondary recovery in oil technologists' jargon.) But much oil remains absorbed on the associated shale and, in the old Czechoslovakia, secondary recovery of oil was successfully enhanced by pumping nutrients for sulphate-reducing bacteria into the well. Unfortunately, as might be expected if one is persuading bacteria to do something in weeks that they hitherto did over centuries, the improvement was modest and transient.

MICROBES IN INDUSTRY

I have discussed the importance of microbes in the formation of the resources of industry. The last paragraph brings me to the question of the deliberate use of microbes in industry. Can any of these processes be made use of today, or do they take so long as to be worthless?

The broad answer is that, at present, there are sufficient coal, oil, methane and sulphur reserves on this planet to last mankind for some time to come and, if a global shortage of any of these basic materials did occur, it would be most logical to prepare them by some industrial chemical process using nuclear or hydroelectric energy rather than to mimic their natural origin. But man, in the mass, is not logical. Because of what seems to be a natural-born parochialism, we are incapable of utilizing our planetary resources on a global scale. Local shortages of raw materials are a chronic disease of our only partly civilized planet, and a classic example of this kind occurred in the world sulphur shortage of the early 1950s. At that time British industry, typical of most Western European industry, was geared to using native sulphur imported from the USA. By 1950 the rate at which existing sulphur domes were being exhausted had exceeded the rate at which new ones were being discovered, the price of American sulphur went up and, starved of dollars as a result of the Second World War, Britain and most of Western Europe found its industrial recovery drastically hampered by a world sulphur shortage. The shortage stimulated further prospecting, and many new deposits were discovered, but the sulphur crisis was only pushed ahead for a decade or so. By the mid-1950s the shortage had eased, but in 1963 output exceeded the discovery of new resources once more and a new sulphur shortage developed. It was less drastic because several major industries had transferred to pyrites and other minerals as their sources of sulphuric acid during the 1950s. Sulphur supplies again eased as the 1970s approached, but a new shortage seemed imminent in the mid-1970s, since when the economic depression has decreased demand. The take-home message, however, is that the world's resources of native sulphur will be too depleted to be useful in a few decades.

As a result of the crisis of the early 1950s, a process was developed in a British government research laboratory for making sulphuric acid using microbes, based on the way sulphate-reducing bacteria function in nature. The late K. R. Butlin and his colleagues, with the present writer interfering at times, showed that one could ferment sewage using sulphate-reducing bacteria and, at least in theory, obtain up to a fifth of Britain's requirement of sulphur by a process that might be called composting sewage sludge with gypsum (calcium sulphate). Actually, the product was not sulphur but hydrogen sulphide, but this was equally useful because it could be converted to sulphur or sulphuric acid as desired by established industrial chemical processes. The sewage sludge, after treatment, had certain advantages as a disposal product (its settling properties were improved, so less water had to be handled to get rid of it) and a London sewage works developed it to a pilot plant scale. However, market forces brought down the cost of sulphuric acid and the work was discontinued; the prospects of Britain becoming even partly self-sufficient in regard to sulphur receded accordingly. Microbiological sulphide production is unlikely to be capable of supplying Britain's sulphur needs, but a microbiological process for producing sulphur from sewage could be useful for countries having a low degree of industrialization and limited resources of foreign currency. A comparable process using industrial wastes was used in the former Czechoslovakia; sulphur farming has been proposed as a possible cottage industry in parts of India such as Masulipatam.

Devices for producing methane by bacterial fermentation of farm waste and sewage are now used to supply power to poorly industrialized areas in Asia and Africa, and installations for running refrigerators on methane generated by bacteria from farm wastes are used in tropical areas. The methane so produced usually contains hydrogen and carbon dioxide, but it is a good fuel. Under the name of biogas, in China, India and several less developed countries it serves as a supplement for more expensive (or less accessible) fossil fuels; the recent world energy shortage has made biogas particularly interesting to technologists. Methane is, in fact, the normal product of one stage of conventional sewage treatment and, in highly industrialized countries, the more sophisticated sewage works use the methane formed in sewage digestion to run their machinery and even, in some instances, to run lorries. Some sewage works supply methane to their country's gas grid. I shall discuss methane production in connection with sewage disposal in Chapter 8.

When the product is a simple chemical such as sulphur or methane, an industrial process based on microbes needs a cheap waste product to work on if it is to be economic. Industrial alcohol, for example, used to be obtained by the fermentation of molasses (as waste product of the sugar industry) by yeasts. Glycerol can be produced industrially by conducting the alcoholic fermentation in the presence of sulphite. Acetone and butanol, both important industrial solvents, have been made by the fermentation of molasses by *Clostridium acetobutylicum*, a process still used in South Africa. Lactic acid, used in the textile industry and in electroplating as well as being an additive to food, is sometimes manufactured by a *Lactobacillus* fermentation. Acetic acid is sometimes produced industrially by the traditional vinegar fermentation. All these products, however, can now be made as easily by purely chemical processes as by-products of the petroleum industry, and though, since the equipment is there, some industries use fermentation processes to make these simple chemicals, it is fair to say that, as industrial fermentations, they are rarely economically attractive. Microbes always form their products in fairly dilute solution, which means that an expensive process of concentrating them has to be undertaken. This, and the fact that their raw material has to remain cheap despite the ever-increasing demands of industry for the product itself, makes the use of microbes for heavy chemicals – those much in demand – a generally uneconomic prospect.

Exceptions to this generalization began to arise in the mid-1970s, when the world energy shortage began to bite. Countries with little or no oil and fossil fuel reserves began to take seriously the microbial production of energy sources. I have already written about biogas or methane, and a comparable product is fuel alcohol. In 1975, Brazil embarked on a government-sponsored programme for the production of alcohol as fuel for automobiles by fermentation, first from sugar cane, more recently from cassava as well. The fuel alcohol programme has been a great practical success: Brazil produced 37 million hectolitres of fuel alcohol in 1979 and over 100 million in 1985. Traffic fumes, I am told, smell quite different in Rio. How economic the process is, is less clear: it appears to have needed constant subsidy; but it has excited interest all over the world, and processes based on corn, artichokes and even treated sawdust are being developed. Ironically, despite its massive oil supplies, the USA does not have enough and plants producing fuel alcohol from surplus corn exist: gasohol, which is petrol with 10 per cent ethyl alcohol, has been available for a couple of decades in the USA. There is a problem: normal petrol engines overheat with alcohol and have to be modified. The world energy shortage is somewhat overshadowed by the end-of-century recession but, as industry recovers, the production of other energy-rich products, such as butanol or sulphur, by fermentation may again become economic.

The future of microbes in industry is secure, however, in the preparation of substances that are for one reason or another difficult for the chemist to prepare on an industrial scale. Citric acid, much used in the soft drinks industry, happens to be an awkward chemical to synthesize, though its molecular structure is simple. Hence it is still made microbiologically on an industrial scale and will probably continue to be. Fumaric acid and itaconic acid are even simpler chemicals than citric acid, but are also not easily made chemically. They have uses in the plastics and synthetic lacquer industries and are produced from sugar by fermentation with moulds of the *Rhizopus* and *Aspergillus* groups respectively. Gluconic acid, a derivative of glucose, has uses in pharmacy as a means of administering calcium to patients (calcium gluconate can be injected safely) and as a component of bottle cleaning and metal-pickling formulations. It is prepared industrially by the action of the mould *Aspergillus niger* on glucose. Complex microbial products have several uses in the cosmetics industry. For example, fermentations by varieties of streptococci are used to produce hyaluronic acid, an elastic and viscous component of normal connective tissue in vertebrates, including ourselves. It is a common component of ointments, creams and make-up (it is also used as an adjunct to surgery); yeasts are exploited to provide substances that have emollient, unguent or moisturizing properties; and several types of bacteria form cyclodextrins, which have the curious property of stabilizing scents so that they do not go stale or evaporate too rapidly. In quite a different economic area, several types of anaerobic bacteria make a storage product which goes by the forbidding name of poly-β-hydroxybutyric acid (PHB for short). It is of considerable interest to the plastics industry because, when extracted, it can be heat-moulded into bottles, sheets, nets, fibres and so on, just like polythene or any other thermoplastic polymer. However, unlike the petrochemical plastics, this 'bioplastic' is environmentally friendly: when discarded it is rapidly and completely degraded. Cultured in appropriate conditions, over 90 per cent of the dry weight of the bacterium *Alcaligenes eutrophus* can be PHB, and it has been produced commercially in this way, but the cost of concentrating the cells and extracting their PHB makes bioplastic goods expensive.

A general class of compounds that are difficult to make synthetically are those that are optically active. This means that they have a particular kind of distorting effect on polarized light which can be detected with appropriate optical instruments. I shall not bother with details of what this effect is here, but its significance is of some importance, because it indicates a subtlety of their molecular structure. For the benefit of non-chemists I shall describe the simplest possible case of an optically active molecule. Consider a carbon compound whose chemical formula is Cwxyz. C is the carbon atom and joined to it are four different atoms w, x, y and z. If you could actually see a molecule of that compound in 3-D it would look like this:

(a)

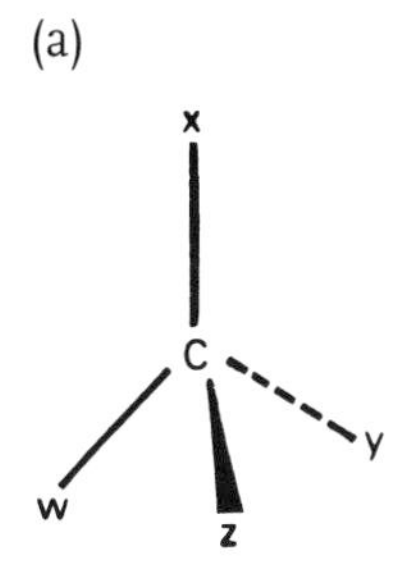

where x and w lie in the plane of the paper, z sticks out above and y below it. (Geometrically, w, x, y and z lie at the points of a tetrahedron, with C at its centre.) This molecule is not symmetrical: if you held up a mirror to it its reflection would look like this:

(b)

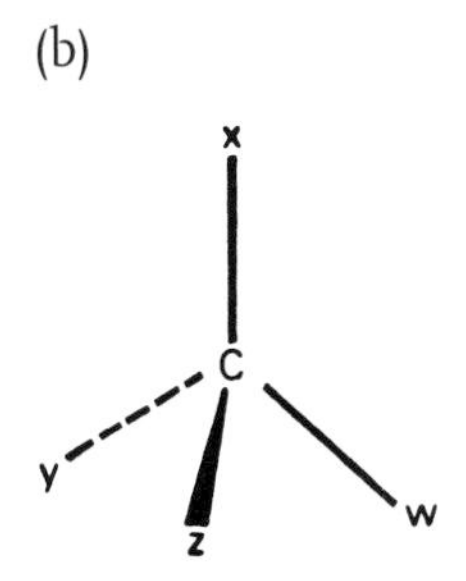

The original molecule and its reflection are different because, however you twist (b) around you could never superimpose it on (a).

It follows that any chemical compound which contains a carbon (or other) atom linked to four *different* atoms (or groups of atoms) can exist in two forms corresponding in structure to these mirror images. But in all their gross chemical properties the two forms behave similarly, only differing in certain fine details such as their effect on light. When chemists synthesize such an asymmetric compound in the laboratory they normally obtain a mixture of the two forms in equal proportions. But when biological systems make or utilize asymmetric compounds, they usually make or utilize exclusively one of the two forms. Indeed, most biological molecules are asymmetrical, and almost all belong to what is known as the left-handed class of molecular configurations. Microbes are used for the preparation of optically active compounds for two reasons: the first is that some microbes will use the left-handed form preferentially, thus enabling the chemist to achieve a separation of the two forms because the microbe leaves one behind; the second is that, if they form a product that is asymmetrical, they usually form only one of the two forms (most often the left-handed form). Why should anyone want to separate optically active compounds if they are chemically so similar? Because the two forms often have different biological effects. For example, pharmaceutical activity often depends on getting the right configuration of molecules that may have not one but several asymmetric centres, and in these circumstances biological, and particularly microbial, methods are the only practical procedures.

Optically active compounds may be needed in pharmacy and research, but they are not the sort of chemicals required by the heavy industries of a nation. They represent a class of fine chemicals for which biological processes will probably always be necessary, but they are far from the sole province of microbes: alkaloids, hormones and many other natural products in the Pharmaceutical Codex are obtained from plants and animals as much as from microbes.

The classical instance of microbes being used to produce something that could not be made otherwise is, of course, in the antibiotic industry. Antibiotics, as I explained in Chapter 3, are substances produced by one species of microbe that either kill other microbes or prevent them growing. Sometimes they are extraordinarily active: penicillin is still one of the most powerful drugs known for use against sensitive bacteria. It is formed by a filamentous mould, rather like that which grows on blue cheese, called *Penicillium* (there is a variety of species that produce substances of this kind, but *P. chrysogenum* is the one used industrially) and, in the pioneer work, the amounts produced were minute.

Despite the existence of resistant bacteria and allergic patients, penicillin has revolutionized medicine and is still one of the most valuable drugs we have. It will continue to be made microbiologically, simply because it is so difficult to prepare chemically. One point of special interest is that the strains of mould now used to make penicillin produce at least three hundred times as much as did Fleming's original strain, and the reason for this illustrates an important point concerning the flexibility of industrial microbiology. I have emphasized often how microbes show great adaptability or, in other words, can adjust themselves to new environments. The process of adjustment involves a process called mutation (which I shall discuss later in this chapter), and just as one can obtain mutants resistant to drugs or able to grow at the expense of unfamiliar materials, so one can obtain mutants that will produce more (or less) of by-products such as penicillin. Maltreatment of strains with substances such as mustard gas or with ultraviolet radiation, γ- or X-rays, increases mutation among those individuals that are not killed, and in these ways mutants of *P. chrysogenum* have been obtained which have the enhanced penicillin productivity just mentioned. All strains used industrially are mutant, and their productivities are closely guarded commercial secrets.

Penicillin itself – or rather the three or four penicillins that are formed by various strains in various conditions – are normally produced by batch culture fermentations. Continuous culture has not so far been of much use. But in the 1960s, Beecham Research

AN INDUSTRIAL FERMENTER. A fermentation vessel of 110,000 litre capacity used by Glaxochem for producing antibiotics. (Courtesy of Dr J. B. Ward, Glaxo Group Research Ltd)

Laboratories in Britain developed a half-microbiological, half-chemical procedure for manufacturing all kinds of laboratory variations on the penicillin molecule, and some of these proved to be exceptionally useful. The penicillin molecules have this formula:

```
                 S
               /   \
R-NH-CH-CH           C(CH3)2
     |  |            |
     C - N ---------- CH.COOH
     ||
     O
```

where 'R' signifies a group of atoms whose precise composition determines which of the penicillins it is. It is possible, using special mutant strains and cultural conditions, to make the mould form penicillanic acid, which has the formula:

```
               S
             /   \
NH2-CH-CH          C (CH3)2
    |  |           |
    C - N -------- CH.COOH
    ||
    O
```

(Notice that it is the same as a penicillin but with an 'H' where an 'R' should be.) This substance is not an antibiotic, but it becomes one if, by chemical manipulation, one of the groups 'R' is attached to it. Now, by attaching groups in the position 'R' that would never have turned up in nature one can make an enormous variety of 'unnatural' penicillins. This has been done by the Beecham's group and a measure of their success is that some of them are active against bacteria that have become resistant to the natural penicillins.

In this instance, a combination of microbiological and chemical processes has been used industrially. One of the antibiotics to be discovered soon after the emergence of penicillin was chloramphenicol, formed by the actinomycete *Streptomyces venezuelae* and active against bacteria (such as the typhoid organism) that penicillin hardly touched. This material has a relatively simple chemical structure, is not difficult to make chemically and is now made industrially without the aid of microbes. Most of the other antibiotics are, however, made by fermentation processes, and though hundreds have been reported in the scientific literature, surprisingly few have proved to be of any real value in pharmacy. It is possible, however, to classify their types, and this I shall do to obtain a synoptic view of those antibiotics that have proved to be of some medical use.

Penicillins: The first, least poisonous and most useful antibiotics to be discovered. Formed by moulds of the genus *Penicillium* and by some species of *Aspergillus*. More than half a million pounds of penicillin are produced each year in the USA. Though penicillin is extremely effective when it does work, its antibacterial spectrum – which means the range of bacterial species against which it is active – is rather narrow. Cephalosporins are antibiotics related to penicillin which have a rather wider range of action and which often attack bacteria which are resistant to ordinary penicillin; they are formed by fungi of the genus *Cephalosporium*. Beecham's semi-synthetic penicillins, as I just mentioned, have an extended range of activity compared with the natural substances.

Polypeptide antibiotics: These are protein fragments, of rather unusual structure, produced by bacteria of the *Bacillus* group and active against other bacteria. They act rather like detergents, damaging the cell wall and, though mostly too toxic for internal use in medicine, they have been used to treat external wounds. Examples are gramicidin, polymyxin and bacitracin.

Tetracycline antibiotics: These are broad-spectrum antibiotics which, though they are rather rough on the normal bacteria that live in association with us, are proving popular with doctors for controlling the secondary bacterial infections that often accompany virus diseases, as well as for coping with penicillin-insensitive bacterial infections. They are formed by both moulds and a group of filamentous bacteria called actinomycetes. A genus of actinomycetes called

Streptomyces has been particularly rewarding in the variety its species make. They include the widely used chlortetracycline (aureomycin) and oxytetracycline (terramycin), which are formed by *Streptomyces aureofaciens* and *S. rimosus* respectively. Over the last two decades Sir David Hopwood and his colleagues at the John Innes laboratories in England have made a detailed study of the genes involved in building up such antibiotics – and related molecules that may have no antibiotic action – in streptomycete cells. They applied this knowledge in the mid-1990s to develop ways of arranging, by genetic manipulation, for genes from different microbes to function together in single strains of streptomycete, which then form hybrid molecules. The way is now open for making novel antibiotics of this kind.

Glycoside antibiotics: Streptomycin, the next antibiotic to be discovered after penicillin, is produced by the actinomycete *Streptomyces griseus*. It is rather more toxic than penicillin and the tetracyclines, but still of great medical value, notably in the treatment of tuberculosis. It attacks a variety of organisms that are insensitive to penicillin and is chemically distinguished by including modified sugar molecules in its structure. Neomycin, related to it and formed by *Streptomyces fradiae*, is too toxic for injection but it is not absorbed from the intestines and is valuable for gut and skin infections. Novobiocin belongs to this group and more distantly related, in a chemical sense, is erythromycin, which has been used to control penicillin-resistant infections. These substances are all produced by species of streptomycetes and, for streptomycin, high-yielding mutant strains have been developed for use by industry.

Polyene antibiotics: Some streptomycetes produce compounds distantly related to vitamin A which are active against fungi. They have been used to treat fungal infections and include nystatin, which is the one most widely used in medicine.

Unclassified: Between 1938 and 1978, when the vigorous search for antibiotics in pharmaceutical laboratories began to slacken off, over 5,500 antibiotics were reported. By 1990 more than a hundred had got as far as being produced commercially, and new ones will continue to turn up or be developed artificially. It is clearly impossible to say much about them here. A group of some special interest is the anti-tumour agents, which cause regression of some kinds of cancer. Actinomycin, the first to be discovered, is intrinsically very poisonous and not of much practical value, but research on less toxic ones such as mitomycin (from the USA) and olivomycin (from the former USSR) has led to some clinical successes. They are all formed by strains of actinomycetes, and they act by interfering with the function of ribonucleic acid (RNA), a component of living cells which controls growth. Other antibiotics that deserve special mention are chloramphenicol, a broad-spectrum antibiotic which, as mentioned earlier, is the only one to be produced chemically. Cycloserine, another *Streptomyces* product, is useful in the treatment of tuberculosis; yet another valuable anti-tubercular antibiotic is rifampicin, made by chemical modification of rifamycin, obtained from *Streptomyces mediterranei*. Griseofulvin should be mentioned, because it is produced by *Penicillium griseofulvum* and also by the microbe that makes streptomycin (*Streptomyces griseus*). It is active against plant pathogens, particularly fungi such as mildews and rusts, and is of considerable value in agriculture. Nisin, produced by the bacterium *Streptococcus lactis*, is in fact an enzyme and has been used in food preservation.

The last two examples show that antibiotics, though normally regarded as wonder-drugs for use on people, also have applications in agriculture and food preservation. They are of obvious importance in veterinary medicine; they have also been used as additives to animal fodder, a matter discussed in Chapter 5. They are certainly the mainstay of industrial microbiology today, in the sense that they are reliable, saleable items which industry can only produce using microbes. Consequently, an enormous amount of money has been spent by industry in the search for antibiotics and in their development; it is really rather remarkable that only some hundred are in fact well suited to use in general medicine and that the best of all, penicillin, should have been the first to be discovered. It is also surprising what a variety of antibiotics is produced by the actinomycetes, notably the genus *Streptomyces*; in 1988 streptomycetes collectively provided about half the monetary value of world antibiotic production, then amounting to some £5 billion. It has been suggested that, since these are slow-growing soil microbes that exist in competition with more active soil bacteria, production of antibiotics may be of selective advantage to them in their natural soil environment. The flaw in this argument is that, in nature, they never seem to produce anything like enough antibiotic to influence their neighbouring bacteria.

Microbes, when used by industry for the production of material such as antibiotics, or the vitamins discussed in Chapter 5, are used by the industrialist as a special kind of chemical reagent. The industrialist uses microbes to convert one kind of substance into something more useful, and it is a small step from this kind of activity to that of using both microbes and chemicals in a sequence of chemical syntheses. I have given examples of this already: the formation of vitamin C using *Acetobacter* on a chemically produced reagent, or the artificial penicillins produced by doing some chemistry on penicillanic acid.

One of the most impressive instances of the use of microbes as reagents in a sequence of chemical syntheses occurs in the industrial production of steroids. These are hormones, and materials related to hormones, which are of importance in pharmacy. The earliest examples were the ergot alkaloids, which can have actions resembling certain sex hormones and which are formed naturally by a fungus called *Claviceps*. This fungus attacks wheat and has occasionally got into bread by accident, causing hallucinations and a variety of other disorders in people who eat it. Ergosterol, related to this group of alkaloids, also occurs in yeast and can be extracted and made into vitamin D. But most spectacular has been the use of moulds of the *Rhizopus* groups to alter the chemical structure of plant steroids and to convert them to pharmacologically active hormones. Most people have heard of cortisone, a hormone of the adrenal cortex which has proved dramatically effective as a palliative in rheumatoid arthritis. It can be obtained in minutely small quantities from the natural glands of, for example, cattle, and in 1949 the only other method of preparing it was to conduct thirty-seven separate chemical operations on one of the bile acids. No wonder it once cost something like $500 a gram! In 1952 research workers at the Upjohn Company in the USA discovered that *Rhizopus* would act on a fairly readily available sex hormone called progesterone to make a product that could be converted to cortisone in only six chemical steps. (Progesterone, originally obtained only from the gonads of animals, could be made from a steroid called diosgenin, found in a Mexican plant called elephant's foot.) Since then a vast literature has developed on the use of moulds, mainly *Rhizopus* and *Aspergillus*, for the transformation of steroids from one chemical configuration to another. Actinomycetes and bacteria such as *Bacillus* or *Corynebacterium* have considerable use in certain reactions called dehydrogenations. In principle the procedure is quite simple: the mould is grown with, say, some glucose and an extract of corn as nutriment, in a medium containing an emulsion of the steroid. (Steroids are inclined not to dissolve in water and must therefore usually be emulsified.) After a suitable time the culture is killed, the steroid material extracted and, in favourable cases, up to 95 per cent of it has been transformed into a new steroid. *Rhizopus arrhizus*, for example, oxygenates progesterone to give an anti-inflammatory drug which is useful in controlling rheumatic and arthritic diseases. Steroids useful in preventing premature abortion and in dealing with disorders of the menstrual cycle, or which are active as oral contraceptives, are now made by processes involving the action of microbes.

Antibiotics, steroids and vitamins might be called the money-spinners of latter-day industrial microbiology, at least as far as the pharmaceutical industries are concerned. There are, however, many more workaday products for medical use in which microbes are involved. Dextrans are starch-like materials formed from sugar which are valuable because they can be used as substitutes for plasma in blood transfusions. They are prepared industrially by letting the bacterium *Leuconostoc mesenteroides* act on ordinary sugar; sometimes the bacterium is killed and the enzyme responsible for the conversion of sugar is extracted from it, because this gives a rather more controllable process. Antisera which are used to protect people who have been exposed to risks of diseases such as tetanus, are prepared by injecting live bacteria into animals and obtaining preparations of the antibodies they form. Vaccines are traditionally prepared by culturing pathogenic microbes in safe hosts, or in cultures, and rendering them harmless by heating or killing with a disinfectant. They may then be safely injected into patients in whom they provoke resistance to the original pathogen. More recently vaccines have been prepared by genetically manipulating harmless viruses so that they resemble pathogens sufficiently to provoke such resistance – more about that later in this chapter.

Higher plants contain all sorts of biologically active substances: poisons, drugs such as quinine, narcotics such as opium or cocaine. It would be surprising if the microbial world did not include organisms which make some such materials, but in fact they are rather few (outside the higher fungi, which are not really microbes). Mould products which lower blood cholesterol or inhibit intestinal enzymes have been reported, but I am not aware of any great use for them.

Enzymes, the biological catalysts that cause biochemical reactions to take place, are an important group of industrially valuable microbial products. They can be extracted from all kinds of living tissue, and microbial tissue is often industrially the most convenient. Pectin, the gelatinous component of fruit that causes jam to set, can be broken down by enzymes called pectinases which are formed by many bacteria, and enzymes prepared from these are used in stabilizing fruit juices. The retting of flax, steeping it to remove pectins and leave the plant fibre, is a traditional procedure that is fundamentally an exposure of the plant to bacterial pectinases; as far as I know, neither pure cultures of bacteria nor preparations of pectinases are used industrially, the traditional retting being preferred. Amylases are enzymes that break down starch, and they are used in laundering and in the paper industry (they dissolve starchy dressings from fabrics which are to be pulped for paper manufacture). They are prepared industrially from the mould *Aspergillus oryzae* among other aspergilli, or from bacteria of the genus *Bacillus*. Proteinases, enzymes that break down proteins, are used to clarify beer, for removing protein stains in laundering, for conditioning dough in baking, for removing extraneous 'meat' and hair from hides prior to tanning and for removing gelatin from spent film emulsions; they have even been claimed to accelerate the clearing of blood clots and bruises such as black eyes. They are prepared from plants or various microbes including fungi of the genus *Aspergillus*. Lipases, enzymes which break down fats, are extracted from certain pseudomonas-like bacteria, or from fungi, for use in laundering and hide treatment. Production of amylases, proteinases and lipases together, which are also common components of domestic biological detergents, has become a huge industry since the 1950s. Cellulases, which are enzymes that break down the cellulose of plant material to form glucose, are formed by certain wood-rotting fungi and pseudomonads; they have been used on a modest scale to make cellulosic waste materials fermentable. An especially heat-resistant form of a group of enzymes called DNA-polymerases, prepared from a thermophilic bacterium called *Thermus aquaticus*, has become so essential a tool of modern molecular biology that its production and marketing is now a multimillion-dollar business operation – I shall tell about its use later in this chapter. Yet another economically important microbial enzyme is invertase, an enzyme that converts cane sugar into glucose and fructose; it is prepared from yeasts. It has been used for making artificial honey, but its most curious use is in making soft-centred chocolates. In this process, a hard sugar fondant containing the enzyme is rapidly coated with chocolate. When it has set and been allowed to stand, the invertase in the fondant breaks down the sugar to the invert sugars, with the result that the fondant becomes partly liquefied.

Enzymes often have the important property of being very specific, by which I mean that they only attack certain types of molecule: cholesterol, glucose, fumaric acid and so on. Therefore they can be very useful in scientific research, for measuring amounts of the molecules they attack, and also in diagnostic medicine, where they can be used to detect and measure substances which either should not be where they are or which are present in wrong amounts.

An impressive development of the last few decades has been the use of electronic monitoring devices called biosensors, which are based on either enzymes or whole cells. Chemists can now make little electrodes, or voltaic cells, which change their electrical output in response to tiny amounts of small molecules such as oxygen, ammonia, hydrogen ions and so on. An example is the 'oxygen electrode', actually a cell whose output depends on the concentration of oxygen in its neighbourhood; it has proved valuable for monitoring oxygen in both research and industry. An oxygen electrode becomes a biosensor if it is combined with, for example, the enzyme glucose oxidase, which consumes oxygen when, and only when, glucose is available. Then its output goes down in proportion to the concentration of glucose present: one has an electrical device which not only senses glucose but measures its amount. Miniaturized devices of this kind were available as long ago as 1970 for measuring glucose in as little as 15 μl of a patient's blood serum. But one does not need to have a purified enzyme in a biosensor; any biological material which reacts with a specific substance to produce or consume a small molecule to which an electrode can respond will work. Living microbes – yeasts or bacteria – immobilized in permeable plastic have been used in conjunction with appropriate electrodes to detect and measure amino-acids; and even the antibiotic cephalosporin. Such devices permit continuous monitoring of production systems, experimental set-ups or hospital patients in ways that were hitherto impracticable. As well as being highly selective, biosensors can be extremely sensitive. An example is a device which uses immobilized luminous bacteria together with a photo-electric cell: the bacteria only produce light if some oxygen is present, but they need very, very little to do so; the combination is nearly 100 times more sensitive to oxygen than the chemical 'oxygen electrode' mentioned earlier.

(I have not mentioned the luminous bacteria before in this book. They are a small group of bacteria whose metabolism causes the emission of light by a biochemical process, similar to that found in fireflies, for which they need air. They may be found free-living in the sea or in the luminous organs of deep-sea fish.)

So far in this chapter I have surveyed the role of microbes in industrial production, industrial microbiology as it was called for several decades, and all the processes I have mentioned were discovered between about 1900 and 1975, or stemmed from discoveries made during that period. However, in the mid-1970s a development took place in our understanding of inheritance in microbes which has introduced an entirely new element into industrial microbiology and (amusingly for some, disconcertingly for others) engendered a change in name. The study of inheritance in microbes is microbial genetics and industrial microbiology combined with microbial genetics is now called biotechnology. Actually, that statement is not entirely correct: it is too narrow. Biotechnology need not necessarily involve microbes at all, but it almost always does; also many industrial microbiological processes which have no genetic component have become absorbed under the blanket name of biotechnology. For example, the biosensors which I have just described are biotechnological devices even though they involve no genetic manipulation. Definitions of broad subjects are always fuzzy at the edges and I shall not try to be more precise about the term biotechnology; the important message for now is that recent developments in microbial genetics have enabled scientists to modify the hereditary properties of microbes in ways which are both spectacular and exciting. The implications for the future of applied microbiology will not be fully realized for many decades, yet they are already tremendous, particularly as regards microbes in production processes. The mundane consequence of all this is, dear Reader, that I must now digress into a brief account of modern microbial genetics so as to tell you about the new frontiers of biotechnology. As with the chemistry which came earlier in this book, I shall keep everything as simple as possible.

MICROBIAL GENETICS

It all stems from a matter I have mentioned several times in this book, the variability and adaptability of microbes and their ability to undergo mutation. How do these variations come about? Can they be controlled and directed?

I imagine most readers are aware of the crucial importance of DNA in heredity. DNA is the technical abbreviation for deoxyribonucleic acid, a chemical found in virtually all living things that acts as a sort of blueprint of what the organism will be like. (It is absent from certain bacterial viruses called the RNA viruses; it is also absent from prions which, as I wrote in Chapter 2, I choose not to regard as living.) The precise chemical composition and configuration of DNA are now known in considerable detail – it is an array of components called bases tied together with molecules of a sugar (called deoxyribose) and phosphate groups. There are essentially only four bases, known as adenine, thymine, guanine and cytosine. Henceforth I shall refer to them by their initial letters, A, T, G and C. In a microbe such as *Escherichia coli*, the DNA is one huge molecule, a very long array of these bases which would cover four centimetres if it could be pulled out straight (remember that *Escherichia coli* is about two ten-thousandths of a centimetre long). In fact, the DNA is coiled and coiled again into a tight, compact bundle called the chromosome. It is rather like that coiled cable which links the handset of a modern telephone to its base and which so readily (and irritatingly) coils up on itself. Actually, DNA is more like what is inside the telephone cable, because it has two strands, like the telephone's two wires, and these strands follow each other coil by coil. The correct name for the DNA coil is a helix, and the molecule which is the *E. coli* chromosome is a double helix. In *E. coli* it is a completely circular structure, as if the two ends of a telephone cable were joined together. The two strands of DNA have an important characteristic: whenever there is an A in one strand, there is a T opposite, and whenever there is a G, a C lies opposite it so:

—GCCATTAG—
—CGGTAATC—

The strands are said to be complementary (brief zones of non-complementarity exist but do not matter for the moment). The abundance and arrangement of the four bases differ from species to species, so that there are as many different DNAs as there are species, some widely different, some only slightly so. The DNA molecule is rather like a sort of coded tape, with four symbols (represented by the bases), which spell out what enzymes and structural components, and roughly how much of them, the organism will consist of. Every species, from bacteria to people, has its special tape which determines what it will be like; indeed, in multicellular organisms every cell carries this tape in its nucleus, and in all but the generative cells the tape is present in duplicate.

The chromosome of *E. coli* is circular, albeit a tightly folded circle, as I said earlier, and so are the chromosomes of most other bacteria, and those of most DNA viruses. A few complications exist: some bacteria have several identical chromosomes; the chromosome of *Borrelia*, although tightly folded, is linear, with a beginning and an end like a knotted up twine; and most bacteria have additional mini-circles of DNA called plasmids (these will prove to be very important later in this chapter). But by the standards of higher organisms such as animals and plants, the DNA of bacteria and viruses is very simply structured. Higher organisms, being altogether more complex creatures, have vastly more DNA, distributed among a multiplicity of differing but paired chromosomes – humans have 46 of them – all housed in the nucleus within the cell. During sexual reproduction parents donate one copy of each of their chromosomes to the offspring, which thus possesses a hybrid nucleus derived from both. The cells of higher organisms also have small extra amounts of DNA in so-called organelles which lie outside the nucleus; organelle DNA does not participate in sexual reproduction and is passed from generation to generation only in the female line. There are many other genetic differences, in both the way in which the information encoded in the DNA is arranged in nucleate organisms, and the way that information is made use of, but it remains true that they all use essentially the same DNA code as bacteria.

An analogy with human languages might be helpful. The DNA code is like a common alphabet in which genetic instructions are written, but nucleate and non-nucleate creatures have their instructions in different languages, with different grammars. The situation recalls the differences between French and German, say, even to the extent that one can discern shared root words, and also linguistic patterns which indicate a common if distant ancestry. Moreover, rather as languages always develop regional dialects, local idioms and even family idiolects, so, despite their shared DNA 'alphabet', every species, every strain even, of plant, animal or microbe has features within the sequence of bases in its DNA that are distinctive. That fact can be exploited for identification and classification – more of that story later.

The nature of DNA, the way in which it is reproduced and the way in which it influences the nature of living organisms, has occupied, indeed sometimes obsessed, biologists for the last five decades. About fifty years ago DNA was recognized as the chemical form of what were then only conceptual elements called genes, which conveyed the hereditary characteristics of living creatures from generation to generation. Biologists had been pursuing the study of heredity – known as genetics – ever since the work of Mendel in the nineteenth century, but in the early 1940s it received an enormous impetus from American experiments on the genetics of a microbe, the bread mould called *Neurospora*. By treatment with ultraviolet light, X-rays or certain chemicals, mutants of *Neurospora* were obtained that had lost certain biochemical abilities – they became, for example, unable to synthesize certain vitamins – and by a systematic study of the progeny of such mutants, duly crossed sexually, the concept emerged that one gene was responsible for the ability to make one enzyme. The discovery that bacteria formed mutants of a similar kind in comparable conditions initiated an extremely rewarding period of research in microbial genetics, as a result of which the chemical pathways were elucidated by which all sorts of components of microbial cells were made. If the biochemistry of the two pre-war decades had been concerned largely with the breakdown of natural products, the war-time and first post-war decades were concerned with their synthesis, and microbes – bacteria and moulds – were the major research material for such studies. As metabolic pathways became clear, or reasonably clear, one after another, interest began to shift in the direction of how these syntheses were controlled: what precise mechanism told the cell what to synthesize, and how much.

By the 1950s the basic importance of DNA in these processes was clear. DNA was the tape bearing all the hereditary information available to the cell; some form of message was transferred from the DNA tape to those centres of the cell (ribosomes) actually capable of synthesizing cell material. Some of the information on the tape remained masked: it passed out no message, until some stimulus of a chemical character removed the mask, permitted release of the appropriate message and initiated synthesis of something new. The procedure by which the message was transmitted, and read, involved a material called ribonucleic acid (RNA), similar in its general chemical pattern to DNA but differing in important details (the bases are not quite the same as those in DNA and the sugar linking them is different). Essentially, one kind of RNA carries the message from DNA to the ribosomes, 'telling' them what protein to make; another kind is actually part of the ribosome; a third kind ferries amino-acids to the ribosomes so they can be joined together as protein. The whole process is very elegant and, naturally, has to be regulated so that things are made in the right order and the right amounts. Several regulatory processes exist: for example, internal feedback processes have been recognized in the pathways of protein synthesis, whereby products of a certain sequence of reactions may slow down and even stop the earlier steps and, in such a manner, ensure that the organism does not make too much of any particular component.

The part played by RNA in reading and using DNA, incidentally, explains the existence of the RNA viruses which I alluded to a few paragraphs ago. They are common and important pathogens of plants and animals; they consist of coiled circles or coiled strings of RNA, not DNA, sometimes as double strands like regular DNA, but more usually as single strands. Many behave rather like wild packages of message-bearing RNA which, after invading a host cell, distort its gene-reading machinery by coding for more of themselves; others oblige the host to translate their message into DNA before causing it to make more virus RNA. The RNA viruses are a genetically strange and quite diverse group of microbes which will crop up again; their nature, modes of invasion and multiplication are an important subject which I must leave for the time being.

I wrote just now that the DNA is a sort of tape which carries the hereditary information. This is true in the sense that the sequence of bases A, T, G and C in DNA does specify precisely which amino-acid shall go next into whatever protein is being synthesized (for example, the trio ATG specifies the amino-acid methionine, ATA specifies iso-leucine). A gene is a long (about 1,000 base) sequence of DNA which encodes the amino-acid structure of the protein which it specifies, and the code is now known: clutches of three bases specify an amino-acid. Arrays of bases lying as relatively short stretches between genes act as punctuation: they tell the reading machinery to stop, start, or slow down. Knowing the sequence of bases in a gene and the start–stop signs, a molecular biologist can tell the chemical composition of the protein it specifies precisely, and most protein structures discovered in the last fifteen years have been elucidated in this way.

Our knowledge of molecular genetics, as it is called, blossomed in the second half of the twentieth century, but it is impossible to go into the details of these advances here – they would occupy a substantial book. Microbial genetics has proved to be the clue to the understanding of the genetics of most living organisms and, in the last forty years, it has led to advances in biology corresponding to the flowering of atomic chemistry in the early years of this century. Just as Dalton's conceptual atoms were shown to have a physical reality in those early days, so Mendel's genes have been recognized as chemical entities and their structure and function understood to a remarkable degree. Enthusiasm among biologists has led to the renaming of parts of the subject: biochemical genetics, molecular genetics and molecular biology have been used at various times as names to describe the research area I have been dealing with. Confusing perhaps, but there is nothing like a new science to attract research grants. Scientists have their fads and fashions, and if ever one justified itself it is the current passion of biologists for molecular biology.

Having painted in some kind of background, so to speak, I can tell something about mutation, which was the question I started out to discuss. A mutation arises simply as a chemical change in the DNA of an organism. The microbe in my example which was exposed to X-rays, ultraviolet light or certain chemicals, had its DNA damaged, because DNA is sensitive to such treatments. It might be able to repair this damage, in which case it would be able to multiply unchanged. It might not, in which case part of the code on the DNA tape would become mis-written. If the code became drastic nonsense the microbe would be unable to multiply and would die; if damage was modest the organism might be able to multiply, but with changed hereditary characteristics. By modest damage I mean an A being changed to a T in the genetic code (so a different amino-acid from normal is specified), or a G changed to a C, or a whole triplet of bases dropped out. Such changes will cause the gene to specify a protein only slightly different from what it ought. Perhaps the protein will then do what it used to do, perhaps not; perhaps it will function, only rather badly. In any case, the microbe will have undergone a mutation, and its progeny will be mutants. The value of microbes, to scientists, has been the enormous variety of mutants one can detect in them and the relative ease with which they can be made and studied.

Mutations actually occur spontaneously. In *Escherichia coli*, about one organism in every ten million is a mutant of some kind. Mutations represent one way in which microbes can change and so adjust themselves to a changed environment. The switching-on of masked information on their DNA tapes is another mechanism of variation. A third process, which arose from the study of mutants, is known as genetic recombination. If mutants of an organism requiring a vitamin (X) are made to mutate twice more, one can get a strain needing, say, X, Y and Z. If one takes a different mutant of the same organism requiring different vitamins (A, B, C) and grows them together in the same culture, some of the progeny may be found to require X and B, X and A, Y and C and so on. Obviously some transfer of genetic information, therefore of DNA, has taken place between individuals of the ABC type and those of the XYZ type. This process most often results from a pseudo-sexual conjugation between individuals in the populations; it has been observed in electron micrographs. Conjugation is rare in most bacteria, but some strains of intestinal bacteria have a high frequency of such recombinations (the so-called Hfr strains). The process has some analogies to sexuality and might be an evolutionary precursor of the sexual reproduction of higher organisms, but bacterial conjugation has several peculiarities, not least of which is the fact that the act of conjugation confers the property of maleness on the female, or recipient, cell. The hereditary factors responsible for maleness in the Hfr strains are incorporated with the rest of the genetic material in the DNA of the chromosome. However, in the majority of conjugating bacteria the maleness, or donor, genes are not on the chromosome proper but on a mini-chromosome called a plasmid. There are many kinds of plasmids and they have become very important in microbiology over the past thirty years, so again I must pause and tell a little more about them.

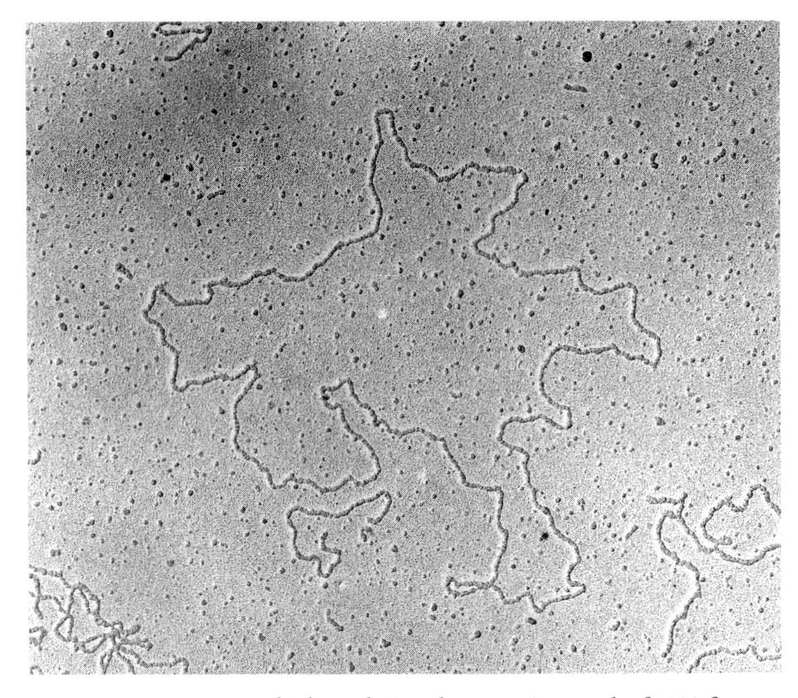

A PLASMID. A high resolution electron micrograph of DNA from a streptomycete. The central 'string of beads' is a molecule of plasmid DNA which has become uncoiled during preparation; it can be seen to be a continuous circle. (Photo by Dr M. Bibb, courtesy of the John Innes Institute)

The majority of plasmids are little coiled circles of DNA which seem to exist alongside the chromosome but to replicate themselves independently of it. They may be anything from a quarter of the size of the chromosome to less than a hundredth; they carry all sorts of genetic information and by no means do they always carry maleness. However, when they do, they can transfer themselves from one microbe to another, carrying other genes with them. Among the better known plasmid-borne genes are those specifying resistance to various antibiotics; some plasmids encode resistance to two or three antibiotics at a time and can contribute to the problems in medication which I mentioned in Chapter 3. Most bacteria carry plasmids, often several at a time, and the function of most of them is not known. However, self-transmissible (sexual) plasmids can carry substantial packets of genetic information into the recipient microbe, so altering its character substantially. A map of an example is sketched in the diagram overleaf. Sometimes self-transmissible plasmids seem to pull (co-transfer is the correct word) other DNA, which might be chromosomal or whole other plasmids, with themselves into the recipient microbe.

Conjugation of the kind I have just discussed is a major means of genetic variation among microbes. Some plasmids can transfer between genera of bacteria so that a drug-resistant *Escherichia coli* can pass its ability to resist a drug to, for example, a *Salmonella*. A worrying example occurs in intensive cattle farming: if calves become infected with a few antibiotic-resistant salmonellae, the resistance can be passed on to their native intestinal microbes without their ever having encountered the antibiotic in question.

I introduced plasmids as the carriers of the genetic information which enables bacteria such as *E. coli* to conjugate, but the variety of biological information which is encoded in the DNA of plasmids is surprising. Resistance to antibacterial drugs causes most anxiety among microbiologists, because so many drug-resistance plasmids have been discovered, but virulence comes a close second. For example, the pathogenicity of the notorious strain O157 of the normally benign *E. coli*, and of a few other nasty *E. coli* strains, is due to genes carried on plasmids, and the genes which convert what would otherwise be a harmless aquatic vibrio into the dangerous cholera agent *Vibrio cholerae* are also on two plasmid-like bodies. But, more positively, a plasmid which makes the soil bacterium *Agrobacterium tumefaciens* into a virulent tumour-inducing plant pathogen has proved to be of inestimable value in the genetic engineering of plants. The amounts of genetic information that plasmids carry can sometimes be substantial: among the rhizobia, the bacteria which form symbiotic nitrogen-fixing associations with leguminous plants, the 'blueprints' for recognizing the right plant, for setting up the association and for making the enzymes which fix nitrogen, are

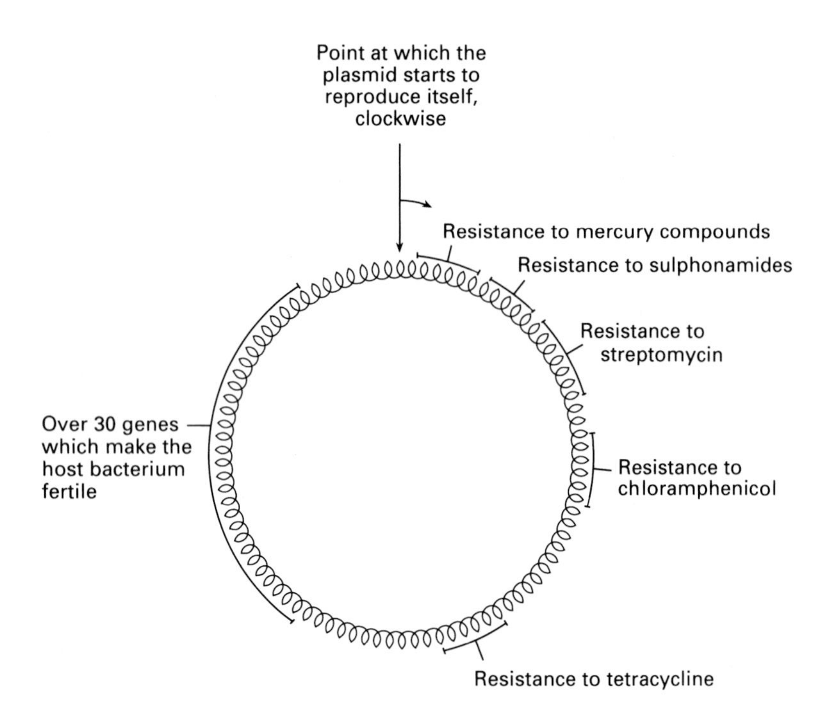

SKETCH MAP OF A PLASMID known as R 100, which makes its host bacteria fertile and also resistant to five different antibacterial substances. The relative positions of the relevant genes on its circle of DNA are known.

usually on plasmids, often relatively large ones. Other properties that have been identified as plasmid-borne in various bacteria include the ability to consume certain organic compounds as food, and the ability to make substances which are toxic to other bacteria. Broadly speaking, possession of plasmids seems to confer certain beneficial properties on the microbe hosting them, though they are not generally the sorts of property that the host is likely to need all the time.

Why do bacteria have plasmids at all, you may ask? After all, bacteria have a perfectly good chromosome which carries the bulk of their genes. It is a good and fundamental question. I mentioned earlier that all cellular organisms, including ourselves, carry small amounts of genetic information away from their chromosomes, in compartments within their cells called organelles. Examples of organelles are the particles called chloroplasts which, residing within plant cells, make plants green; or tiny structures within animal cells called mitochondria, which play a fundamental part in generating energy. Biologists usually regard the separate packages of genes that such organelles possess as being left over from the way in which the host creature's genetic machinery evolved: as relics of ancient symbionts which, long ago in evolutionary history, became incorporated into the cell – I shall return to that story in Chapter 10. Something of the sort may explain the abundance of plasmids in bacteria: some plasmids may have arisen from bacterial viruses which happened to have acquired genetic information useful to the host, and to which the host became adjusted. For example, the plasmid-like structures which make the cholera vibrio virulent are actually more like viruses than conventional plasmids. But however they originated, plasmids seem now to have become a kind of 'gene bank' for groups of bacteria: a reserve of genetic elements which carry packages of biological information which is not needed as a rule, but which can be called upon if circumstances change in a way that would make the information they carry useful. Being mobile, plasmids can be exchanged among strains, species and even genera of bacteria – they give bacteria, unusually for a group of living things, a degree of emergency access to each other's genes. This is what makes plasmids so valuable in biotechnology: some of the properties on plasmids are potentially useful to man, and in addition scientists can now modify and exploit plasmids to alter the genetics of microbes, and of other organisms, in useful ways. More about that shortly – I must return to the topic of genetic variation among microbes.

A second mode of microbial variation is called transformation. It arises because sometimes bacteria can actually absorb DNA from their surroundings without damaging it. Therefore, if DNA from a bacterium (P) is added to a culture of another (Q), a proportion of the Q culture takes up the P-DNA and acquires P-like characteristics. The DNA-added can come from the chromosome, or else, in special circumstances, whole plasmids can be taken up.

Mechanisms of variation in bacteria

Mutation. By some accident, the microbe's DNA becomes altered. The change is usually lethal, but when it is not, it becomes hereditary.

Conjugation. A 'male' bacterium donates some of its DNA, usually as a plasmid, to a 'female' recipient, giving her new genes which become hereditary. These include genes that change her into a 'male'.

Transformation. A bacterium in a suitable state (e.g. subject to abrupt cooling) takes up raw DNA from its surroundings and builds some of it into its own DNA. The new DNA becomes hereditary.

Transduction. A bacterial virus carries some DNA from a previous host into a new host, and the latter survives to use the imported DNA, which becomes hereditary.

Finally, there is a process called transduction that can lead to variation among microbes, and it involves participation of bacterial viruses or bacteriophages. Many bacteriophages kill their hosts, but some, called temperate 'phages, do not. They appear to live peacefully within their hosts and only multiply and damage them under the influence of some external stimulus (ultraviolet light is an example). They can then re-infect new hosts and, when they do so, they may carry some of the hereditary characteristics of their previous host into the new one.

I have collected the main modes of bacterial variation together in the box above. It is the understanding of microbial genetics, of how their processes of inheritance work at the molecular level, which has given the new impetus to biotechnology. The particular event which did more than anything else to push things along was the discovery of ways of taking isolated DNA, in the test-tube, chopping it up with certain enzymes, reassembling it in new combinations and transferring it back to living microbes, in which it would persist. It became possible to extract plasmids, for example, open up their DNA chains, insert DNA from other organisms – not necessarily microbes, as I shall tell – and then to close up the plasmid DNA circle, transform it into *E. coli*, say, and thus confer a completely new inheritable property on the recipient microbe.

How is this done? The crucial materials are two types of enzyme, both obtained from microbes. Restriction enzymes are enzymes which nearly all microbes possess, which they use to protect themselves from invading DNA (e.g. from a virus). A restriction enzyme will recognize alien DNA and split it, and eventually the cell will excrete the fragments. These enzymes cut the chain of bases which makes up DNA at certain specific sequences (e.g. between the G and A in the sequence GAATTC); sometimes they cut across the two DNA strands symmetrically, sometimes unsymmetrically. For example, a much-used restriction enzyme called EcoRI recognizes the base sequence I have just mentioned and cuts at the GA junction in both complementary strands of DNA, leaving fragments with single-stranded tails, thus:

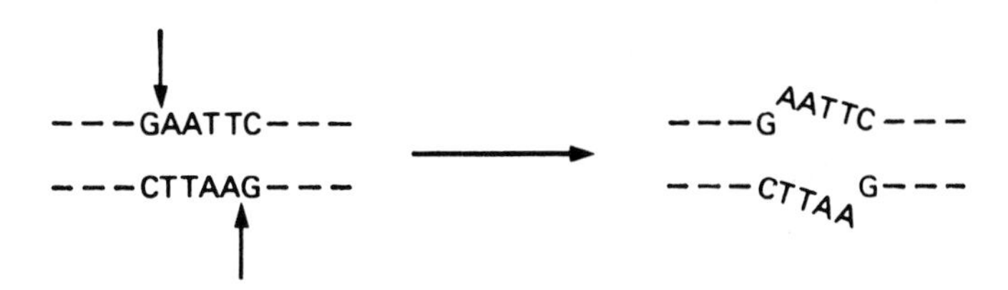

Sites at which these enzymes cut can be quite rare – more than a gene apart – and some plasmids have only one site for a given enzyme. Restriction enzymes are used to cut DNA into fragments, many of which will carry whole genes, and also to cut, for example, a small plasmid or the DNA of a virus.

That is half of the story. The other half is that enzymes exist which enable one to stick the cut ends together again. These enzymes are those which microbes use to make their DNA, or to repair damaged DNA; they are called DNA ligases and can be used, still in the test-tube, to stick the cut ends of restricted DNA together. For example, a plasmid with a single restriction enzyme site can be reassembled with a ligase, although some funny by-products also turn up when a cut end of a plasmid does not link with its own tail but instead hooks on to the cut end of another. If, however, you take a plasmid which has been cut in one place, mix it with quite different DNA (from anywhere – man, mouse, plant, microbe) which has been cut with the same restriction enzyme, and then treat the mixture with an appropriate DNA ligase, you will get (given certain conditions concerning the nature of the DNA and its cut ends) not only reconstituted plasmids but also chimaeral plasmids, new plasmids into which fragments of alien DNA have been incorporated. Your test-tube will contain all sorts of chimaeras, depending on how many fragments of alien DNA you added. If you now use that DNA to transform *E. coli*, for example, each individual will pick up on an average one plasmid – chimaeral or reconstituted – so a population of several thousand will have representatives of all the chimaeric plasmids which were formed in the mixture. The alien DNA in such chimaeral plasmids is said to be 'cloned'.

Easy, is it not? Well, it is not *quite* as easy as I have written it. How can you distinguish recipient *E. coli* which have taken up plasmids from those which have not? Answer: start with a plasmid that has a gene for, say, penicillin resistance on it and spread your population on a medium containing penicillin; only those with the resistance gene, and therefore carrying the plasmid, will grow. How can you distinguish *E. coli* cells carrying plasmids which have alien genes in them? One way is to choose a plasmid with the target site of the restriction enzyme in a different drug resistance gene – one for tetracycline resistance, for example. If an alien gene is spliced into the tetracycline resistance gene, that gene will no longer specify whatever substance it is that makes the cell resistant to tetracycline (to be technical for a moment, the tetracycline gene will no longer be expressed), so the *E. coli* cells which have become penicillin-resistant but sensitive to tetracycline are the ones which contain chimaeral plasmids. How can you tell precisely which genes have been spliced into the many chimaeral plasmids you have probably made? Enough! I am sure you do not want to know, at least not now. Rest assured that there are all sorts of tricks that molecular biologists can and do use to discover what kind of DNA they have cloned into their plasmid. More important, for this chapter, is what can be done with such cloned DNA.

However, before I return to that theme I must mention a few more genetic manipulations which will be relevant. I indicated earlier that the precise sequence of bases along any cloned fragment of DNA can be worked out. To explain how that is done would take me into a labyrinth of detail; the important message is that it can be done. A couple of decades ago it took months of slow and laborious work to work out the base sequence of a single gene – be informed that *E. coli* has about 4,000 genes – but today, with the help of automatic analysers and computers, DNA can be 'sequenced' quite quickly. In 1997 the first complete sequence of the DNA comprising a genome, that of a small virus, was elucidated in Cambridge, UK; by mid-1998 the genomic DNA sequences of about a dozen bacteria had been published, and those of some 60 more microbes, including yeasts, were being sequenced by groups working in various laboratories. Moreover, a grandiose international programme was under way to clone and read the whole human genome: i.e. the base sequence of all the DNA that constitutes our chromosomes, plus a small amount of DNA which exists outside these. All is expected to become known in the early 2000s. What can be done with such knowledge? To start with, one can identify genes within those DNA sequences and deduce what sorts of proteins they specify; in about half of the genome sequences so far worked out this information has given researchers a good idea of what the genes are doing. Moreover, once the base sequence of a stretch of cloned DNA is known, it can be modified, altering one or a few bases deliberately, so that the amino-acid composition of the protein specified by a gene can be changed (a new technology called 'protein engineering', which I shall return to later, exploits this principle). Genes within a stretch of cloned DNA can also be inactivated, or removed altogether, and then returned to their host; in that way one can, for example, make an unpleasant bacterial pathogen harmless by inactivating or deleting a gene which is important for its pathogenicity. One could also do the reverse: make a normally harmless microbe into a nasty pathogen, a prospect that has not escaped the attention of those concerned about biological warfare. On a different tack, one can recognize and clone stretches of DNA which, while not genes themselves, carry the switches that regulate a gene's function. One can then put quite different genes under the control of those switches, thus causing those genes to make more, or less if one so chooses, of their usual gene product – or to make it in response to some trigger such as age, a chemical, or a stress.

Finally, although I have largely centred my discussion on what can be done in the way of introducing alien genetic elements into the bacterium *E. coli* and then manipulating them, cloned DNA need not be conserved in microbes: there are now ways of transferring clones to animals or plants, to make so-called 'transgenic' organisms: creatures that possess new genes, copied from those present in entirely different organisms. But microbes, most often *E. coli* but also yeasts or viruses, remain the vehicles that are used to move genes around.

I am well aware that my account of the science and techniques which underlie what is now called genetic engineering has been narrow and imcomplete, but this book is not a manual or monograph; I excuse myself from writing more because good popular accounts are nowadays available in books and magazines. I hope I have conveyed something of the flavour of genetic engineering. In its earlier days such genetic manipulation was rather a hit-and-miss process, but now, with the tremendous advances in our understanding of gene structure and function brought about by the underlying basic research, genes from all kinds of organisms can be cloned and manoeuvred with remarkable precision, with some surprising and exciting practical consequences. Let me return to those.

BIOTECHNOLOGY

As far as its public image is concerned, genetic engineering got off to a rather poor start in the mid-1970s. In the first place, it generated a certain amount of public alarm – more about that in Chapter 9 – and secondly, among those in the business community and general public who cottoned on to its exciting possibilities, it aroused over-optimistic expectations. The non-scientific world heard of these marvels through the journalistic media and, as always, expected them to happen next week, or next year, but it is in truth a long slog from the clever laboratory experiment to the market. It took a little over a decade for the first genetically engineered product to become available commercially: human insulin.

Human insulin is a good example to dwell on briefly. Insulin is a hormone essential for diabetics, of whom there are known to be more than 60 million in the world. It has brought a once lethal organic disorder under control. Normal insulin as used by diabetics is the pig hormone (for the obvious reason that the human hormone has not been available) and it works. However, injecting pig protein into people always carries a slight risk of side-effects and several genetic engineering companies set about cloning the gene for human insulin and persuading a bacterium (our familiar *E. coli*) to make and release it. The story is a fascinating one, not least in that the gene had to be sandwiched into another *E. coli* gene to disguise the product, to prevent the cell from recognizing that it was making an alien protein and from destroying it as it was formed. The upshot, however, is that human insulin made in this way became available for clinical test in 1980 and was cleared for marketing in the USA and UK in 1982.

Other products of medical value which are being prepared in genetically engineered microbes are the pituitary growth hormone, essential for treating pituitary dwarfism and hitherto available only in minute amounts from human cadavers, and certain other hormones. Interferons, the materials involved in immunity, can be made in this way and are exciting considerable interest because of their potential value in the treatment of virus diseases and (possibly) certain types of cancer. They have been very scarce indeed, even for clinical trial, since their discovery, and the successful cloning of interferon genes has at least made more material available for testing. But it will take a lot of time to sort out the pharmacological actions of the several kinds of interferon; no really startling results have yet emerged. Human genes have been cloned for diagnostic reasons. A few people suffer from disorders which are hereditary: they arise from genetic defects within certain families. Examples are cystic fibrosis and a kind of muscular dystrophy. The genes responsible for these examples, and some others, have been cloned and the actual defects within the genes have been identified. It is now possible, using a tiny blood or tissue sample, to tell intending parents, for example, whether they are carrying a defective gene and if they are likely to pass it on to their offspring.

In principle it ought to be possible to cure, or at least alleviate, the simpler genetic disorders, and attempts to do this are already in hand. Cystic fibrosis provides a case in point. Its primary symptom is excessive fluid secretion by the mucous membranes lining the intestines and lungs, which leads to episodes of intestinal obstruction, inadequate breathing and a proneness to lung infections likely ultimately to culminate in respiratory failure; sufferers require lifelong medical attention and regular physiotherapy, and they die young. This tragic disease afflicts some 50,000 people worldwide. It has nothing to do with microbes; it is hereditary and has been traced to mutations in a single gene which specifies an enzyme concerned in mucus secretion. The healthy form of that gene has been cloned and sequenced; if it could be restored to the cells lining the respiratory apparatus some of the worst symptoms of cystic fibrosis ought to be prevented. In principle there are two ways of doing this. One is to find a virus which will infect the mucous membrane cells but not damage them, then engineer the healthy gene into that virus's genome, then infect the patient's respiratory tract with it, when the healthy gene ought to find its way into the host's cells as the virus infection is combated. The other way is to extract the good DNA from whatever vehicle it was cloned in and pack plenty of it into some sort of particle that mucous membrane cells can absorb, and hope that some of it displaces the defective DNA in those cells as the alien particle is broken down. Both procedures are being tried. The infective organism chosen is an adenovirus; it ordinarily causes coughs and sore throats but the genes which make it pathogenic have been disabled and a healthy copy of the cystic fibrosis gene has been cloned into its DNA. The particle is a natural subcellular organelle called a liposome, which can enter cells without damaging their membranes, and to which DNA sticks rather firmly. Patients receive suspensions of either in an inhalant spray. Preliminary experiments on animals were encouraging. Though only a small number of tests on humans have so far been reported, they, too, have yielded promising results. The interim position at the time of writing is that the idea is working, but that the virus package produced some unwanted side-effects and the liposome package was not always effective. But it is early days and more trials are needed.

In the USA two children who suffer from defective immune systems because their version of the gene coding for an enzyme called adenosine deaminase (ADA) is faulty have responded to gene therapy: repeated injections of leucocytes into which a good ADA gene has been engineered are apparently controlling the condition. There are also reports that a kind of cancer has been caused to regress by introducing a gene, cloned into an adenovirus, that acts against that particular sort of cancer. Gene therapy is of immense interest to the pharmaceutical industry, because it will most likely become a very valuable adjunct to medicine, and a lot more research has been done than has been published; secrecy is one of the ethical issues raised by the applications of genetic manipulation to which I shall return in Chapter 9.

In another area of medicine, the vaccines used to protect humans and animals from microbial diseases have traditionally been either dead preparations of the pathogen itself, or live relatives of the pathogen that are either non-pathogenic or so nearly harmless that they cause no serious illness. But for many diseases such preparations are either not available or are unsafe. Genetic engineering has now provided a route to safer vaccines. When, for example, people or animals become immune to a virus it is usually because their immune systems recognize one or a few of the proteins which form part of the surface of the virus. The immunity-provoking proteins are called antigens. (With bacteria much the same happens, though the antigens in their coats may include carbohydrates as well as proteins; carbohydrates are mainly responsible for the diverse antigenic patterns of salmonellae which I mentioned in Chapter 2.) In the 1980s a biotechnology company, Biogen, developed a vaccine against the hitherto intractable hepatitis B virus by cloning the gene responsible for making its coat protein into yeast; preparations of the protein extracted from the yeast are now widely used to confer immunity to hepatitis B without exposure to the virus itself. Earlier in this book I mentioned that baited vaccines have been used to control the spread of rabies in Europe. The most widely used anti-rabies vaccines were in fact live but virtually non-pathogenic strains of the rabies virus, but these caused anxiety because there is always a risk that a weakened strain might spontaneously regain or develop pathogenicity in the wild. More recently the gene coding for an antigenic coat protein of the rabies virus has been cloned into the harmless cowpox virus, providing a safer vaccine which, in field trials, has proved very successful.

Cloning antigen genes from pathogens into harmless microbes, and using these to provoke immunity to the nasties, is a new and seemingly effective way of obtaining relatively risk-free vaccines. But why stop at microbes? The ability of biotechnologists to clone genes from plants or animals in microbes and have the microbes express those genes, or to clone genes from microbes and have them expressed in higher organisms, has already led to useful and marketable products. The next step, obviously, was to clone plant or animal genes in microbes, and then to transfer those genes back to plants or animals, and in recent years this approach has revealed further vistas of practical possibilities. Genes for numerous complex biological substances of animal origin and medical potential – proteins, hormones, enzymes, antigens – have been cloned and expressed in experimental mice; for practical exploitation it is often desirable to have an engineered gene expressed in a tissue culture, or in regulated secretions from which the product can be harvested easily. Milk is the favoured secretion, with urine second. (The latter source is not as bizarre as it may sound: female sex hormones for clinical use have long been prepared from the urine of pregnant mares.) Mice are too small for practical exploitation on any substantial scale, so large farm animals such as sheep, cows or goats are preferred. Among the earliest developments in this direction was the cloning of the gene for an enzyme inhibitor, α-1-antitrypsin, which can be used to help cystic fibrosis sufferers. A copy was transferred into a ewe, the antitrypsin gene being put under the control of the ewe's lactation machinery so that its product appeared only in the ewe's milk and nowhere else. An anti-blood clotting agent called antithrombin 3 has been cloned in a comparable manner to appear in nanny-goat's milk, and I have met and patted on the head an amiable transgenic ewe which secretes human Factor IX, a protein used for the treatment of haemophilia, in its milk. The transfer of the gene into the right part of the host animal's genome is still a haphazard and difficult process; nevertheless, such products were under clinical trial in the late 1990s.

However, in the more distant future tissue cultures and the secretions of mammals may well be replaced, as sources of cloned biologicals, by organisms which are more amenable to subsequent fractionation: plants. I shall be more precise about how one moves alien genes into plants a little later on – either a plant virus or the bacterium called agrobacterium can be used as the vehicle. Whole plants or tissue cultures of plant cells can be hosts for alien genes; even more attractive are the fruits, seeds or other reproductive bodies of plants, because their formation involves the timed expression of genes which are normally silent, as does lactation in mammals. Transgenic leafy plants such as tobacco or a legume called cowpea are already being tested as sources from which antigens could be extracted. Tomatoes carrying antigens to hepatitis B virus, and potatoes with antigens to a troublesome diarrhoea agent called Norwalk virus, are among several such projects which were being developed by biotechnology companies in the late 1990s. When such plans first received publicity in 1995, the press seized upon the idea of 'vaccinating' children against diseases by offering them bananas genetically engineered to contain antigens – a nice thought, but therapeutic bananas are not yet with us. As hosts for genes which will express such biotechnological products, plants, albeit slow-working, have enormous economic advantages in terms of husbandry costs – but, as with all this kind of work, the risks of spreading unwanted alien genes among normal plants will need to be watched most meticulously.

The innovations in vaccine production and the prospects for future developments brought about by genetic engineering are thus impressive, and the biotechnologists involved are justly proud. However, there is always a joker in the pack. In 1996 some American researchers reported briefly that there was probably no need to go through the hassle of growing a mass culture of the microbe carrying the cloned antigen gene, and then extracting the antigen for purification and use. They found that they needed only to isolate the plasmid into which the relevant gene had been cloned and, provided they had put that gene under the control of a powerful on-switch, they could inject the whole plasmid and it then induced the intended immunity in experimental animals. Received with scepticism at first, because it seemed incredible that plain DNA could induce an immune response to a protein, there is now increasing evidence that the procedure works, and clinical trials on volunteer patients have been started. How could it work? The bacterial DNA comprising most of the plasmid would be rapidly destroyed by the muscle cells' normal defences against alien nucleic acids, but presumably enough of the DNA constituting the antigen gene, together with its switch, escapes to reach the nucleus and be expressed. Compared with a protein antigen, plasmid DNA is remarkably stable; it can be stored frozen, even freeze-dried. If so-called 'DNA vaccines' sustain their apparent success in trials, they could revolutionize vaccination in remote and under-privileged parts of the world.

In a wholly different direction, genetic manipulation has proved to be spectacularly useful in police and forensic work. Most readers will be familiar with the fact that fingerprints are unique: that no two people leave identical prints. There are zones of DNA in people's chromosomes, zones which are not actually genes, which are also unique to the individual and which can be used in a similar way. As with genes, their DNA can be cloned in microbes, extracted and analysed. The sort of analysis used in these tests involves fragmenting the cloned DNA with a restriction enzyme and using an electrical device (electrophoresis) to arrange the pieces in a pattern of sizes, and in the mid-1980s Professor Sir Alec Jeffreys realized that these patterns were a kind of 'DNA fingerprint'. It transpires that the pattern of such DNA in a child, for example, is a hybrid of its parents' DNA patterns, but even in the same family the children's DNA patterns differ slightly from each other. 'Fingerprinting' of this kind (it is not fingerprinting, of course; the only fingers that actually come into it are those of the scientist) first showed its value in cases of disputed paternity and in establishing kinship.

For DNA fingerprinting, the scientist needs a few milligrams of sample (tissue or body fluid, for example) from which a few micrograms of DNA may be extracted and characterized. However, today only tiny traces of DNA are necessary and a few hairs, scrapings of skin and so on will provide sufficient. The reason so little is needed stems from a technical development made by a US biotechnology company, the Cetus Corporation, around 1985, by which a fragment of DNA can be copied again and again in the same vessel until the concentration of that fragment has increased, many million-fold if necessary: the fragment has been 'amplified'. The details cannot detain us here; in principle the process exploits a DNA-synthesizing enzyme (called a DNA polymerase) which is caused to undergo cycles of heating and cooling, and during each cycle it doubles the

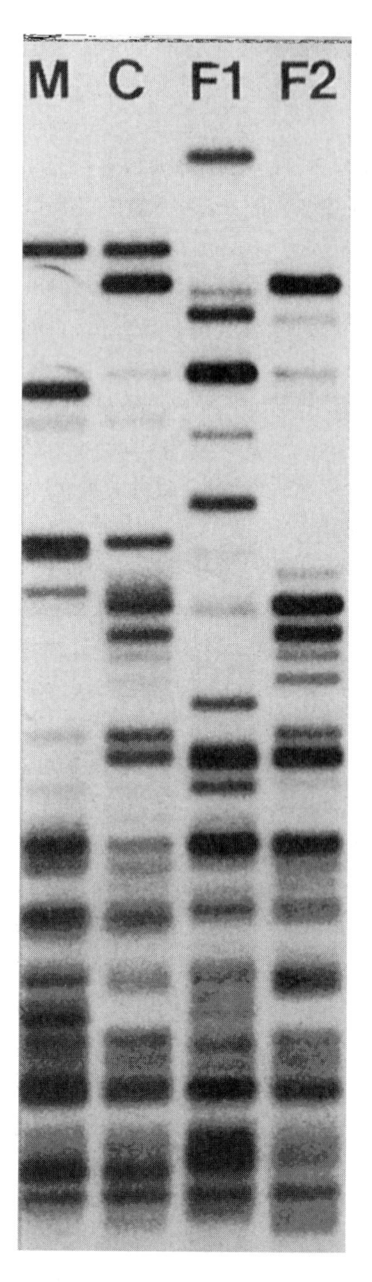

DNA FINGERPRINTS RESOLVE A QUESTION OF PATERNITY. M and C are 'fingerprints' of the mother and child. The bands are fragments of the same zone of their DNA sorted electrically according to size and made visible. F1 and F2 are comparable 'fingerprints' from the same zone of DNA samples from two possible fathers. The child clearly has many bands in common with F2, who must therefore be the father. (Courtesy of Professor Sir Alec Jeffreys and the Royal Society of London)

number of copies of a fragment of DNA initially provided by the operator. Most such polymerases are destroyed by the heating step; the trick was to use one extracted from a thermophilic bacterium called *Thermus aquaticus* which is heat-stable. Automated devices for doing the amplification are now widely available. With so sensitive a technique, operators have to be extremely careful that fragments of their own hair, scurf, dried nasal mucosa and so on do not get into the system – luckily, errors due to inadvertent contamination soon become glaringly obvious.

The amplification procedure is called PCR (standing for Polymerase Chain Reaction) in molecular biologists' jargon. A variant of the procedure enables scientists to amplify RNA instead of DNA; that variant is now used, for example, to detect HIV virus (an RNA virus) in patients who have not yet developed AIDS. Genetic material is much more easily cloned, whether for study or practical use, if it is plentiful, so amplification of DNA or RNA by PCR has become a routine step in genetic research and genetic engineering, used whenever the nucleic acid of interest is in short supply.

A refinement of DNA fingerprinting using PCR has been developed by Jeffreys. It is called 'DNA profiling' and has revolutionized forensic science. Criminals can now be identified readily if only the tiniest traces of their hair, saliva, blood, or a body fluid such as semen, have been found at the scene of the crime. Its first practical use was in Britain in 1986; the Leicestershire police were able to catch a rapist who had murdered two 15-year-old girls. There was no satisfactory conventional evidence against him, and there was another suspect who had made a confession. However, that suspect had the wrong DNA profile and was exonerated. Samples from 5,000 males living locally were then tested. The true criminal had actually arranged for a substitute to give a sample in his name, but the subterfuge was detected; in due course an authentic sample was obtained and then, four years after the crime, the true killer was arrested and convicted.

DNA profiling is now widely used throughout the world. In 1995 the British police were allowed to set up a data-base of DNA profiles, from criminals and volunteers, analogous to and supplementing their older data-base of real fingerprints. The procedure is extremely reliable because the chances of fortuitous matching are considerably lower than one in a million – provided, as defence lawyers will insist vigorously, that samples are properly taken and handled: neither muddled nor contaminated.

Comparable DNA analyses have been used to identify unrecognizable bodies after catastrophic accidents such as aeroplane crashes or fires. One remarkable use in the mid-1990s was to test the putative bodies of Tsar Nicholas II and his family, who were murdered during the 1917 Soviet revolution and rumoured to have been buried at Ekaterinberg in the Russian Republic. That rumour proved to be correct. In contrast, in 1998 a long-standing rumour that Martin Bormann, a close associate of the German Dictator Adolph Hitler, had escaped to South America at the end of the Second World War was refuted by profiling DNA from a skull which had been found near Hitler's suicide bunker in Berlin and preserved: its DNA matched samples given voluntarily by Bormann's children. The skull, with another found nearby, had had the remains of a glass cyanide capsule in its teeth: Bormann had clearly attempted escape but, sensing failure, had followed his *Führer*'s example and committed suicide. A remarkable use of the procedure was reported in Canada in 1996: the DNA profile of a cat's hair found on a blood-stained jacket led to the conviction of a murderer, because it was identical with the DNA profile of his family cat.

PCR has been used to amplify the DNA remaining in samples from a preserved mammoth and from ancient cadavers, but one must remember that the actual DNA left in such samples is not the long chain-like molecule that existed in the living cells; over the years DNA molecules slowly break up of their own accord, into increasing numbers of shorter and shorter chains; it is these fragments, or some of them, that are amplified successfully. Even after a thousand years it would be unlikely that a whole intact gene, for instance, could be amplified out of residual DNA, and after fifty thousand years almost all of the gene would have disintegrated. Reports of stretches of DNA isolated from insects which became fossilized millions of years ago in amber, or suggestions that dinosaur DNA might be cloned, are inconsistent with the chemistry of DNA as it is known at present. They should be regarded with reserve unless they are confirmed by several laboratories, and most scientists attempting confirmation have only succeeded in amplifying contemporary DNA contaminating the samples. Would that things were otherwise! *Jurassic Park* was a fascinating film, was it not?

DNA amplification, and to a lesser extent RNA amplification, are novel research tools that continue to have an impressive impact on the whole of biology and medicine. In the context of general microbiology, DNA amplification can now be used to detect specific microbes, using characteristic stretches of their DNA, when very low numbers are present in samples taken from patients, food, or the natural environment. For example, a single *Listeria* cell in a glass of milk can be detected in this way. The procedure provides scientists with a means of detecting pathogens, including biological warfare agents, much more quickly than by conventional culturing. Such research has also revealed hitherto unknown stretches of bacterial DNA in nature, deep beneath the sub-soil, for example, indicating the existence out there of types of microbe that have never been cultivated in the laboratory.

As a result of modern techniques for manipulating DNA, whether in the test tube or as mobile genetic elements transferred between living cells, it is now quite easy to transfer properties among microbes – to make new strains of antibiotic-producing *Streptomyces*, to alter the genetic character of yeasts, to enhance microbial solvent production. A protracted fuss took place in the USA over one such microbe in the 1980s. Frost can seriously damage strawberry crops in California, the trouble being initiated by strains of a bacterium, *Pseudomonas syringae*, which lives harmlessly on and around the plants and which happens to have proteins in its coat which seed ice-formation, rather in the way that chalk fragments attract frost in the

British winter. This ice-formation causes the plants to be attacked by frost at temperatures around −4 °C, at which it otherwise would not happen. In the early 1980s, ingenious microbiologists in California produced an 'ice-minus' strain of *P. syringae* which lacks the ice-seeding protein, and they proposed, in 1982, to release copious numbers of these on an experimental strawberry plot, to out-compete the natural strains and thus to protect the crop. They brought upon themselves the wrath of local environmentalists, anxious because unnatural microbes would be spread around, worried that they might have unforeseen effects, and not reassured when informed that natural 'ice-minus' strains of the organism actually exist in nature. (They do, but they are in a minority.) After much public debate, legal and scientific, permission to release the pseudomonads was given, but even then the first field experiment was vandalized by activists. Proper tests were delayed for several years but in the end they worked. This was an instance of excessive over-reaction, by a minority of laymen, against a minuscule risk, but it makes an important point. The capacity biotechnologists now have to produce all sorts of hitherto unknown life forms will sooner or later entail risks, and these risks must be thought out, and guarded against convincingly, well before release of such organisms is contemplated. More about that in Chapter 9.

On a more positive theme, agriculture seems likely to benefit considerably from new biotechnological advances. Genes from microbes have successfully been transferred into plants, making use of a special plasmid, called the Ti plasmid, which exists naturally in a bacterium pathogenic to plants called *Agrobacterium tumefaciens.* This plasmid is able to transfer itself from the microbe into the plant, where it integrates itself into the plant's nucleus. Often it causes tumours on the plant – not usually serious ones – but biotechnologists have developed strains that do little or no damage. Microbial genes for antibiotic resistance have been cloned into the Ti plasmid and introduced into plants, where they work; desirable genes can be cloned and introduced along with an antibiotic-resistance gene.

A TRIUMPH OF GENETIC ENGINEERING. Two tobacco plants are shown. The one on the right has been given a gene from *Bacillus thuringiensis* that causes it to make a protein poisonous to caterpillars; the other is the same variety of plant without the bacterial gene. Both were deliberately infested with caterpillars 11 to 12 days before the photograph was taken. The plant with the bacterial gene is unharmed, because it makes its own insecticide; the other is devastated. (Courtesy of Marc van Montagu, University of Gent)

There are ways of ensuring that the Ti plasmid is then largely or completely eliminated from the plant. I mentioned on p. 147 the successful transfer into plants of the gene which makes *Bacillus thuringiensis* toxic to insects, so that the plants themselves became toxic to their pests; the 'vehicle' which imported the toxin gene was the Ti plasmid and the result was little short of spectacular, as the tobacco plants in the illustration show. The Monsanto company has engineered a comparable 'Bt^{+}' strain of potato, which resists the dreaded Colorado beetle, and three varieties of Bt^{+} maize are available in the USA.

Plants bearing *B. thuringiensis* toxin could revolutionize pest control in arable farming and it is worth considering what risks their use might entail. If such plants were grown in an open field, could the toxin gene become transferred in pollen to related wild plants? Might it then devastate, perhaps eliminate, local insect flora, including beneficial insects? Might birds become starved of insect food? There are ways round these problems: cultivars of sterile plants which form no pollen could be used, or the toxin gene could be engineered into organelle DNA, which is not transferred sexually. However, a more subtle risk exists. The concentration of toxin in the leaves of a transgenic plant is quite low, and living things acquire resistance to toxic material most easily when (a) there is not a lot of it about but (b) some is present for a long time. Insect variants which are resistant to *B. thuringiensis* are rare, but they do exist; the probability of toxin-resistant variants of sensitive insects turning up is higher if the pest receives a steady but low exposure to the toxin from transgenic plants, than it is if it receives a suspension of *B. thuringiensis* itself, or a solution of its toxin, sprayed on a crop, because with the sprays the initial toxin concentration is high and the material which misses its target soon washes away or decays. There are at least three kinds of precaution that could be taken to minimize that risk. One is to plant, alongside your toxin-forming crop, a patch of the same crop cultivar which does not form the toxin; this will support a reservoir of non-resistant insects which, by mating with any resistant ones that turn up, will lower the chances of resistant progeny establishing themselves in the local population. A neater precaution rests with plant molecular geneticists: one can arrange for expression of the toxin gene to be under the control of the farmer, not the plant, by putting the Bt gene under the control of a genetic switch that responds to a chemical added from outside – dilute alcohol, for example. In those circumstances there would be no risk of resistance developing until, under threat of infestation, the farmer chose to 'switch on' his plants' poison with a spray of weak alcohol. But perhaps the most practical precaution is to adopt a procedure known as 'integrated pest management' in which both insecticidal chemicals and transgenic plants are used, sometimes together, sometimes in succession. These considerations illustrate the sorts of problems that must be foreseen and scrutinized closely when releases of genetically engineered organisms are contemplated.

Cauliflower mosaic virus, a DNA virus which can be modified to do no damage to its host plant, and into which alien genes can be cloned, was used to introduce into tomato or tobacco plants a gene – not from a microbe this time but from another plant, a petunia mutant – which makes the plant resistant to a priopríetary herbicide. This development inspired several companies to confer herbicide resistance on important crop plants, including rape, cotton and soya beans, so that the farmer can then spray fields of them with the herbicide, killing weeds without affecting the crop he wants. The agrochemical industry, which then sells lots more herbicide, is more enthusiastic about this development than those who enjoy the wild flowers of the countryside.

In comparable ways plants can be made resistant to natural plant virus infections and, in theory, the nutritional value of plant products could be up-graded. For example, plant protein is often deficient in the amino-acid lysine and a microbial (or other) gene for making more of this might be introduced. In the USA the shelf-life of the fruit from a variety of tomato has been prolonged by silencing the gene which makes the ripe fruit soften, so it stays firm longer; it has been a great commercial success under the name 'Flavr Savr'. And, by introducing a gene from *E. coli*, potato plants have been constructed which make tubers having a substantially enhanced starch content. Why? Because they absorb less fat in cooking, and so make healthier chips – sorry, French fries.

In all the instances I have mentioned the new property manipulated into the plant is coded for by a single gene. This is an important point. It is not difficult for a higher plant, for example, to read and express a single alien microbial gene. But some properties which one might wish to confer on plants involved the coordinated activity of several genes. If these come from other plants or even fungi, difficulties might arise, but they ought not to be drastic. If the clutch of genes came from a microbe, however, the problems would be very serious, because microbes read and translate their genetic tapes in quite different ways from higher organisms. To use my earlier linguistic analogy: to put a bacterial gene cluster into a plant and to expect the plant to use it is like giving an English-speaking engineer instructions in Hungarian – though the alphabet is recognized, it will make no sense. There are ways round this problem, involving detailed knowledge of the reading mechanisms in the two types of creature, but success is still some way away in the future. The problem is real because, for example, if one could confer the property of nitrogen fixation on crop plants, thus relieving them of their dependence on nitrogen-fixing bacteria or artificial nitrogenous fertilizer, one would remove one of the major restraints on world food production. This will, I am sure, be done one day, but the hard problem is how to get plant material to cope with the nitrogen fixation genes, of which there are twenty with a complicated regulatory system. To date weak expression of just one of these genes in plant material is all that has been achieved.

Moving genes around among microbes and, by way of microbes, to and from plants and animals, is now commonplace in research, and its impact is changing the face of industry. But why stop there? Why not make entirely new genes in the laboratory and put them into living things? For the genetic code is understood and chemists can not only make the component molecules of DNA, but can also string them together into gene-like chains. Well, that thought has not only been thought already, but put into practice. As long ago as the early 1970s, H. G. Khorana and his associates in the USA synthesized chemically two small genes; they were concerned with moving amino-acids around within yeast and *E. coli* respectively. Today, however, it is easier to clone an existing gene, purify it, and alter its chemical structure by removing stretches of DNA and putting back synthetic ones. Indeed, one can now buy computer-controlled machines which will synthesize lengths of DNA to specification for that purpose. The great majority of genes code for proteins. Therefore, to be anything more than a scientific exhibit, a synthetic or partly synthetic gene has to be spliced into a plasmid and introduced into, say, *E. coli*. Provided it has been spliced in at a place where the plasmid DNA has 'read me' signal (this is not difficult to arrange), the *E. coli* will dutifully make whatever protein the new gene specifies. Things do not always go smoothly – the new gene product may not be good for its host – but the strategy works so well in general that the new technology mentioned earlier called 'protein engineering' has arisen: the design and construction of entirely new proteins. Among the earliest products of such work are enzymes with altered specificities towards their substrates – they are useful for elucidating how the enzymes actually work – and it will soon be possible to create enzymes able to attack entirely new materials, substances which have not hitherto been substrates for biological systems. This ability could have all sorts of industrial implications in the future, and microbes will play a central part in the necessary manipulations.

Modification of enzymes by way of the relevant genes, then, opens vistas of new and useful catalysts; modification of plants by genetic engineering has already happened; modification of animals is under way; and modification of man, actually curing genetic disorders by gene therapy, is just over the horizon – all these in addition to new vaccines, new pharmaceuticals, new ways of detecting and indentifying biological entities ranging from viruses and bacteria to corpses and criminals, plus a myriad of other ingenious applications, most of which I have had to omit. In all these activities, microbes are indispensable tools. The pay-off may have been slow in coming, but now it is all happening – and the prospects are still marvellous. In my all-too-brief survey I may have seemed to emphasize matters of health and food, but are these not the primary interests of mankind? Certainly industry has no illusion over where the money lies, as I shall have cause to report in the next chapter. The way present research is now developing has imposed on this chapter an emphasis on production, on money-making processes. Indeed, whole textbooks have been written on applied or industrial microbiology which deal with little more than production processes. Well, what is good for industry is often good for the public – but not always. Let me now turn to the use of microbes for the public good.

7 Deterioration, decay and pollution

I am the proud father of a family, one which has grown up and has, indeed, extended our lineage into a new generation. So my own children are encountering a result of the noble estate of parenthood which, to me at least, was quite unexpected. But it will become familiar to them, and to readers who find themselves in a comparable position: it is the amount of junk that accumulates in the house. (I hasten to add, lest one of my grandchildren should read this, that I know it is not junk to them: every celluloid duck, plastic brick, fluffy rabbit, coloured felt-tip, pop record, decayed transistor, pop-art wall poster is known personally to each offspring, complete with history and ownership.) When I was young, toys had a finite, almost predictable lifetime: a clockwork train, for example, would break, be repaired a couple of times, then find its way to a dustbin – within weeks if one was unlucky, in months on an average, in years if one was careful or the toy was particularly good. Today, it seems, toys are indestructible. Bouncer, my forty-four-year-old daughter's brushed nylon washable cuddly-dog is two years younger than its proprietor, having survived for five average lifetimes of the pre-war teddy bear and, albeit a trifle battered, is well set to accompany my granddaughters into adult life. Well, as far as Bouncer and my descendants are concerned, I am happy for all of them; it is the great number of objects of uncertain function, usually fashioned from some kind of plastic, bent, with screws missing, that leaves me with the impression that, in another decade, whole rooms of people's houses will be given over to the storage of beloved toys belonging to beloved offspring!

What has this to do with microbes, you ask? Ah! I see you have taken the point. But I shall spell it out to make sure, anyway. Just as a small family collects a sort of sediment of indestructible matter, so civilized man on this planet accumulates a mass of artefacts, many of remarkable durability, made of wood, iron, stone, concrete, brick, plastics, tin, glass, pottery and so on. Moreover, they discard clothes, remains of food, husks and residues of vegetation, the bodies of their fellows, corpses or pets and domestic animals, excreta, paper, hair and nail parings into the biosphere of this planet. What prevents us from being knee-deep in our own detritus?

Microbes, you reply. Quite so. Micro-organisms in soil, water, sewerage systems and refuse dumps transform the detritus of human society, converting it to materials that can be re-used, or that are at least innocuous. It is here that microbes provide their most valuable function for mankind, for picture what the world would be like if wood did not rot, corpses did not decay, excrement and vegetation lay where it fell and so on. An impossible situation, of course, because the biological cycles of the elements would have come to a standstill millennia ago, but one which is instructive in considering the economic value of microbes. Microbes (helped a little by fires, natural or man-made) return materials that man has withdrawn from the biological cycles – or carbon, nitrogen etc. – that keep life going on this planet. Decay, deterioration and destruction are the reverse of growth, synthesis and production, but are quite as important for the terrestrial economy. Regrettably, however, they are not obviously important to the economy of industrialists, so I shall have to point out, as I survey these processes here, that there are gaps in our knowledge of the microbiology of these processes. They arise from the relatively moderate research effort that has been put into understanding them, which, in its turn, has been determined by the availability of funds and laboratories for basic research in economic microbiology. But let me not grind an axe; let me look at what microbes actually do on the return side of this planet's economy.

Deterioration, decay and disposal are three names for what, microbiologically, are similar processes. People use the name deterioration for something they wish did not occur; decay is on the whole neutral; disposal is to be encouraged. I shall discuss disposal processes in the next chapter; in this one I shall look upon the worst side of the picture and note the corrosive, destructive and generally obstructive parts microbes can play in mankind's dealings with the inanimate world.

As usual, I shall think of our stomachs first, and consider the spoilage of food. As everyone knows, food goes bad if it is kept around too long, unless it is pickled, sterilized, dried or deep frozen. The process of going bad occurs when microbes grow on or in the food, altering its consistency, taste and smell; the processes used to preserve foods are those that delay or prevent microbial growth. It will be reasonably clear to anyone who has read Chapters 3 and 4 that almost any form of food is a good medium for bacterial growth. A meat stew left open in a warm kitchen for a day or two will collect all the airborne microbes that happen to fall into it, plus those coughed or sneezed about the place by passing humans and pets, those scattered by the wings of insects, those falling off the hair and clothes of the cook. Imagine that the ingredients have been prepared, put together but not yet cooked, so that, in addition, they are liberally infected with organisms from the cook's hands and have a modest infection of miscellaneous mouth and other contaminants on the cooking vessel, left over from when it was dried with a contaminated cloth last time it was washed up. A depressing prospect, it may sound; but in fact the preparation at this stage is perfectly wholesome. The microbes are mainly dormant, few if any are multiplying and most of them are harmless – though if the cook has cut a finger that is going septic, a few potentially nasty pathogens may be present. Even so, the mixture is harmless, because of the small number of microbes present. The meat and vegetable tissue is largely intact, as it was in the living animal or vegetable, the water is pure enough and the salt and flavourings are not much use as nutrients for the microbes. If it were to stand in a warm place for a few hours, the meat and vegetable tissues would begin to break down, partly by the action of the microbes, partly by intrinsic chemical processes, and more nutrient would become available for the microbes to multiply. But for the while the mixture is quite safe. Then the cook boils it for some hours, in either a casserole or a saucepan, and all the microbes are killed. Unless the cook is very unlucky, all the spores are killed too. Assume the cook is preparing a casserole stew: if it were removed from the oven at the end of three hours and served hot, nourishing and, one trusts, delicious food would be provided to which the microbes I wrote about made an undetectably small contribution. Now assume there is some left over. It has cooled, so that airborne and hair-borne microbes start to fall in it again, and now, because it has been cooked, all the most nutrient juices and substrates have been extracted from the ingredients. The microbes find a perfect, warm culture medium and they start to multiply.

Supposing, as an illustration, ten staphylococci got into it from someone's thumb as it was carried out at 8 p.m. after dinner. It is covered up, put on one side and forgotten. The kitchen is warm – there is a boiler at the other side of the room – so the staphylococci start to multiply. By 9 p.m. there are twenty, by 10 p.m. there are forty, by midnight one hundred and sixty. Now assume that the organisms divide every hour – and in a really warm place they can divide four times as fast as this – then by next day at noon there will be something like 600,000 staphylococci in that stew. It will be beginning to smell a bit, but it will look all right still (the population of microbes has to reach about 100 million in each thimble-full to *look* bad). However, some rather depressing chemistry will be taking place in it. Amino-acids, components of the meat and vegetable proteins, are being transformed into substances called ptomaines and rather toxic products of the growth of the microbes are being formed. Suppose the cook does not notice, but warms it up in the oven for lunch. The microbes will be killed, but the ptomaines will remain, with one of three consequences. Whoever eats it will have an upset tummy, but it will probably be over quite soon. Or they may just find it tastes a bit off but does no further harm. Or they may not notice. Which of these things happens depends really on how warm it was when the food was stored: by a stove it could become quite toxic overnight; in a cool larder it might last a day quite safely; in a refrigerator the staphylococci would not have grown at all. However, psychrophilic (low temperature bacteria) could grow slowly and cause additional ptomaine formation, although this would take several days.

Now, just imagine what would have happened if it had been not a reheated stew, but a pie that had been intended to be eaten cold. Whoever ate it would have ingested a jolly good dose of live bacteria, and if those had happened to be pathogenic, they could have caused a nasty infection of the mouth and intestines. This is how most cases of food poisoning happen: pre-cooked food has been stored in too warm a place and has not only gone bad faster than it ought to have done, but has grown pathogenic bacteria picked up from someone who handled it during preparation. This is why preservatives are put into prepared foods: they are in fact disinfectants that have a negligible effect on man but keep the microbes at bay. Personally, I should often prefer to do without the foods than bear with some of the preservatives that are in common use, but that is a matter of taste.

Though chemical preservatives are widely used and unavoidable, there are many traditional processes available for preserving food from microbes. Pickling, which is steeping the food in acetic acid (vinegar), preserves food by making it too acid for bacteria to grow. Sugaring also preserves, as in jams and syrups, because few bacteria can grow in strong sugar solutions. Yeasts and moulds can grow in sugar preserves, but they usually do no harm or else become so obvious that no one would think of eating the food. Salting is a method of preserving meats and fish that depends on the fact that most putrefactive bacteria cannot grow in a strong brine. If the brine contains potassium nitrate (saltpetre) or sodium nitrate, a halophilic microbe called *Paracoccus halodenitrificans* grows and converts the nitrate into a preservative substance called nitrite. This forms a red compound with meat protein which thus becomes much less susceptible to ordinary microbial attack; such meat is said to be cured. Bacon is red because of this curing process. It is not really necessary to grow the microbes to cure meat: the chemical sodium nitrite will have a similar effect, but as it is slightly poisonous its use is controlled by law in most countries. (But, often, one can use as much nitrate and *P. halodenitrificans* as one likes!) This is why one so often sees sodium nitrite as one of the ingredients of canned meats: it is a preservative and curing agent.

Deterioration of canned and bottled foods can occur if they are inefficiently sterilized: *Desulfotomaculum nigrificans* is a thermophilic bacterium (meaning it grows in hot water, remember?) which forms spores that resist prolonged heating. Being an anaerobe, it positively welcomes being canned; being a sulphate-reducing bacterium, it produces the evil-smelling gas hydrogen sulphide. It is obvious why this kind of spoilage of canned foods (such as canned corn) is called sulphur stinker spoilage. Happily, this particular food spoilage was understood in the 1930s and modern canning procedures make it very rare. *D. nigrificans* also enjoys strong sugar solutions. Molasses, which is unrefined treacle, is heated to quite a high temperature when it is processed so that it will flow easily, and it is then hot enough to kill most bacteria. But *D. nigrificans* grows quite well in this environment and is a source of constant nuisance to the sugar industry. Another spore-forming anaerobe, *Clostridium thermosaccharolyticum*, can generate gas in canned foods and cause explosive effects when the can is opened; but one of the most dangerous is *C. botulinum*, which appears occasionally in canned or potted foods, most often meat or fish though it has caused trouble in honey. The organism itself is only pathogenic in exceptional circumstances, but the toxin it forms is one of the most poisonous substances known. (It has been proposed as an agent for biological warfare.) Botulism, usually a fatal condition, results from eating food made poisonous by this organism.

Moulds spoil foods such as bread, cheese and so on, and are usually obvious and fairly harmless. But there are more drastic effects. Moulds of the group *Aspergillus* can spoil grain that has been stored insufficiently dry, and the solution to this particular problem is to damp-store the grain in hermetically sealed chambers, where the grain produces so much carbon dioxide by its own respiration that the mould is prevented from growing. *A. fumigatus* is a mould that is pathogenic to poultry which grows through the shells of eggs, forming spores on the inside, and the chicks develop a lung infection when they hatch. Pigeon-pluckers in France have developed a pseudo-tuberculosis from plucking infected birds. One of the most spectacular cases of spoilage by moulds emerged in connection with groundnut (peanut) production in the 1950s. *A. flavus*, another mould, may contaminate harvested nuts and, when it does so, it forms within the nuts poisonous materials called aflatoxins. These were first detected when poultry, fed on groundnuts, developed liver damage; alarm increased considerably when aflatoxins were found to be capable of producing cancer in people and animals and to be present in some batches of food intended for human consumption – peanut butter, for example. Though the situation is now under control, there was a period when the possibility of carcinogenic matter in peanut preparations was a cause for anxiety.

Spoilage of foods by microbes is familiar to everyone. It could be said that the whole distributive and catering trade in civilized communities is based on procedures, traditional or modern as the case may be, designed to delay or arrest microbial deterioration of the product being purveyed. Think of the problems in the distribution of fish, for example, and the manners in which they have been overcome. A whole technology of food microbiology has grown up concerned with the understanding and control of deteriorative, infective and protective processes in the food industry. The examples given so far in this chapter and in Chapter 5 have been illustrative rather than comprehensive because, as with almost every chapter in the book, a proper survey of the field would require a book on its own. For my synoptic view of microbes and man, it is more interesting now to turn to the destructive action of microbes on materials that are not foods.

Have you seen a pair of old shoes, gardening shoes, for instance, that have gone mouldy? Or observed the efflorescence of mildew over the walls and ceiling of a derelict house? These are two examples of microbes attacking and damaging materials that one expects to have a reasonable degree of permanence. In fact, however, I have chosen those two examples rather carefully, because in neither case is the basic material itself being attacked. Leather, even in tropical countries, is remarkably resistant to microbial attack, and insects and worms are its most serious destructive agents. But the dressings and conditioning agents used to polish or improve leather can be attacked by microbes. It is generally these that the moulds use as food when they grow on leather, but having grown, they form pigments, erode the surface of the leather and generally make an unsightly mess of it. Similarly, the growth of mildew on ceiling plaster or walls is not really because it can use the plaster itself as food, but because fining agents, paper and, often, the paste used to stick on wall and ceiling paper, can be used as substrates for growth. Most decorating materials contain microbicides to prevent growth of moulds, but where a house is excessively damp – being very new or derelict, for example – the microbicide may leach away and moulds will grow. A less severe instance has become common as domestic central heating has spread throughout Britain: window panes act as condensers for the moisture in the warmed indoor air, so water droplets collect at their edges depositing minute traces of nutrients from cooking, breath, sweat and so on; moulds then grow and a greeny-black stain duly appears. Such stains, so infuriating to householders, are usually the pigmentation of the spores of the otherwise barely visible moulds. In the tropics, moulds can cause enormous damage. Lacquers, resins and the insulating layers of electrical equipment, for example, all contain materials that can support growth of moulds. *Aspergillus restrictus* and *A. glaucus* are notable in that, when they grow, they produce substances that can etch glass, and during the Second World War they damaged lenses of cameras, binoculars and such by growing as a film on the glass.

Paper pulp, which is a warm, wet mash of pulped wood, recycled paper, other cellulosic materials and diverse additives such as white clay, mineral sands and starch, is host to a wide variety of microbes, especially bacteria. They are inevitably introduced with the primary ingredients and the majority do no harm. However, those that form slime can interfere with the paper-making process, and one or two types could be hazardous to the health of operatives. Microbicides are used to control them. The foxing of paper, only too familiar to those who handle old manuscripts, artwork or prints, seems to involve filamentous microfungi, but whether their presence is a cause or a consequence of the damage is not clear.

Wood is a fairly resistant material, but anyone who has encountered dry rot domestically will realize the expense and trouble that fungal attack on wood can cause. In this instance, the wood itself, and not any kind of dressing, is the substrate for growth of the microbe. There exists a wide range of wood-rotting fungi, ranging from the huge beef-steak fungi one sees in woods or on fallen logs – the beef-steak is its fruiting body – to the rare but very active *Myrothecium verrucaria*, invisible except when it forms spores. Wood conditioners such as creosote protect against wood-rotting fungi for a time, for years even, but the really effective treatment is to keep the wood dry. Even in a country as damp as Britain, roof beams will last for centuries if damp is avoided.

I indicated earlier that destructive microbes do have their positive aspects: they play an important part in the recycling of the biological elements, and the wood-rotting fungi are valuable in nature because they bring the carbon of wood back into biological circulation. Cellulolytic bacteria, as they are called, break down the cellulose of wood, paper and plant material to simple fatty acids which other microbes can use. They even live in the guts of wood-boring insects or molluscs and help to digest the wood. Fungi require air to grow, so a mound of garden refuse, for example, tends to be broken down more by bacteria than by fungi, and the interior of a compost heap usually consists of a mass of various anaerobic bacteria all living on the products formed by cellulolytic bacteria from cellulose. As I wrote in Chapter 6, methane (natural gas) is one end-product of this process, and it is probable that vast natural compost heaps of this kind developed during the Carboniferous era and ultimately formed coal. On the horticultural scale, of course, the process does not even go to the stage of peat formation because it is interfered with, but even on that small scale the compost can get quite hot (some of the energy generated by the microbe's metabolism is released as heat – just as you and I get hot if we run) and one can then understand why thermophilic bacteria are so widespread on this planet. They come into their own during large-scale natural fermentations.

Wool is attacked by microbes, though householders in temperate countries are perhaps more familiar with attack by insects' grubs, called woolly bears by my parents. But in damp environments and, particularly, soil, both moulds and bacteria decompose wool quite quickly. Since wool is fundamentally an animal protein – keratin – its breakdown releases nitrogenous matter. I recall another parental belief: never throw away an old woollen garment, bury it in the rhubarb bed and you will get better crops of rhubarb. I am sure it works, but for myself I soon have enough rhubarb.

Microbes can degrade paints and, here again, as with leather and wall plaster, it is the additive rather than the pigment itself that they use. Oleic acid and related materials such a linseed oil are widely used to support the pigments used in paint manufacture and, particularly in tropical areas where the paint may be exposed to warm and humid conditions, bacteria and fungi may attack these materials and destroy paint rapidly. An interesting side-issue, if one may call it such, to all this is that until the 1930s arsenic compounds were used as pigments in some paints and wallpapers. Many of the more primitive moulds, belonging to genera such as *Aspergillus, Mucor* or *Penicillium* can, when growing on other materials in the arsenical pigments, convert the arsenic to the gas arsine. This has a garlicky smell and is intensely poisonous: deaths have occurred because people breathed air containing arsine formed in this manner over a long period, the last of this kind in England being recorded in 1931.

Rubber is usually regarded as a fairly stable material, but it is in fact attacked by a particular species of actinomycete. Dr La Rivière of Holland showed in the 1950s that rubber gaskets and washers all over the world act as enrichment cultures for this particular actinomycete, which can be found anywhere. This organism attacks the actual polymer (latex) that constitutes rubber. But there is a second way in which rubber can be corroded by microbes, which depends on the fact that natural rubber, before it is used, must be vulcanized. Vulcanization involves adding sulphur to the rubber; if the rubber is wet, the sulphur-oxidizing bacteria *Thiobacillus thio-oxidans* grow at the expense of this sulphur, converting it to sulphuric acid. This acid attacks the rubber and any fabric associated with it: during the Second World War considerable damage was done to National Fire Service fire hoses for this reason. The remedy was to dry out the hoses adequately, which is why fire drill is so insistent on this seemingly trivial detail. Cases of a similar kind leading to destruction of rubber gaskets sealing bottled fruits and other materials have been described. In all such instances, the breakdown of the rubber is associated with the formation of sulphuric acid. This is not the only time that I shall have cause to describe thiobacilli behaving destructively as a result of their ability to form sulphuric acid.

Some synthetic rubbers (the chlorinated rubbers and the silicones) and some plastics (the fluorinated hydrocarbon polymers) are, as far as we know, immune to microbial attack. Chlorinated plastics such as polyvinyl chloride were once thought to be attacked by bacteria, but this seems not to be true: some components or additives sometimes support the growth of fungi or bacteria, but the plastic itself seems not to be attacked. Fortunately, both burn, or decompose in light to compounds which microbes can attack, so the plastic rubbish of present-day society does get back into the biosphere in the end. Substances which microbes can decompose are called biodegradable and materials which are not attacked by microbes at all are called recalcitrants. Among compounds of the carbon atom remarkably few are recalcitrant: the synthetic polymers

I have just mentioned, carbon itself and some laboratory chemicals. Some compounds are broken down only very slowly: the humic acids in soil, formed from decaying vegetable matter, and some of the complex organic compounds in coal tar. Interestingly, and logically if you think about it, the spore coats of those microbes which form spores are usually very resistant to microbial attack. But they do go, or we should be surrounded by them.

Many microbes can consume naturally-occurring hydrocarbons. The commonest are the methane-oxidizing bacteria, organisms capable of growing while oxidizing natural gas, but bacteria, moulds and yeasts capable of oxidizing oil hydrocarbons are also known – I introduced both methane-oxidizing bacteria and yeasts able to utilize oil hydrocarbons when they appeared as possible sources of food protein. Hydrocarbon-oxidizing microbes occur naturally in oil deposits and their presence around seepage areas has been used in oil prospecting. It is when they get involved in stored petroleum products that they become a nuisance, because they can spoil the fuel. Petroleum and kerosene are stored in huge tanks at the bottom of which is usually a layer of water. This water bottom is normally unavoidable. If the storage tank is by the sea, the fuel has usually been pumped into the tank from an oil tanker and the pipe along which it was pumped started full of sea water. Inland tanks do not get wet in this way, but petroleum dissolves an appreciable amount of water and releases it on cooling, so that, even inland, water tends to accumulate as a layer in the bottom of the tank. In this water, mainly at its interface with the petroleum, hydrocarbon-oxidizing microbes grow.

It is important to emphasize that these microbes grow in the water, not in the oil, petrol or kerosene as the case may be. I have seen quite learned accounts of microbes in oil technology in which this point is not clear, so perhaps I should be more emphatic: microbes crop up in many aspects of oil technology, including the case of spoilage that I am now discussing, but in none of these do they grow in anything but water. If you will forgive the digression, the same principle explains why, in the same cupboard or refrigerator, butter lasts

– it goes rancid only slowly – but cream goes bad quickly. Butter is an emulsion of water in fat, and such bacteria as get trapped in the droplets when it is made are confined: they can only multiply in the water; they cannot spread. Cream is an emulsion of fat in water, and a single microbe, multiplying, can spread throughout and ruin the lot.

To return to oil technology. Growth of microbes in water beneath stored fuels always occurs and, generally, it does little harm. A moderate sludge of microbes appears in the water layer and an infinitesimal amount of the fuel is consumed, but little damage is done. However, if the sludge gets too thick and the turnover of fuel is slow, the water can become anaerobic (because the microbes consume all the dissolved oxygen) and this is when trouble starts. The sulfuretum gets established because sulphate-reducing bacteria grow, reducing sulphate dissolved in the water by means of organic matter made available by the hydrocarbon-oxidizing bacteria. Some oil technologists have long believed that sulphate-reducing bacteria can oxidize hydrocarbons themselves, using sulphates, but for many years most scientists were doubtful, because pure strains demonstrably able to attack oil hydrocarbons proved elusive. However, the technologists were probably right: in the early 1990s a species aptly named *Desulfobacterium oleovorans* was discovered which can do this. Hydrogen sulphide is formed and contaminates the fuel, becoming at least in part converted to free sulphur, and rendering the fuel corrosive to certain parts of the fuel injection system of aircraft. This problem occurs particularly in tropical and subtropical areas: in 1952 and again in 1956 portions of the RAF were grounded at politically awkward moments because of bacterial spoilage of fuel in storage tanks. The symptom is an increase in the copper strip test, a test based on measuring the speed with which sulphide in the fuel blackens a strip of bright copper. Once spoiled, there is no remedy but to use the fuel in less sensitive engines such as motor cars (to downgrade it); prevention is mainly a matter of cleaning out the bottom waters regularly, though certain chemicals active against sulphate-reducing bacteria are also effective.

The usual consequence of microbial spoilage of petroleum is a simple financial one: the fuel becomes less saleable than it would have been and its owner loses money. But instances of more serious damage are known. The iron sulphide, formed as a result of bacterial sulphate reduction, becomes oxidized on exposure to air and, on rare occasions, it may become hot enough to ignite petroleum vapour when the tank is being cleaned. Two serious explosions of petroleum tanks occurred in Britain in the 1930s for this reason.

When metal tools and parts are manufactured industrially, the metal is often cut on a lathe, and the cutting edge has to be cooled and lubricated simultaneously. This is done with a jet of an emulsion of oil in water, a cutting emulsion. Such emulsions are a marvellous habitat for bacteria, particularly during periods when the factory is closed and the tank of emulsion is stagnant. Once hydrocarbon-oxidizing bacteria have got started in the emulsion others can grow and a noxious, even infectious, brew can be generated which no longer serves its original purpose. Standards of hygiene in this context used to leave something to be desired, and in the mid-1950s an authority on the topic wrote: 'workers should not be allowed to spit, urinate or throw portions of their lunches into the emulsions'. I believe awareness of the problem has improved matters; certainly, disinfectants are now added routinely to cutting emulsions.

Petroleum and oil are just two examples of many hydrocarbons of economic importance. Asphalt and bitumen are mixtures of carbon and hydrocarbons and both are used in road surfacing. In wet and warm climates, both can be decomposed by soil bacteria and, in the Southern States of the USA, this process causes appreciable damage to roadways. It probably happens elsewhere, but this particular cause of deterioration is still not widely recognized. Bitumen coatings have been used to protect buried pipelines and, again, soil bacteria limit the lives of such coatings by feeding on them. An extreme case of microbial attack on hydrocarbons may occur in the spontaneous ignition of coal heaps: even in temperate climates stacks of coal may become mysteriously warm and sometimes catch fire spontaneously. One possible explanation is that bacteria oxidize the coal, or components of the coal, and, as in a compost heap, some of the energy is released as heat so that, in special conditions in which the heat is not readily dispersed, a cycle of oxidation leading to ignition can take place. It must be admitted, however, that the existence of bacteria able to do this has never been convincingly demonstrated, nor does it seem likely that bacteria could generate enough heat to reach the flash point of coal. But Will-o'-the-wisp exists as a precedent and there is no better explanation at present.

One might expect iron or steel pipes to be immune to microbial attack, but this is in fact not so. Iron pipes – and all other iron structures that are not protected in some way – rust in damp air. This fact is familiar to most people, and the fact that both water and air are necessary for rusting is also well known. If one immerses an iron nail, for example, in pure, air-free water and seals it against access of air, it will remain shiny and bright for years. Admit air and it rusts rapidly. Iron pipes, buried in soil, are pretty well protected from air, particularly if the soil is waterlogged and there are plenty of microbes around to consume any air that penetrates to the pipe. Yet, in these circumstances, iron pipes can corrode faster than they would in air, and the cause of this corrosion is now known to be the bacteria which have featured so often before in this book, the sulphate-reducing bacteria. Underground corrosion of iron pipes was estimated to cause a loss of between 1.6 and 5 billion dollars to the USA in 1990. It is a serious and expensive process, so I shall respectfully devote a little space to studying its subtleties.

If you took a lump of pure, unrusted iron and put it in water, it would react, splitting the water molecules so as to form hydrogen and iron hydroxide. In chemical terms:

$$Fe + 2H_2O \rightarrow Fe(OH)_2 + H_2$$

Normally, if nothing but water were present, the reaction would no sooner start than it stopped, because the hydrogen would stick to the surface of the iron and prevent any further reaction taking place. If, however, air is present, oxygen from it would react with the hydrogen, turning it back to water, so the process can go on indefinitely until the iron has rusted away. (Readers with some knowledge of chemistry will remark that rust is not $Fe(OH)_2$. No matter. The process I have described is the first step in rusting and, though all sorts of further reactions take place, iron would not rust if this first step did not happen.) The sulphate-reducing bacteria, I repeat, do not use air for respiration but reduce sulphates instead. To save turning back, I shall write the reaction again, using calcium sulphate as an example:

$$CaSO_4 + \text{food} \rightarrow CaS + \text{oxidized food}$$

They make calcium sulphate into calcium sulphide while oxidizing whatever material is available as food. Most of them, especially the genus *Desulfovibrio*, also have the property of being able to use hydrogen for this reaction:

$$4H_2 + CaSO_4 \rightarrow CaS + 4H_2O$$

and though the hydrogen is not strictly speaking a food (it contains no carbon), the reaction provides the bacteria with energy and enables them to use such carbon-containing food as is available more economically. Confronted with an iron pipe, with its protective film of hydrogen, they tend to use this hydrogen for sulphate reduction, converting it to water. So the iron corrodes. As a further reaction, the sulphide reacts with some of the iron to form iron sulphide, so one can always recognize underground corrosion of this kind because the corrosion product contains iron sulphide. It is black instead of brown and often rather smelly.

Underground corrosion of iron pipes, as I have indicated, is one of the most expensive kinds of microbial corrosion, and a fair amount is known about how it happens and how it may be prevented. It attacks water and gas mains, drainage pipes and, because sulphate-reducing bacteria thrive in sea water, it destroys marine installations and damages the hulls of ships. But there is no easy cure, and the basic

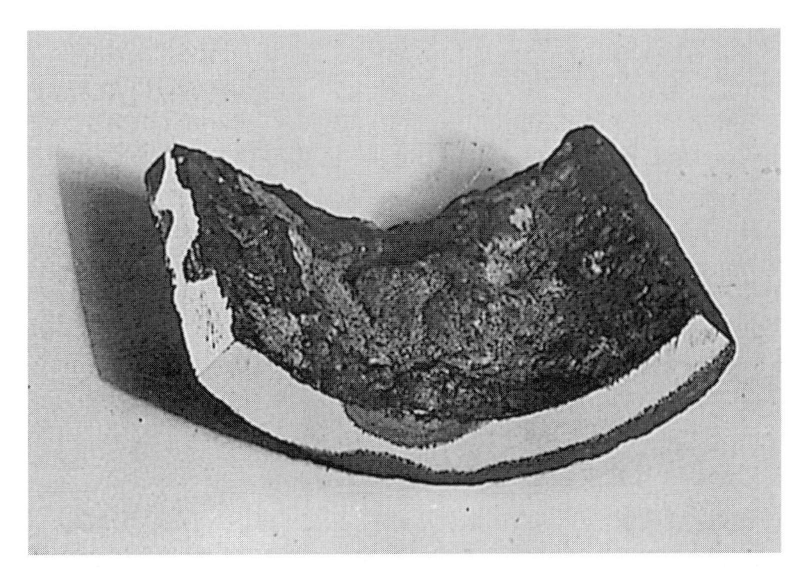

BACTERIA CORRODE IRON PIPES. The photo shows a fragment of an iron water pipe cut to show corrosion by sulphate-reducing bacteria. It encroaches from both outer and inner surfaces. The encrustations to be seen on the inner surface were probably caused by iron bacteria; they served to screen the anaerobic sulphate reducers from air dissolved in the water.

principle remains: do not bury iron pipes unless there is nothing else you can do, and if you *do* bury them, either see that air has free access to them or coat them with so thick a cover that bacteria cannot penetrate to the metal. (Remember that cloth, bitumen, wax, many paints and plastics are decomposed by soil bacteria; remember, too, that when you have found a suitably impenetrable coating, it needs only one careless workman with a pickaxe accidentally to make it penetrable again. There are electrochemical methods of protection that are expensive but probably worth it in the long run.) Even hot-water systems are not immune to corrosion of this kind, because some strains of sulphate-reducing bacteria (*Desulfotomaculum nigrificans*, which I introduced when discussing food spoilage, for example) are thermophilic; in the right circumstances they can even corrode copper piping in domestic hot-water systems, because they grow in the cooler parts of the circulation system and the sulphide they form diffuses throughout the pipes, attacking the copper by converting it to copper sulphide. In the USA I have taken hot showers which smelled like the spa at Bath and, therefore, noted sadly that my host's domestic water system was due for breakdown in a year or two. A curious case of 'should a doctor tell?' I have only once had the courage to do so, and my host was so upset to be told that his hot water smelled of bad eggs (it had developed so slowly that he and his family had grown accustomed to it) that it would almost have been kinder to leave him in ignorance. (Fortunately, he was a microbiologist and recovered his morale quickly. His solution? He sold his house and bought another. What, you may ask, is the difference between a scientist and a second-hand car dealer? Forgive me if I duck the question.)

One can fairly ask this question. If sulphate reduction by sulphate-reducing bacteria is the basic process that causes underground corrosion, why not get rid of the sulphate, at which point the whole process should stop? Indeed it would, but it is in practice impossible to remove the sulphate. The hardness of ordinary tap water is due mainly to calcium sulphate; all soil waters contain sulphates; plaster and other building materials contain lots of calcium sulphate. Thus three inescapable materials of everyday life, water, mud and dust, are sources of sulphate for these microbes and there is almost nowhere on this planet that remains wholly deficient in sulphate for long. The bacteria, after all, require very little sulphate and they are in no hurry: the fastest recorded corrosion of a water main took about three years. Even the sulphate-deficient soils I shall mention in Chapter 11 probably contain enough sulphate for those bacteria to grow on; it is with plant crops, which need much more sulphate, that the soil deficiency shows.

Underground corrosion is not the sole manner in which microbes attack metals, but it is the most important one. (I should, for completeness, mention that corrosion *can* occur underground without the intervention of bacteria, but the reasons are usually rather special. Underground corrosion is something of a misnomer, though widely used, and a better term is bacterial corrosion.) Ordinary microbes growing in films of water on metals can accelerate the normal corrosion process by altering the electrochemical character of the metal surface in ways that I cannot discuss here (specialists must be content to be told that they form differential aeration cells). Acid-forming bacteria, such as the thiobacilli, can generate enough acid to destroy metals and machinery. Moulds, as I have told, can attack the coatings of cables and wires, and, where they do so, some of the products they form from the coating materials attack the metal; both lead and zinc can be corroded in this way. In these instances it is a product of the microbe's action that is corrosive.

Fungi can also erode stonework, stone facings and concrete. They do it by making organic acids from hydrocarbons which, originating from traffic fumes and the smoke of fires, get deposited on the stone. It is these acids which attack the stone. In a rather similar way, research workers have recently observed that the nitrifying bacteria, the microbes which so usefully make ammonia into nitrates in soil, can cause extensive damage to buildings by converting atmospheric ammonia into corrosive nitric acid. But perhaps the classical example of corrosion by a microbial product is the case of stone and concrete disease. The temple of Angkor Wat in Cambodia is apparently a glorious relic of early Malayan architecture, antedating most European buildings (I have only seen pictures of it). It is slowly decomposing, overgrown in the jungle. The reason for its breakdown, according to work by Dr Pochon and his colleagues in the 1950s at the Pasteur Institute in Paris, is that sulphide soaks up the stone from the relatively polluted tropical soil on which it is built. This sulphide, the reader will by now hardly need telling, is formed by sulphate-reducing bacteria. At the surface of the stone, the sulphide becomes oxidized to sulphur and sulphuric acid, largely through the agency of thiobacilli, and it is this acid that is destroying the stone. Angkor Wat is a particularly tragic case of microbial corrosion of stone, and its protection has not been helped by the terrible wars, strife and privation seemingly endemic to that area. Many similar instances are known, nearer home, as it were. Stone statues in Paris have been found to corrode in a similar way. Sewer pipes are often made of concrete and may corrode rapidly for analogous reasons: in this instance the sulphide comes from the sewage itself, diffuses as hydrogen sulphide to the roof of the pipe, and here the thiobacilli convert it to sulphuric acid. A characteristic of this type of corrosion is that the roof of the pipe caves in. Many industrial effluents contain sulphide, so cooling towers, concrete pipes and concrete manhole covers have been known to corrode for similar reasons. Do not assume, however, that all corrosion of buildings is bacterial: sulphur oxides are major components of atmospheric pollution and in large towns these chemicals can have quite as big an effect as microbes. Westminster Abbey, for example, is being corroded as a result of a process that includes sulphuric acid formation, but thiobacilli play little if any part and the main source of corrosive acid is London's polluted atmosphere.

I could go on looking at microbial deterioration indefinitely. I have said nothing of how cyanobacteria can damage newly disclosed Stone Age cave paintings such as those at Lascaux in France; of how oil wells can become clogged by sulphate-reducing and associated bacteria; of how iron bacteria and algae can obstruct water supplies and filters, causing what water engineers refer to rather charmingly as water calamity.

But I ought to say more about the general topic of water pollution, because anyone who has bathed in the seas and rivers of this island knows something of the damage such pollution causes. I shall deal with pollution of the sea by oil in the next chapter; generally speaking, if any organic matter – leaves, paper, food, manure and so on – gets deposited in water, microbes grow and the water becomes polluted. The sea or a fresh running river can take a fair amount of pollution, because there is plenty of air available and the microbes can

FISH ARE KILLED BY BACTERIAL SULPHATE REDUCTION. In 1971 Lake Palic, Yugoslavia, became badly polluted. Aerobic bacteria flourished, consuming all the dissolved oxygen, and this enabled the anaerobic sulphate-reducing bacteria to multiply abundantly. The hydrogen sulphide they produced poisoned great numbers of carp. (Courtesy of Dr R. Vamos)

oxidize the organic matter to carbon dioxide and products such as lignin and humins, which are almost immune to microbial attack and merely settle harmlessly as sediment. Serious pollution occurs, however, when the water is so stagnant that the microbes use up all the air available. Not only do anaerobic bacteria start growing, and producing putrescent smells, but fish and plants die, making the pollution worse. Sooner or later the sulphate-reducing bacteria start growing too, and since the hydrogen sulphide they form, as well as smelling particularly nasty, is toxic to most living things, they augment the pollution even further. Hence microbial water pollution is self-perpetuating: it is far more difficult to stop once started than to prevent.

Natural lakes and ponds usually have polluted zones near the bottom where free sulphide is present. Normally, just above this layer, there is a zone of sulphide-oxidizing bacteria making use of this sulphide: since many sulphide-oxidizing bacteria are photoautotrophs (i.e. they need light to grow), the depth of this layer depends on how clear the water is and how far into it light penetrates. Above this layer, fish, algae and plankton grow, and the whole lake is a stable system with the sulphur cycle progressing quietly in its lower reaches. Lakes, canals and even seas (such as the Black Sea) are like this. The drawing illustrates a typical system:

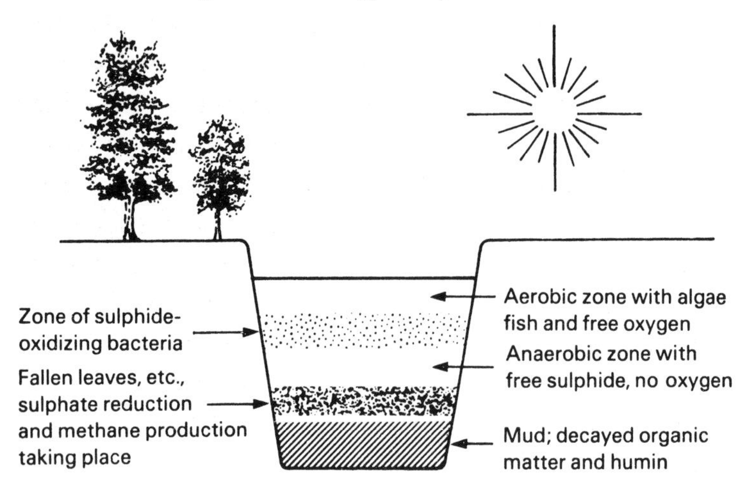

Artificial pollution with, for example, sewage or industrial wastes can have a catastrophic effect on the natural balance, causing the anaerobic zone to spread and to comprise the whole water system. Sometimes a sort of transient condition occurs, when a whole lake or stream turns red owing to the growth of coloured sulphur bacteria – I once encountered an ornamental lake to the west of London which looked as if it consisted of red paint – but this situation depends on a rather fine sulphide concentration being maintained and rarely lasts long. Nevertheless, do you remember '. . . and all the waters that were in the river were turned to blood. And all the fish that was in the river died; and the river stank, and the Egyptians could not drink of the water of the river . . .' (Exodus 7: 20, 21)? Moses may well have been aided by coloured sulphide-oxidizing bacteria of the genus *Chromatium*, because the Wadi Natrun in Egypt, which is rich in these microbes, is traditionally associated with the first plague. Many coloured waters are due to sulphur bacteria, and characteristically they smell of hydrogen sulphide. However, this is not always so. In East Africa there are certain soda lakes which are so alkaline that little hydrogen sulphide escapes; a distinctive species of red sulphur bacteria inhabits the lakes and turns them bright red. Apparently flamingos browse these bacteria and the red pigment keeps the birds' feathers pink.

Not all blooms of coloured water are due to sulphur bacteria: brown algae, green algae, small aquatic plants and coloured protozoa can also cause such colours. The reddish tint of the Red Sea has been attributed to a filamentous cyanobacterium called *Trichodesmium*. Episodes of abundant growth of coloured cyanobacteria occasionally occur in ponds, lakes, slow-moving rivers and sometimes over considerable areas of the sea. They are accompanied by micro-algae and various protozoa, and they form a coloured bloom over the water surface. Such blooms are generally symptoms of some kind of pollution from elsewhere: in the sea it might be an up-welling of cold nutrient-rich water or an influx of man-made pollution; in lakes, rivers and ponds, run-off of fertilizer from farm land is a common cause. Rarely, but disastrously when it happens, the blooming cyanobacteria might include species which form toxins lethal to fish and animals: cultured oysters have been rendered poisonous to humans in this way. In any event, in a few weeks or months the bloom generally out-grows its nutrient supply and the microbes die off. In the sea they will disperse, but in more enclosed fresh waters they can generate a worse pollution as they decompose, while other bacteria multiply at their expense and deplete the oxygen supply, higher organisms become asphyxiated and die, and our old friends the sulphate-reducing bacteria take off and set up a sulfuretum.

The lower reaches of the Thames were, in the 1950s, a good example of a polluted river, too toxic for fish, with a fair level of dissolved sulphide in most circumstances. Ships' hulls corroded in it, paintwork darkened and there was generally a nasty smell around the place. The situation was so bad that corrective legislation was enforced in the 1960s and, by the 1980s, improvement had been so spectacular that there was hope that salmon, a most pollution-sensitive fish, would return. Industrial waters are often polluted, despite government action intended to control such pollution. Though industries were often responsible for initiating it, the larger manufacturing concerns are now usually reasonably responsible about not polluting inland waterways. It is the small producer, the houseboat or the drain that everyone has forgotten, that tends now to keep the trouble simmering gently. Sewage is only discharged into rivers after highly controlled treatment and is usually innocuous; its discharge into the sea is less carefully regulated and we are fortunate that most sewage microbes are killed by the salinity of sea water. Estuarine waters are often very polluted: the black sand, familiar at estuarine resorts, is black because sulphide, formed by sulphate-reducing bacteria, reacts with iron salts in the sand, forming black iron sulphide. On exposure to air the sand turns brown again because the sulphide becomes oxidized back to brown iron oxides. In warm weather, particularly in tropical climates, such polluted sand and water can develop blooms of luminous bacteria, so that, at night, each footprint or swirl of water glows with light. The bacteria usually responsible, *Vibrio fischerii*, glow when oxygen reaches them, stirred into the environment by the pressure of a foot or the turbulence of an oar. In nature, the effect can be both dramatic and romantic; it is a pity that the smell is usually less conducive to romance.

Venice, one of the most beautiful cities of the Western world, is a haven of pollutant microbes. My colleague, the late K. R. Butlin, answered the question why the gondolas of Venice are black. He pointed out that, due to pollution of the canals with hydrogen sulphide, they would turn black soon enough whatever colour they started out. Where does the sulphide come from? Why, from sulphate-reducing bacteria, of course. Water pollution of this kind can occur on a dramatic scale in nature without the intervention of mankind. Walvis Bay, an area off the south-west coast of Africa, suffers from periodic eruptions of hydrogen sulphide from the sea bed which kill fish for many square miles. The sea breezes then carry so much sulphide that they tarnish metalwork and paint in the inshore town of Swakopmund and blot out the face of the town clock, and, in the evocative words of a local reporter, 'Sharks come gasping to the surface on the evening tide'. I have seen pictures of the beach near Swakopmund nearly three feet deep in dead fish as a result of one such 'disturbance' in 1954: the smell of rotting fish apparently added piquancy to the general sulphurous smell as putrefaction set in. This is a most dramatic consequence of the actions of the sulphate-reducing bacteria and one that is quite uncontrollable.

Of course, the late summer smells of the canals of Venice and Bruges, or of a blackened, fishy harbour at low tide, or again, of a ripe refuse tip, are examples of atmospheric pollution brought about by microbes. They arise from hydrogen sulphide, amines and other volatile substances formed as microbes decompose whatever organic matter is around. As environmental problems they are rarely more than transient nuisances, dwarfed by those generated by mankind's industrial activities. Microbes do not always tolerate our smells, either. The yeast *Sporobolomyces roseus* lives on the leaves of trees and can easily be counted because it forms pink colonies on agar. It is very sensitive to sulphur dioxide, a common and harmful pollutant of urban air. Microbiologists from Trinity College, Dublin, have used the numbers of these yeasts on the leaves of ornamental trees to measure sulphur dioxide pollution in cities, including Aberdeen, Belfast, Hamburg, Lyon and Brussels.

I could continue indefinitely on my general theme, but the examples I have given ought to be sufficient to illustrate the extraordinary ramifications of microbes in what one might call the negative side of mankind's economy. Before I look at the reverse side, at how we can make use of these proclivities, is there any moral to be drawn? I think there is. If one compares the amount of research being put into production of pharmaceuticals by genetic manipulation or the search for new antibiotics, for example, with the amount of effort expended on, say, sulphate-reducing bacteria and water microbiology, one reaches a melancholy conclusion. Research that produces pounds sterling or dollars for a company can be sure of support: research that shows prospects of profit nowadays stands a good chance, because most industrialists are fairly science-minded. But research that promises only economies, or furthers only the general public good, with no obvious reward to any particular person or group, attracts no-one. The alert reader will have noticed how rarely molecular genetics and the new biotechnology have featured in this chapter compared with the last. They will reappear when I come to disposal processes, but the degradative side of the processes which sustain the biosphere seems to lack charisma: research on them only gets done because things get out of hand, so that someone darned well has to have a go at it. As a result, not to put too fine a point on it, some thoroughly bad work is performed and published and, what is really more serious, particularly good research is wasted because one scientist with a half share in an assistant is plodding on with some fragment of the field, in some backwater of the mainstream of scientific advance. The problem is essentially an administrative one, for there is no scientific reason for this neglect. The questions raised by the curious microbes concerned in the destructive processes I have discussed are of enormous intrinsic scientific interest, because they represent the chemical fringes of living things; the extremes, as it were, of terrestrial biochemistry.

8 Disposal and cleaning-up

In the last chapter I looked at the destructive effects that microbes can have on materials. I noted at the start, however, that these destructive effects represent an important function that microbes perform in the natural economy of this planet: they remove the detritus generated by higher plants and animals so as to recirculate the biologically important elements contained in it. Deterioration, corrosion and pollution, when brought about by microbes, are simply special cases of this general function, and so are disposal and cleaning-up processes, in which microbes are deliberately used to get rid of unwanted matter and pollutants.

The most important example of a microbiological disposal process is sewage treatment and, since it is fundamental to the health of all civilized societies, as well as being intellectually a most satisfying form of applied microbiology, I shall spend a little time discussing it.

When the population of the world was small, sewage disposal presented few problems. The Greeks and Romans had hygienic systems, often building their baths and lavatories over or near running water. The Romans, in particular, built their baths and sewerage systems to last, and there is a pleasing irony in the fact that these sewerage systems are often all that remain in excavated Roman communities. Standards then fell, and descriptions of life in the Middle Ages and during the Renaissance tell us that human habitations must sometimes have resembled pigsties: steps and odd corners were used as lavatories, refuse was thrown into streets, chamber pots were emptied into the streets and people rarely bathed. The widespread use of scents and nosegays becomes understandable. By the mid-nineteenth century, with the populations of towns increasing as a result of the industrial revolution, it became obvious that it was dangerous, as well as disagreeable, to behave in this manner. Diseases such as typhoid and cholera were widespread, and the Thames, by 1860, had become a vast open sewer, with the rain-washed refuse of London flowing into it. Civic action in sewage collection and treatment was initiated in Britain by the Public Health Acts of 1845 and 1875 and a Royal Commission of 1898. By the early twentieth century sewage was collected and piped in most urban centres, though often the sole treatment it received amounted to discharging it on to municipal land (which was cropped for produce such as tomatoes – the origin of the term sewage farm). The water usually became purified as it percolated through soil strata, and the soil became incidentally fertilized, but it soon became evident that the purification was a haphazard process and that, without detailed knowledge of subterranean water flow, there was a serious risk of polluted water reaching drinking wells. Though a few sewage farms still exist (and in too many coastal areas raw sewage is still discharged directly into the sea), on the whole sewage technology has made enormous strides in the last half century and, in a modern sewage works, sewage treatment is a highly automated and efficient process.

Some idea of the magnitude of the problem can be obtained by quoting a few figures. In the old West Middlesex area of London the population uses over 50 gallons of water a day per head, all of which washes detritus to the local sewage works. An installation serving over one and a half million people must handle more than 75 million gallons of raw sewage a day, which it collects through a local network of pipes running from drains, sinks, baths, lavatories and industrial effluent conduits (the sewerage system). This sewage represents something like five thousand tonnes of organic matter; something has to be done with it before it gets into the rivers and seas, or it would cause unimaginable pollution as aquatic microbes recycled its carbon, nitrogen, sulphur, phosphorus and so on. In effect, what a sewage works does is to allow these processes to carry on in controlled conditions, so that the water which carried the sewage is purified and the solid components of sewage are rendered innocuous. This is easily done by modern sewerage techniques: the processed solids reach a state in which they can be sold as soil conditioners or fertilizers and the treated water is so pure that, at the Mogden works west of London, for example, the staff will demonstrate the purity of their effluent water by drinking a glass for visitors. (The visitors are unaware that they do something of the sort themselves daily: the water economy of this country is such that quite a lot of purified water finds its way back into the drinking reservoirs. I used to wonder how often an average glass of water had been drunk by someone else before I consumed it. Then I learned from Professor Hutner of New York that London water has passed through an average of seven sets of kidneys when it is drunk. Now I am wondering how the calculation is done.)

Sewage from a typical city, as I have told, consists mainly of the rinsings from sinks, lavatories and bathrooms, mixed with some industrial effluents and a certain amount of natural drainage water. A little thought makes it obvious that its main component is human excreta, supplemented with hair, paper, food debris and detergents. It is a suspension of solid matter, rich in bacteria, in a rather strong solution of organic substances; it is an extremely satisfactory medium for the growth of bacteria. The way in which it is treated can be best described by considering an imaginary sewage works, into which I shall introduce the main sewage treatment processes used today.

The sewage, then, flows into settling tanks, in which the solid matter settles as a sludge to the bottom. The settled material is called settled sludge, and I shall describe what is done with it shortly. The liquid part flows into special ponds where it is stirred and aerated vigorously, so that aerobic bacteria grow in it and oxidize much of the organic matter away to carbon dioxide. This escapes to the atmosphere, and one step of the purification process has thus occurred. However, more bacteria have grown, so the sewage needs to settle once more, to yield a sludge of bacteria. This second kind of sludge is called activated sludge, and some of it is collected and added back to the aerated ponds to accelerate the original formation of CO_2: the whole process is a sort of aerated continuous culture in which some of the microbes produced are returned to the original culture vessel (represented by the aerated pond). The rest of the activated sludge is either added to the settled sludge or packaged and sold as fertilizer. This whole process is called the activated sludge process, and after such treatment the water is considerably purified and can often be released into a river. If not, it is sprayed over beds of a porous material such as coke, often by the rotating sprays that one sees in the grounds of sewage works, and allowed to trickle through a metre or so of this. Here films of moulds, streptomycetes and bacteria grow on the coke, removing the last traces of organic matter and yielding an effluent of almost sweet water. Sometimes particles are removed by filtering through sand, but the water is pure enough to be discharged forthwith into a river or the sea.

The settled sludge presents a rather more difficult problem. Though it has settled, it is still more than 90 per cent water, so it can be pumped into huge vats called digesters. These are continuous cultures of a different kind: they are not aerated, so that mainly anaerobic bacteria grow. Sulphate-reducing bacteria produce hydrogen sulphide from the sulphates dissolved in the water, cellulose bacteria destroy paper and similar materials, but the main microbes to grow are the methane bacteria. These, aided by the other anaerobes, break down the organic matter, mainly to CO_2 and methane, both of which are gases. Now, this methane is the same gas as marsh gas or natural gas: it is a valuable source of power, and though some old-fashioned sewage works still burn it away, most modern ones collect it and use it to drive their machinery. Sludge digesters need to be stirred slowly and the methane is used to drive the stirrers and air-pumps; it may also be compressed into cylinders and used to drive lorries; sometimes it is sold. The methane fermentation of settled sludge is therefore a useful source of power to a modern sewage works.

A digester may have a capacity of as much as a million gallons of sludge. Every day between 5 and 10 per cent of the treated sludge is added, so the digester is an anaerobic continuous culture whose contents are replaced every ten to twenty days. The fermentation can be so active that the fermentors have to be cooled. The digested sludge still consists largely of water, and is again held for a while in tanks so that the bacteria and recalcitrant solids (those attacked only slowly or not at all by microbes) will settle out. Usually this process – called dewatering – is rather inefficient, because some residual methane production continues and prevents the solid from settling well; settling agents that inhibit methane production can be added at this stage. After settlement, the water is run off into activated sludge plants, thus ultimately finding its way into a river or the sea. The settled, digested sludge has to be carted away and disposed of somehow; it is fairly innocuous and can be spread on soil as a soil conditioner – it has some value as a fertilizer though most of the useful soluble elements (nitrogen, sulphur, phosphorus) have been extracted from it. For this purpose it must usually be dried and, despite the ready availability of methane to heat it with, this process is rarely economic. In Britain it is often carried out to sea in barges and dumped: by law it must be carried some twenty miles offshore and so it contributes negligibly to the pollution of coastal areas.

A small sewage works may use only some of these processes, but most modern installations use them all. However, one must record dismally that, in 1997, there were still many sites around Britain's coasts where local water companies discharge untreated sewage into the sea; one in ten British beaches still failed to reach the European Union's minimum standards on sewage pollution.

Trouble comes when the in-flowing sewage contains materials that overload the plant or poison the microbes. Abattoir effluents and diary wastes, for example, are examples of waste fluids that are so rich in organic matter that they must be diluted with great quantities of water before the average works can handle them. Industries dealing with such materials on a large scale are often obliged to set up their own plant for dealing with their wastes: biological effluent treatment, the process is called. Chemical industries and the coal gas industry have comparable problems, because their effluents are poisonous to ordinary sewage microbes. But, as I told in Chapter 2, there exist bacteria that can metabolize a wide variety of curious chemicals, and it is possible to set up biological disposal plants using them. The Monsanto Chemical Works at Ruabon, Clwyd, produces effluents containing various phenols, most of which act as antiseptics towards normal microbes. By establishing populations of phenol-oxidizing microbes in activated sludge plants and trickling filter systems, they can so purify their effluents that the water can be released into the River Dee without further treatment. The coal gas industry produced effluents that contain phenols, cyanides and other poisonous substances, and biological effluent plants have been devised to treat these. The paper industry produces effluents rich in organic matter, extracted from wood, and containing sulphites, which are used in pulp preparation. This effluent is a particularly noxious brew, because, though the sulphate is toxic to many microbes, it is received with delight by sulphate-reducing bacteria (it is as good as sulphate for their metabolism): given careless handling, it causes the worst kind of pollution at once. Again, it must be extensively diluted before it can be accepted by a normal sewage works, or a special population of microbes must be developed to deal with it in a special plant. Methods have been developed for using sulphate-reducing bacteria to remove the sulphite in a special kind of trickling filter, but they have not, to my knowledge, been adopted by the industry. On the other hand, these bacteria have been used successfully to remove toxic metals from industrial effluents. For example, the mining industry produces discharges and washings which are often rich in iron, particularly if the working includes strata of iron pyrites. If such an effluent is passed through a biological filter bed in which sulphate-reducers are the dominant microbes, the sulphide they form precipitates the dissolved metal as black iron sulphide and the liquid can be accepted by a conventional sewage or drainage system.

Nitrates in drinking water have caused some anxiety in the latter part of the twentieth century. The problem arises because nitrogenous fertilizers, chemical or manure, and nitrogen fixed by bacteria in and around the roots of plants, release nitrogenous matter into the soil. This is rapidly converted to nitrate, which plants can use, by soil bacteria. However, the plants rarely capture more than half of the nitrate; the rest gets washed by rain into rivers, lakes and underground water reserves. Increased agricultural activity, particularly as regards the use of fertilizers, has led to a gradual rise in the nitrate content of drinking water in much of Western Europe, including Britain, and in the USA, a rise which is still going on, because it takes some time for run-off or leached nitrate to reach the reservoirs. If the process continues in this way, a health problem may arise, because nitrates in unusual amounts could be harmful. The European Union has therefore set upper limits to the permitted nitrate content of drinking water. Indeed, in hot, dry summers such as Europe enjoyed in 1979 and 1990 some British water reserves exceeded this limit, but the water boards coped by mixing waters from various sources. All of which leads up to a pleasing process developed in Holland for removing nitrate from water. In principle, the water flows through a column, like in a water softener, which contains an ion-exchange resin, a material which traps the nitrate by swapping it for bicarbonate. Later the nitrate is released (swapped for bicarbonate, so the column can be re-used) and piped to a vessel where denitrifying bacteria convert it to atmospheric nitrogen. These bacteria have to be fed, so they are given a little methyl alcohol. Thus, at the cost of a little methyl alcohol, which is quite cheap, microbes help to render drinking water fresh, sparkling and nitrate-free.

A pollutant that causes considerable public nuisance is sea oil, the tarry material which accumulates on beaches as a result of the discharge of oil by ships at sea. Despite international legislation, accidental and partly accidental contamination of the sea regularly takes place, and normally the spilled oil is oxidized away by marine bacteria of the kind I mentioned as contaminating the bottom waters of petroleum tanks. Unfortunately, the sticky, tarry components of crude oil are oxidized only slowly, and the tarry material that today dirties one's clothes and children on most European beaches is the residue of such pollution, still slowly undergoing microbial decay.

The nuisance occurs because more oil is discharged on the sea than the natural microbes can dispose of before it is washed up on the beaches. World consumption of oil has increased relentlessly throughout the twentieth century and, as more and more is shipped about on the world's oceans in giant tankers, spillages of a catastrophic kind have become inevitable. The public was first alerted to this hazard when a spectacular disaster of this kind occurred in the spring of 1967: a huge oil tanker, the *Torrey Canyon*, was wrecked off the south-west coast of England and thousands of tons of oil were released to contaminate the beaches of England and France. That disaster was dwarfed in March of 1989 when the *Exxon Valdez* spilled 11 million of its 60 million gallon cargo into Prince William Sound off the coast of Alaska, producing a slick which spread over 100 square miles of sea. Even more serious pollution has originated directly from oil wells or storage tanks. In 1980 a blow-out in the Ixtoc well released some 150 million gallons into the Gulf of Mexico; substantial escapes into the Persian Gulf occurred as a result of military action, first during the Iraq–Iran war in the mid-1980s, then, early in 1991, as a deliberate release by Iraq during its dispute with the United Nations (the military objective in 1991 was to clog the Gulf desalination plants which provide drinking water for much of Northern Arabia). Environmental disasters of this kind kill thousands of sea birds and sea mammals, and damage fishing and crustacean industries, as well as spoiling beaches; they call for crisis measures. Microbes work only slowly and the damage must be contained until they can act. Floating booms are used to contain the oil slick; sawdust has been used to help the oil sink and to give microbes a good surface from which to attack the oil. Detergents are used to clear beaches and to help to disperse floating oil, but they are of limited use: the next tide will often return oil to a seemingly cleaned-up beach and, in any case, many detergents are disinfectants which delay the action of microbes. They also add to the damage suffered by other life in and on the sea.

The aftermath of such spillages can last for several years around Britain or in the cold Alaskan seas, where all microbial activity is slow. Recovery was much quicker, a matter of months, in the warmer waters of the Persian Gulf, into which oil seeps naturally and ensures that appropriate bacteria are already present and active. However, prospects for microbiological control of oil pollution have improved: chemical fertilizers supplying nitrogen and phosphate appear to have hastened recovery after the Alaskan spill and strains of microbe have been generated in recent years by genetic manipulation which act relatively quickly and can be introduced in emergencies. For example, Dr Chakrabarty in the USA observed that, in certain bacteria of the genus *Pseudomonas*, the ability to decompose oil and other hydrocarbons is specified by genes which reside on plasmids; by genetic manipulation he has been able to construct plasmids which enable these bacteria to perform faster than usual.

Oil is not the only contaminant of the natural environment that could be treated with microbes. I mentioned in Chapter 2 the field in Smarden, Kent, which became contaminated with fluoracetamide, a powerful poison used as a pesticide; emergency measures were taken to remove and dispose of the contaminated soil and only later did it become obvious that bacteria exist able to decompose fluoracetamide to harmless products. It is probable that treatment of the soil with microbes adapted to decompose fluoracetamide would have provided a quick and effective remedy. A problem was that it took a couple of months, even with good luck, to isolate a microbe capable of decomposing fluoracetamide and, in a disaster situation, one cannot wait that long. A comparable catastrophe took place in July 1976, at Seveso, North Italy, when a very poisonous intermediate in the manufacture of disinfectants and herbicide escaped from a chemical works into the town and its environs. The chemical is known by the acronym TCDD (2,3,7,8-tetrachlorodibenzo-*p*-dioxin, if you really want to know; it was somewhat incorrectly referred to as dioxin by the press). It killed cattle and vegetation and caused a nasty skin condition (chloracne) in many of the population as well as causing deformities in unborn children. Elaborate earth-moving measures were taken to decontaminate the neighbourhood. Microbes capable of decomposing TCDD exist, but they were not available then. They would, of course, have been of little use for decontaminating people, but they could have been invaluable for cleaning up the environment afterwards. Selective herbicides, insecticides and such materials disappear from soil as a result of microbial action, usually by mixtures of microbes. The herbicide 2,4-D vanishes in three or four weeks, but the still widely used and stronger 2,4,5-T (made from TCDD, actually) can persist for up to a year. Populations of microbes able to degrade it have been developed. The herbicides I have just discussed are all organic compounds with chlorine atoms in their molecules, and this class of chemicals often includes substances that microbes find difficult to degrade. The insecticide DDT is a notable example. It has been marvellous for the control of disease-bearing insects and agricultural pests, but it is not degraded by microbes and it persists in the environment, entering the food chains of all kinds of creature. Its use has had to be banned in many countries because it has undesirable ecological side-effects. Another group of persistent chloro-organic compounds is the polychlorinated biphenyls (PCBs), once widely used as insulating fluids, which present a troublesome disposal problem. However, in this case there is hope: bacteria that attack them have been found in Hudson River water and they or their genes could be recruited for a disposal process.

War gases present especially awkward disposal problems. How does one get rid of old stocks when one cannot incinerate them, or treat them chemically, without grave risk to the operators and the neighbourhood? And the certainty that wherever one might bury, dump or store them, someone or something is likely to disturb them sooner or later? It is a problem that tends to spend a long time in the 'pending' trays of administrators. However, bacteria which degrade nerve gases and which aid the breakdown of mustard gas in water have been isolated, and fermentation processes using these are being explored in the USA.

There are impressive possibilities for the deliberate construction, by genetic manipulation, of microbes able to remove unwanted toxic materials from the natural environment and, conscious of the threat of environmental legislation, industries concerned with herbicide and pesticide production are sponsoring fascinating biotechnological research in these directions. The deliberate use of appropriate microbes, either as pure cultures or as mixed populations, to remedy local incidents of industrial pollution is now called 'bioremediation'. It has been a success when applied to oil and petroleum spillages, but with more exotic pollutants it is still largely at an experimental stage. In practice several difficulties arise. In the first place, bacteria which have been trained or manipulated to consume a certain pollutant, such as a pesticide, often still prefer to consume such ordinary nutrients as they may find in soil or water, and variants then appear which have dispensed with their ability to decompose the pollutant. The remediation process thus tends to peter out. In the second place, if the remedial population works well, it then flourishes and multiplies, and in consequence predators such as protozoa and nematodes, which live naturally in the soil or water, soon discover an abundance of food and begin to eat the remedial bacteria, multiplying in their turn and bringing the remediation process to a standstill. To cope with these two problems, it is wise practice to add bioremedial populations in as active a state as possible, and in pulses rather than continuously. If it is fine soil that needs treatment, it can sometimes present an additional, non-microbiological, problem: should the pollutant have crept into sub-microscopic cracks and fissures in soil particles, or become bound tightly to their surfaces, the remedial microbes may simply be unable to reach it. Bioremediation shows great promise, but it still has teething troubles.

An important problem in sewage technology is how to deal with materials that are not attacked by microbes at all. Wastes from chromium-plating industries, for example contain the chromate ion which can upset the microbial population of a sewage plant and thus put the whole process awry. Since the industries producing such effluents are generally known, their discharge into the sewers can usually be regulated sufficiently to avoid trouble. What is more troublesome is the domestic use of recalcitrant materials. Some of the detergents in use a few decades ago used to interfere with sewage treatment, and this could happen in two ways. In the first, the microbes may have been capable of handling the detergent chemically, but it produced such a froth in the activated sludge plant that the access of air to the sewage was restricted and the whole purification process slowed up. I recall the Director of a local sewage works in the early 1950s telling me that he could tell when a detergent firm was having a sales campaign in his area: his activated sludge plants disappeared under the mound of froth. Expensive use of anti-foaming agents was then necessary. Such treatment is not always successful, and detergent foam could pass right through a sewage works and contaminate rivers – I have seen rivers afroth with foam both in London and in the Midlands, but this problem is less common today.

A more insidious problem occurred with certain non-foaming detergents, which were used in the catering industry and in domestic washing-up machines. Some of these were recalcitrant: no microbes were known which attacked them at all rapidly, with the result that they might have got through the sewage treatment process quite unaffected and, in fact, get into drinking water supplies. By 1960 it was known that minute traces of these substances had reached several reservoirs of drinking water in the UK and, though they caused no obvious harm, the amounts were clearly going to increase and possibly become harmful. Though some success was obtained in developing strains of microbe that attacked them, the solution has lain more in the direction of altering the chemical character of the detergents so as to make them susceptible to microbial attack: biodegradable, as the specialists call them. Legislation regarding the marketing of non-degradable detergents was proposed, here and in the USA, but I believe that the detergent industries largely abandoned hard detergents voluntarily.

Modern sewage treatment is usually highly automated: flow, settlement, charging of digesters and so on are generally directed from a central control area by push-button mechanisms of which sewage engineers are justly proud. The unsavoury nature of physically handling sewage has contributed in part to this high degree of automation, but an important additional factor has often been the fact that sewage works can be self-sufficient as regards energy. The methane produced by anaerobic sludge digestion, as I told earlier, is usually more than sufficient to power the pumps and machinery used in sewage processing, and several sewage works have been able to sell excess methane to the national gas grid. I also mentioned that small sewage plants for making methane from domestic and farm residues have been devised to power refrigerators and domestic machinery in tropical countries such as India.

The productive nature of waste treatment has caused scientists to consider what useful products other than methane might be obtained from sewage, and a number of interesting projects have arisen. Sulphur, as I told before, is an element which is becoming scarce, at least in its reduced form. I mentioned the method of making sulphur from sewage sludge developed during the 1950s by the late K. R. Butlin and his colleagues: sewage sludge was composted semi-continuously with gypsum (calcium sulphate) and sulphate-reducing bacteria converted this to calcium sulphide. Sewage gas (methane plus carbon dioxide) was used to remove the sulphide by converting the calcium sulphide to carbonate and releasing the sulphide as H_2S. As a result a net purification of the sewage took place. Butlin calculated that a North London sewage works which processed about a million gallons of sludge per day could be adapted to produce 5,000 tons of sulphur per day – but it would then have no methane. In practice a balance between sulphide and methane fermentation would have to be reached, not only because the methane is useful to carry off the H_2S but also because the CO_2 that accompanies it is needed to displace the H_2S. The process, as I also mentioned, proved to have an additional virtue from the sewage engineer's point of view: the sulphide-digested sludge settled more efficiently than conventional methane-digested sludge. The reason was that sulphate-reducing bacteria competed with methanogens, and stopped gas bubbling and stirring up the settling sludge. Disposal of the digested product was thus a much more economical process, because less water needed to be transported with it. Although conceived as a means of alleviating an industrial sulphur shortage, Butlin's process seems now to have more promise as a waste disposal technology, with sulphide as a saleable fringe benefit.

Comparable processes have been developed in India, in the USA and in the former Czechoslovakia; in the latter country sulphur fermentation has been used successfully to pre-treat strong wastes – effluents from yeast and citric acid manufacture – that are too rich to be handled by conventional sewage processes: a preliminary sulphate fermentation downgrades the effluent sufficiently to make it acceptable to a normal sewage works and yields sulphur as a bonus. I mentioned earlier the particularly noxious character of wastes from the paper industry which contain sulphite and organic matter and told how sulphate-reducing bacteria can be used to pre-treat these also, but the yields of sulphur are, according to Russian workers, too small to be economically worth collecting. American workers have used paper wastes to grow yeasts which could then be used as animal fodder, and a project was developed to grow mushroom spawn on paper and woody wastes, to make packaged mushroom soups. I believe the flavour did not come up to standard. Sewage sludge itself is a useful source of vitamin B_{12}, though at the present time I am not aware of any commercial exploitation of this source.

Sewage also contains compounds of a variety of heavy metals. Iron salts, for example, are present in all kinds of washings and discharges; when they reach the sewerage system, some of the iron reacts with sulphide in the sewage and becomes concentrated as a black deposit of iron sulphide, and some sticks to the solid organic matter of the sewage: to the microbes and general detritus. Copper, zinc and lead compounds are also components of ordinary sewage, because they are normally present in small amounts in pipes, dust and all sorts of domestic materials; they, too, tend to become concentrated in the sludge, partly as sulphide deposits, partly because they stick to the cell walls of sludge microbes such as fungi and bacteria. Surprisingly, sludge also contains trace amounts of compounds of valuable rare elements such as zirconium, germanium, gallium and selenium. There have been proposals to extract them for commercial use, but they have not to my knowledge got beyond the planning stage.

The fact that heavy metals become concentrated by sticking to the surfaces of sewage microbes led to the idea of using microbes deliberately to free effluents of toxic metal compounds. Copper, zinc and cadmium compounds, which can be very poisonous at quite low concentrations, are present in a diversity of industrial discharges; lead salts are also toxic and occur in lead mining and smelting discharges, and in washings from the manufacture of batteries. The use of microbes, which may be bacteria, fungi or both, to absorb such toxic matter from effluents is called 'biosorption'. It works and, with the right mix of microbes, it can be very effective because microbial cells often prove to be excellent scavengers, whether they are dead or alive. But biosorption is a relatively new technology and it is not yet widely used – at least, not intentionally.

Microbes can also be used to treat one of the most awkward of effluents, a type which has arisen only in the last half-century: that which emerges from industries, laboratories and hospitals which use radioactive materials. For example, the nuclear fuel industries in both the UK and the USA sponsor research into the use of bacteria such as *Citrobacter* to remove the last traces of uranium from their effluents. In one process, the bacteria are immobilized, in a filter cartridge for example, and they entrap the metal around their cell surfaces as crystals of an insoluble uranium phosphate. Many types of radio-isotope can be concentrated spontaneously by a variety of microbes, a fact which can be exploited for bioremediation. However, it can sometimes be inconvenient because even if an industry dilutes a radioactive effluent to a harmless level, one cannot be confident that the radio-isotope will stay that way; that it will not become concentrated again somewhere by naturally-occurring microbes. Moulds, algae and bacteria, as well as plants, may concentrate radioactive isotopes and there seems to be no particular rhyme or reason in whether or not a given species will concentrate a given substance. For these reasons such effluents must be segregated carefully and are not normally accepted by ordinary sewage works; the use of microbes and plants deliberately to extract useful isotopes from such effluents has been proposed but not, to my knowledge, developed.

The main bulk products available from sewage treatment are methane and sulphur (or, rather, hydrogen sulphide). The other product (besides water) is the digested sludge and, though this can be used as a fertilizer and soil conditioner, it is often rather unsuitable, because of its content of copper, zinc, lead and trace elements which are not good for plants. One disappointing aspect of conventional sewage procedures is that they lead to loss of inorganic constituents that would be useful to agriculture. Potassium, phosphates, sulphates and nitrates tend to be removed from sewage during the treatment, becoming diluted in the purified water and eventually finding their way to the sea. Quite a lot of nitrate is lost by bacterial denitrification to nitrogen gas during the activated sludge process and subsequent settling. Thus there is a net loss of useful agricultural elements from the land to the sea and air. One of the long-term problems of civilized communities is that of returning these elements to the land: in the old days, when sewage could be spread on land and sewage farms could be operated, this drainage of intrinsic fertility occurred on only a small scale. Today the deficit must be made up with chemical fertilizers and careful husbandry.

Talking of the land, picture, if you will, this rural scene. A stream winds lazily along a valley, edged with lacing of reeds and tussocky grass. On either side verdant fields slope gently down to the sparse gallery of trees and shrubs which mark the stream's course, and healthy cattle graze on the abundant grass, a testimony to their owner's care for the land and its fertility. Perhaps the stream itself has a little sedimentary mud at its bottom, and the odd decaying branch or clutch of leaves, but the water is clear, the water plants flourish, and it is home to a thriving community of fish, insect larvae, water birds, amphibians and water voles. It is a pleasant, peaceful spectacle, is it not? But if you pause and think about it, surely a constant stream of pollutants is running, day and night, into that stream – groundwater flow bringing in the excreta of cattle, surplus fertilizer, agrochemicals, and the effluent of decomposing vegetation? This in addition to pulses of dust, dirt and surface detritus which are periodically washed or blown into the stream by rain and storm (not to mention contributions from heedless passers-by)? Indeed it is. So why does the stream remain so clean? The answer is that, at its edges, the reed beds and, to a lesser extent, the roots of the trees, form an efficient biological filter: films of bacteria and fungi grow over the network of plant material, just as they do on the bed of a sewage percolation filter; these consume, decompose and oxidize the pollutants in the drifting groundwater. They multiply, and become food for a population of protozoa (which, incidentally, eat up pathogens from the animals' excreta). Nematode worms, insect larvae and small fish graze upon the surplus microbes which emerge with the now highly purified groundwater leaching into the stream. It is a natural self-cleansing system which works very well, and woe betide the landowner who decides to dredge up that ribbon of rushes which so irritatingly interfere with his fishing or boating! Many streams in Britain became greeny-black, smelly and almost lifeless before the value of their gallery vegetation, and the microbes it harboured, was widely understood. Of course, as with a sewage works, gross pollution can overwhelm the system, but in ordinary conditions it works well, and filter beds based on plants and their root microflora are sometimes used deliberately to contain small-scale effluents.

To return to sewage: an interesting and quite different way of exploiting it goes back to a very early pattern of sewage disposal. Small communities can, in some circumstances, discharge their sewage into ponds or small lakes, or sometimes a series of such pools (called waste stabilization ponds), allowing the natural microbial degradation processes to take place such that the water purifies itself, bacteria, worms, water beetles, plants – the whole flora and fauna of the pond – flourish and the ultimate beneficiaries are fish and the water birds that feed on them. Algae grow particularly well in such oxidation ponds (so called because the algae, by photosynthesis, make oxygen, which aids purification of the water) and can be good fish food. Deliberate use of sewage to fertilize ponds as a means of fish farming has been proposed and, I believe, it has been used in developing countries such as Indonesia.

Sewage treatment deals with a waste material that is fluid: it can be pumped around a sewage works and handled like a bulk liquid, despite its content of solid matter. But a lot of urban, agricultural and domestic waste is solid or semi-solid. While much of this can be burned, there is much that cannot be, and to dispose of this many local authorities use processes that are basically similar to the gardener's composting. Urban refuse has a high content of vegetable matter from paper and food residues, and, after removing useful items such as tin cans (the tin can be recovered and sold), many municipal refuse works bulldoze refuse into huge compost heaps where microbial degradation sets in. The interior gets so hot that thermophilic bacteria grow and cause very rapid breakdown of the organic matter. Development of insects and multiplication of rodents at the surface of such refuse dumps has to be controlled, but in a surprisingly few years quite fertile soil may so be formed. Composting is a cheap and safe means of disposing of urban refuse, but it requires the separation of biodegradable waste (food residues, plant material and paper) from glass, metal, plastic and so on. In Germany, where in many communities households, shops etc. are obliged to keep such wastes separated, millions of tonnes of biodegradable urban waste are composted and sold for agriculture and land reclamation.

Land reclamation with composted material is fine, but historically, in Britain and elsewhere, raw urban refuse has been used, relying on the filled site to behave like a huge compost pit. Raw refuse indeed makes an admirable material for in-filling old pits, quarries and the like, notably where there is pressure for new building. However, the disposal pit must be monitored for the gaseous products of decomposition. Enthusiastic disposal agencies have been known to cover such disposal heaps with soil and hard core, to bring them into use as building sites, too soon: before the composting process is properly over. Methane, still being generated from residual refuse by our old friends the methanogenic bacteria (some of which are thermophilic), then accumulates beneath the ground. It may diffuse through the top soil harmlessly but, obstructed by building foundations, it may also erupt, or leak in substantial quantities into buildings when, in the worst episodes, it has caused serious explosions. In fact, this feature of composting can be managed, and even turned to advantage: if a well impacted compostable landfill site is lined with sheets of a non-degradable plastic, the methane can be piped off and either used as an energy source or burned off. A medium-sized pit will continue to yield methane for several years, and in some countries new landfill sites must by law be enclosed and controlled in this way. The plastic lining also controls leaching of polluted fluid into groundwater supplies.

A different approach to the problems of disposal of town wastes and the reclamation of waste land was taken in the area around Staines and Twickenham, on the outskirts of London. This area was scarred with waterlogged old gravel pits, great artificial ponds which were useless because most of the exploitable gravel had been removed, and the pits abandoned. Because of the intense local demand for building land, the idea was to reclaim them by filling them in with urban refuse. Yet to tip raw refuse into a waterlogged pit is to court disaster: a most glorious pollution will develop within weeks and the local authorities will be deluged with complaints, injunctions and legal actions caused by the resulting smells and damage to paint and metalwork. Mr A. S. Knolles, Borough Engineer of Twickenham in the 1950s, developed an ingenious procedure for containing such pollution and yet reclaiming land: the clinker from the combustible part of urban refuse was used to divide the pit into lagoons, each of which was filled with raw refuse rapidly before pollution could get established. Provided the lagoon walls could be built ahead of the influx of raw refuse, whole lagoons could be filled in and recovered for building without nuisance. The clinker contained sulphate and the refuse contained organic matter, so the ingredients for massive pollution by bacterial sulphate reduction existed, with the consequent risk of the most noxious kind of nuisance but actually minimizing the production of methane. However, provided the process was conducted rapidly, with an understanding of the microbiology involved, the pits could be filled and the land recovered with nothing but benefit to all concerned.

Some ingredients of urban refuse can be burned, but it is not always desirable to do this. Plastics of the chlorinated hydrocarbon kind (polyvinyl chloride, for instance) are widely used and disposed of today and, if these are burned, hydrochloric acid is released and damages the furnace and flues as well as producing noxious fumes. There have been claims that bacteria exist which can degrade these materials, and a composting process for disposing of them would, if it worked well, be more useful than burning in practice. Used tyres are an on-going disposal problem. I am impressed by a project to render their rubber re-usable by powdering the tyres and using thiobacilli or other sulphur-oxidizing bacteria to remove the vulcanizing sulphur: a fitting exploitation of the deterioration of fire hoses brought about by such bacteria. I do not know how well or how economically it works.

From the point of view of disposal, then, microbes are essential to the social organization of civilized communities and, as I have already pointed out, if it were not for their activities we should all be up to our necks in an appalling morass of the detritus of human activity. The character that makes microbes so valuable in this context is their extraordinary chemical versatility: there seem to exist microbes capable of destroying and degrading almost any material mankind can produce. Scientists understand but little of the biochemistry of these processes and, indeed, have only a vague knowledge of the microbes involved. This ignorance arises because, as with the corrosion and deterioration processes discussed in the previous chapter, fundamental research has lagged behind practical experience in these areas of knowledge. The fragments of knowledge scientists have concerning the roles of methane bacteria, sulphate-reducing bacteria, detergent- and plastic-degrading bacteria make it clear how rewarding a sustained and basic scientific investigation of microbiological disposal processes could be. The problem is, who will pay for it?

I have, in the last three chapters, tried to show how basic to our economy the microbes are and the reader has, I hope, noticed occasional references to genetic manipulations of a worrying kind, to environmental hazards and even, as in the story of the ice-minus *Pseudomonas*, to militant antagonism generated by microbiological research and its applications. Let me briefly, step aside from my main theme and look at these matters. It will be a melancholy spectacle, but it will not take long.

9 Second interlude: microbiologists and man

I grew up in an area when science and technology, inseparable in most people's minds (including my own), were wonderful things. I suppose one has to be tapping on towards eighty to remember the time when an aeroplane was a thing to be stared at, when the crackling wireless was a miracle, when funny bug-eyed cars edged horses to the side of the road, when one or two of one's friends had tuberculosis, and telephones looked like black daffodils made of weird material called Bakelite, originating from coal-tar. As the 1920s segued into the 1930s the seeds of modern communication, travel, medicine and plastics were germinating and scientists were leading society to a new and glorious dawn; mankind would live well-nourished, well-cared-for lives in a happy technological Utopia, fulfilling its creative potential in arts and science, untroubled by war, deprivation, hunger and cruelty, for there would be enough of everything for everyone. Bigotry, prejudice and vindictiveness would vanish as scientific rationalism prevailed, putting flight to tribalism and mysticism (dignified by the uninformed as patriotism and religion). H. G. Wells was our prophet and, though we may not have followed his ideas in every way, none of us doubted that science, technology and the well-being of mankind went hand-in-hand, with man himself most in need of shaping up.

How innocent we were! For today there is a positive surge of feeling against science. Science, to many, has generated the threat of atomic holocaust, destroyed the environment, released new poisons, carcinogens and illnesses on an innocent and unsuspecting public and has done nothing to alleviate the perennial plagues of society: unemployment, poverty, drug abuse and violence. It is useless to point out that science does not do these things and that real living standards have improved immeasurably throughout the world. Few people get the point that it is what people make of science that causes the trouble. That is a boring thought. It is easier to blame the scientist, so the net effect is that an anti-science attitude has spread throughout the lay public. Science is seen not to have fulfilled the promise of its halcyon days but rather, some would say, to have made things a lot worse.

Why has this happened? It is an important question, one that applies to all of science, not just to microbiology. It is also a subtle question which it would be distracting to tangle with in an overview such as this because it has several partial explanations rather than a single clear-cut answer. But one line of thought is relevant here, which the story of microbiology illustrates. It is this. Scientists can explain and discuss points of interpretation and exploitation of research findings among themselves, knowing that their colleagues will understand the inconsistencies, uncertainties and possible errors involved in any kind of laboratory or field research. All participants will be aware that nothing in science is absolutely certain; research is a matter of checking up on scientific knowledge and, if necessary, amending it. That is the way science works; in a philosophical sense there are no immutable scientific truths, only statements of overwhelmingly high probability. However, most lay people expect truth. They find it incredible that the science which underlies the tremendously complex technology of modern civilization, which we all take for granted, is an edifice of probabilities, not certainties, and is subject to constant revision. A few therefore conclude that all science is unreliable, which is palpable nonsense. But informed scepticism is fair enough. Our understanding of the microbial world and its effect upon ourselves and our societies is constantly changing, generally to the benefit of all concerned; I hope the preceding chapters have given readers some idea of when to worry and when to relax in the face of our invisible companions.

Medicine, often regarded as a science, has suffered less from the anti-science outlook, because its many benefits are obvious. But even here its failures are leapt upon by some, and today fringe medicine has never been so flourishing – nor so dangerous. As a parable for today, I recall a musician friend who was convinced that something had gone seriously wrong with Western medicine because nearly everyone dies of cancer these days. (Not a statistical truth, I am told, but widely believed.) It had not occurred to him that modern medicine, by its success in keeping many of the older microbial disorders at bay, allows us all to live longer, so cancer gets more of us. Nevertheless, reform is certainly called for in medical matters; I shall return to that topic very soon.

Microbiology came to suffer from the anti-science attitude rather later than, say, atomic physics or pharmaceutical chemistry. The crunch was brought about in a quite remarkable way by a group of microbial geneticists who, in 1974, publicly announced that they thought their research was dangerous. It was obvious to competent microbiologists that they were wrong: nothing they were doing posed a threat in any way comparable to that of handling a natural dangerous pathogen, nor did it require such elaborate precautions. They were in fact engaged in the excellent fundamental work which underpinned all the genetic engineering that I have already discussed: cutting out and transferring genes from one organism to another. Yet for reasons of their own, they announced to a bewildered public that they were not sure that they might not create ghastly new pathogens, microbes which might escape and decimate humanity. Although there was not the vaguest chance of this, the foreseeable results occurred: the public was seized with panic, whipped up by press, radio and TV; environmentalists, activists and politicians joined the bandwagon, watch committees and biological safety committees were formed, scientific debate was replaced by political posturing, lots of time, money and mental energy were wasted – and the instrument manufacturers made a packet out of selling unnecessary microbiological containment facilities to bewildered researchers. The furore lasted about five years, until Dr Sidney Brenner of

attacked
l closed

'Frankenstein' project given go-ahead in US

From Peter Strafford
New York, Feb 8

The scientists at Harvard and the Massachusetts Institute of Technology won a victory in Cambridge, Massachusetts, last night when the city council voted unanimously to allow them to carry out advanced genetic experiments in the field of what is termed recombinant DNA research.

This means that molecules of the genetic material DNA (deoxyribonucleic acid) from different species are combined and transplanted into living cells. Traits and capabilities of one species, such as humans, could be transferred to other forms of life, such as bacteria.

The decision came after months of controversy and public hearings, in which scientists, environmentalists and others expressed fears of where the research might lead.

Opponents said today that they intended to carry on the battle to prevent the research in Cambridge and elsewhere in the United States. They say it might lead to the creation of some new organism which humans would have difficulty in controlling or resisting.

Mr Alfred Vellucci, the Mayor of Cambridge, has talked of "some sort of Frankenstein" emerging from the laboratories. Other opponents talk of a pathogenic agent which would cause disease, or else argue that scientists have no right to embark on experiments which could lead to "an absolute biological catastrophe".

A board appointed by the city council recommended approval of the research, provided there were regulations to control it, and this has now been adopted by the council. A Cambridge Bio-hazards Committee is to be set up to keep watch on the research.

[illegible] of vaccinations for diphtheri poliomyelitis and tetanus ove over the past three year Whooping cough vaccinatio had dropped by nearly 60 p cent. Mr Ennal's appealed parents not to turn their bac on vaccination and said th gains greatly outweighed th risks.

Our Health Services Corre pondent writes: After M Ennals's Commons statemen and a press briefing later, M Jack Ashley, Labour MPP fo Stoke-on-Trent, South, who ha been campaigning for brai damaged children, said nothin had been done to help them They are estimated to numbe three hundred.

He welcomed the move to give doctors and nurses the most recent information. But he added: "In view of the very clear conflict in the medical profession about the vaccine and the deep public anxiety, I think that he has to have an independent inquiry."

The minister had a clear responsibility to provide compensation. There was no reason to await the report of the royal commission.

Parliamentary report, page 6

s Japan offers Europe **Moderates' leader**

moderates are looking to Mr

Leader page, 15
Letters: On the Bullock Report, from Mr John F. Phillips and Mr Michael

GENETIC ENGINEERING – SHOCK! HORROR! Even *The Times* of London (February 9, 1977), in those days a model of sober journalism, could not resist a dramatic headline when reporting on the regulation of research with recombinant DNA.

Cambridge put a stop to it by calculating some of the risks and demonstrating that they were minuscule. The microbiological world relaxed and got on with research; the public and media found new things to worry about; a couple of Nobel prizes were awarded.

What is the lesson? Do not cry wolf when what you see is a puppy dog. But some good came of it all. At the time when the major fuss was taking place, almost all the research was being done with a variety of *Escherichia coli* called the K12 strain, one which is totally harmless to humans, and also prone to die rapidly away from its cosy laboratory cultures. The biochemists and geneticists doing the work treated it most casually; they had no need of, indeed rarely knew about, the techniques for asepsis which I outlined in Chapter 4. Their procedures for handling their cultures, for disposing of their wastes and for avoiding contamination of themselves and their surroundings were usually horrifying to microbiologists familiar with pathogens. Well, many sloppy laboratories were compelled to tighten up their routine procedures; nothing but good there, but to panic the general public was an extravagant way, to put it mildly, of engendering such reforms.

Yet fear of the new technology remained, not only in the wider community but among research workers themselves (not to mention their technicians, who not unreasonably wished to be clear about what risks they might be running). In the earlier 1970s official Health and Safety bodies in various countries began to look closely at the possible risks which various kinds of genetical research might entail and to prescribe the appropriate containment requirements. Molecular genetical laboratories were required to set up Biological Safety Committees, which would include non-scientists to reassure the wider public, and to appoint Biological Safety Officers, to assess and guard against possible hazards from their research and development programmes. New research and development projects had not only to be approved locally but to be submitted to National committees, such as Britain's Genetic Manipulation Advisory Group, with legal penalties for failure to do so (to the consternation of industrial laboratories, who felt their commercial security would be breached; they accepted official reassurances, albeit grudgingly). Much of this bureaucracy proved to be tedious over-reaction, but it was valuable in compelling research and development workers in the general area to think seriously about risks, genetical or other. And it had a less parochial value, too. As the risks came to be seen more objectively, and to recede, the tight watch on the more basic kinds of research eased up during the 1970s; but by the early 1980s the products of all these genetic manipulations began to reach the stage of practical application. Industries and field workers wished to release genetically manipulated materials, such as new plants or vaccines, into the natural environment. A new set of possible hazards needed to be thought about: risks not so much to people as to the environment. For example, I mentioned in Chapter 6 a couple of ecological problems that could arise from the widespread use of plants which had been genetically engineered to carry the Bt gene, the one which enables them to produce the insecticidal toxin of *Bacillus thuringiensis*. One is that the Bt gene might spread to other plants. This consideration applies equally to any other genes that might be manoeuvred into plants. A clear example would be the genes responsible for resistance to herbicides, which could be a terrible nuisance if they were passed on to weeds. I also suggested ways of overcoming such risks. However, there are more subtle scientific problems. As a general principle, genetic engineering of plants involves the use of genes which act as markers. These genes are not necessarily, nor even usually, part of the essential genetic information being introduced into the plant: they are genes that, being easily recognizable, are attached to the essential genes being introduced and manipulated alongside them. Tests for the marker genes tell the geneticist whether or not the whole package of genes being manipulated has arrived in their new hosts successfully. For example, a widely-used marker is a bacterial gene which codes for resistance to the antibiotic ampicillin. In an actual genetic manipulation one may have several hundred colonies of manipulated cells which *might* have taken up the alien DNA and several million which have not. How is one to tell them apart? If the colonies are tested for ability to withstand ampicillin, which is very easily done, only a few will resist the antibiotic, and these will be the ones which also possess the alien DNA.

Marker genes are thus invaluable; but they can and should be removed from the genome of any genetically engineered product before it is released for general use. However, their removal can be a tedious and time-consuming process. Commercial competitiveness being what it is, firms have cut corners, and already agricultural products carrying a marker gene have been released in some countries. The 'Flavr Savr' tomato carries a marker gene for resistance to the antibiotic kanamycin, and an American variety of maize which carries both the Bt gene and a herbicide-resistance gene has caused much concern within the European Community because it has also an ampicillin-resistance marker gene. When regulatory bodies permit the release of such constructs they do so because the particular marker involved presents a trivial hazard. For instance, the argument goes that a human or animal pathogen is infinitely more likely to acquire ampicillin resistance through careless use of ampicillin itself than from any step in the intricate chain of post-harvesting events to which a genetically engineered crop would be subjected, such as digestion, cooking, food processing; even just leaving the crop around or composting it. Nevertheless, it is sloppy technology to leave marker genes, and other non-essential DNA, in place; customers may reasonably worry about what other genetic determinants might have been left behind.

Other sources of anxiety about genetically engineered crops tend to be social rather than scientific, though they are none the less important for that. For example, if genetically manipulated plants are rendered sterile, to avoid unwanted spread of their introduced gene or genes, the farmers cannot save a quota of seeds from their harvest for next year's sowing and must perforce buy new seed from the supplier – if they can afford it, a serious problem to small farmers. Again, crop plants engineered to carry resistance to industrial herbicides undoubtedly give improved yields of food, fibre or whatever the product may be. So in principle they can feed or otherwise sustain a larger population for a given unit of agricultural land. And they benefit the manufacturers of the herbicide financially, as well as those farmers who can afford to buy both seed and herbicide. However, developments such as these bind agriculture to the agrochemical industry more tightly than ever, which suits that industry very well. In fact this situation is nothing new: chemical fertilizers have benefited manufacturers, wealthier farmers and consumers, and tied farming to the agrochemical industry, for over a century, and today over one-third of the world's population is alive and fed by virtue of the Haber process, an industrial procedure whereby nitrogenous fertilizer is made from air. But the ultimate economic effect of all this high-tech agriculture is to concentrate wealth even further in the hands of a decreasing minority of farmers and industrial conglomerates. In developing countries this trend can lead to serious social problems. It is a pity that resentment of this fact, rather than perception of serious scientific risks, underlies much fringe activism against genetically engineered crops.

A comparable socio-economic shift, wealthy farmers becoming wealthier at the expense of others, took place when hybrid short-stemmed rice saved much of Asia from starvation during the 'green revolution' of the 1960s and 1970s. I suppose one can be somewhat reassured by evidence that, since most societies adjusted to the green revolution, they will adjust anew to the agro-genetic revolution. Certainly, given the still expanding world population, high-tech agriculture, be it genetical or agrochemical, cannot be uninvented.

Happily some of the more distasteful developments can be discouraged, even prevented. There is an animal gene which codes for bovine somatotropin, a natural hormone which influences milk production. It was cloned in *E. coli* in the 1980s, isolated, and duly licensed for marketing to the dairy industry to enhance and protract milk production in cows. It worked, but cows so treated proved to be prone to mastitis, and public distaste for the whole idea, coupled with a declining market for milk, has inhibited its widespread use, at least in Europe. A comparable hormone from a cloned gene which promotes fast growth of lean pigs is no longer used at all, because it has caused serious animal welfare problems among the unfortunate pigs.

These are important problems in which microbes play a part, and they have to be addressed. Unfortunately, the element of tendentiousness in the response of protest groups and activists to genetic engineering and the release of its products has led to some bizarre non-science being put about. There is no doubt that the new technology needs to be closely and intelligently monitored and regulated, in agriculture, in medicine, indeed wherever it can be applied. But a balance has to be struck. Continuing my theme of genetically manipulated crops, agriculture and forestry have historically engendered mankind's most devastating effects on the natural environment, and they continue to do so. So should we not welcome new technologies which promise to moderate their environmental impact by increasing productivity, controlling disease, decreasing agrochemical use or fertilizer inputs? Of course – provided they do so as safely as can be foreseen. It is very sad that, at the time of writing, market forces rather than the public good are setting priorities – crops that thrive in saline water would have been far more useful to the developing world than strains which sell more herbicide – and I do not doubt that new anxieties and conflicts will arise in the future, but that is how market forces work. The fact is that the official safety committees and regulatory bodies which exist in most parts of the world usually do a good job and act quite wisely.

However, I am straying too far from microbiology. So far genetic engineering has done a lot of good and no obvious harm to anyone. Ironically it is in more conventional biotechnology that urgent problems are appearing. BSE is a clear example; it involved no genetic engineering at all. At the time of writing a public inquiry is under way into the origin of the epidemic in the UK; as I have recounted, it is more than likely that the feeding of cooked animal matter to creatures that were naturally herbivorous initiated this unforeseen disaster. I should be foolish to try to second-guess the inquiry's findings, but it is already clear that political considerations, vested interests represented by the farming and animal feeds lobbies, fear of public panic, fear of hierarchical wrath, obsession with economy (the government's research on scrapie was being run down), and simple bureaucratic myopia, combined to cause administrators of goodwill and integrity to make stupid, even catastrophic, decisions – or perhaps just to cross their fingers and hope for the best.

BSE, with its attendant nvCJD, is the most high-profile of the microbiology-related problems that are now looming. Another is food-borne infections. If the media are to be believed, the public is anxious about its food, including its quality, its nutritional value, the nature and effects of additives, flavourings and preservatives used in processing, the possible presence of pesticides and other chemical contaminants, not to mention that dreaded possibility that alien DNA from genetically manipulated crops might be around (as if people had not been eating alien DNA for ever!). All these are legitimate concerns, yet only rarely has sickness or physical disorder been unequivocally associated with any of them – except perhaps when poverty or cult diets have lead to malnutrition. In contrast, as I have earlier pointed out, microbial diseases caused by food-borne pathogens have caused some nasty epidemics in the last couple of decades. They were a normal hazard of life in the nineteenth and early twentieth centuries, but became impressively rare as a result of the judicious use of preservatives, sterilization, pasteurization and hygienic handling. Now they are on the increase. Why? Forgive my boring you with repetition; recall that it is because new technological processes – intensive animal farming, centralized food processing, chilling, deep-freezing, gas and vacuum storage, have all combined to make convenience eating the modern way of life. But standards of hygiene, at every stage from farm to kitchen, have not kept up with the pace of innovation.

Why is that? I shall offer you a reason soon, but let me first return to medicine, where a comparable thing is happening.

In the eighteenth and nineteenth centuries hospitals were dangerous places, to be entered at your peril. Your immediate illness or injury might be treated sensibly, but you were as likely as not to catch something else from the staff or from fellow patients. As the twentieth century advanced, progress in understanding how microbial diseases spread became translated into hygienic medical practice, in-house infections declined impressively, and hospitals became the havens of health the older among us still expect. Those were the days of the Dragon Matron, a formidable lady in charge of the ward. She would inspect the nurses' finger nails, skin and clothing regularly, would ensure that floors were neither brushed nor dusted, but swept with disinfectant-damped brooms, and that other surfaces were wiped with similarly damped dusters. She was obsessive about the disposal of wound dressings and food residues, and about the cleanliness of bathing and lavatory arrangements. She ensured that patients for operations were checked for cleanliness, and shaved in appropriate places, which were then swabbed with acriflavine or alcohol. And woe betide anyone (even a Royal Princess, as happened in the mid-1990s) who entered the theatre with a coif of hair uncovered! Those unsung heroines of hospital life understood about skin-borne, cough-borne, dust-borne and airborne pathogens.

Then along came sulphonamides, penicillin and the rest of the antibiotics. Many traditional practices began to seem unnecessary, for it was a simple matter to give a patient a shot of a broad-spectrum antibiotic at the first sign of a secondary infection – or even without such a sign, simply as a prophylactic. Almost imperceptibly standards of hospital and clinical hygiene slipped. Here and there antibiotics were used carelessly, dosage courses were not completed, or inadequate amounts were given – perhaps through ignorance of the elements of microbial genetics, perhaps because dosage was left to feckless or absent-minded patients, perhaps simply because of human error – doctors and nurses are as human as the rest of us. So strains of bacteria began to emerge which were resistant to antibiotics or drugs. In Britain in 1997 infections acquired in-house played a part in some 20 thousand deaths in hospital, the infection itself being directly responsible for a quarter of them. Many were due to antibiotic-resistant pathogens which, in the last decade or so, have become a worldwide medical problem. Diseases such as tuberculosis, which we thought were conquered, have begun to reassert themselves in drug-resistant forms.

What have food poisoning and drug-resistant pathogens in common? Once again ordinary practice did not keep up with modern technology. Elementary principles of microbiological hygiene became unnecessary for a couple of decades and were allowed to lapse, sometimes forgotten. The reasons why have been various and only too familiar: overwhelming demand for the product or service, technologically ignorant management and operatives, exhausted or confused staff, preoccupation with profit or economy cuts, blinkered bureaucracy in all sectors – pick-and-mix among those as suits the case. The upshot has been that new generations have simply been inadequately trained.

I must not overstate the position. Some professionals are only too well aware of the situation. Hygiene and monitoring of victual-borne pathogens are perennial topics at gatherings of food microbiologists. And, reporting on hospital-acquired infections to a meeting of the British Medical Association in 1998, an occupational health consultant shocked his audience when he spoke of surgeons entering the canteen in theatre clothing, and of a nurse who, for fear of theft, wheeled her wet bicycle through the recovery room of a cardiac unit! The tip of an iceberg, I fear. The world of microbial pathogens demands constant vigilance. Nothing that I have written is not in official or parochial reports already in the UK's public domain, and doubtless it is all available to the governmental machinery of other countries, too. There is a crying need for a renewed input of ordinary microbiology into both food technology and medical practice.

My chosen examples have been in areas which bear upon the health and well-being of large numbers of people in obvious and immediate ways, and which can be dealt with by intelligent application of existing knowledge. But BSE and its attendant nvCJD were something wholly new and unexpected. No amount of training would have forearmed us against their emergence. As I made clear earlier, scrapie in sheep had done no consumer any harm for several generations, and it had not spread spontaneously among animals. Several decades' research on the disease had been slow-moving, and the nature of its mysterious agent seemed to be of little interest except to a few academic enthusiasts. Laboratories studying scrapie were being closed down and the fact that anyone could find finance to work on scrapie at all was largely a left-over from the halcyon days of the mid-twentieth century, when governments still accepted that curiosity-motivated research would sooner or later pay off.

Today research on transmissible spongiform encephalopathies is reasonably well supported. But it took the BSE epidemic and its consequences to force government to find the money. Yet who else should find it? The cattle feed manufacturers? The butchers? The brow-beaten farmers? How were they to know? No matter how

Treasuries, Chancellors and politicians may wriggle, there are large areas of civilized society which simply have to be supported out of public funds. We accept that the police, education at all levels, the legislature and defence are examples of such areas. Also among them is basic scientific research.

Notice that word 'basic'. I have chosen it carefully. In these hi-tech days few major industries can get along without doing, or sponsoring, scientific research, but their research, quite reasonably, needs to bear upon the interests of that industry and its shareholders: there has to be a reasonable prospect of a practical pay-off within a few years in the form of economies, or new processes or products. In times of affluence, companies, especially newly-formed companies looking to exploit the latest scientific advances (in silicon chips or biotechnology, for example) may embark upon relatively basic research projects, usually with a flourish of publicity, but come a recession or severe competition, then costs have to be cut – and basic research is the first to go. It has happened time and again; only giant multinationals, which have swallowed up smaller fry as markets tightened, can sustain relatively long-term research programmes of a basic character, and these are, of course, still constrained by their commercial interests. Well, that is fine: it is good that industry should do its own research in its own way. But market-directed constraints not only limit the choice of research topics, they also imply commercial secrecy. This means that a company's researchers may not discuss their progress with scientific colleagues from outside lest they say something which might benefit a competitor or influence the company's quotation on the stock market, and any research findings which might have broad scientific value are slow to reach the scientific community because their publication is delayed, sometimes indefinitely, for similar reasons. In the later 1990s a telling example of this situation became public – a few biotechnology companies had elucidated the DNA sequences of the complete genomes of several pathogenic bacteria. Those sequences could have been of immense academic value for advancing our understanding of bacterial genetics and pathogenicity, but they were also invaluable for the design of new pharmaceuticals. So reluctant have the industries been to release their data – they wished to 'protect their investment' – that in 1998 Britain's Wellcome Foundation, a charity, set aside £7 million to fund the sequencing of the same genomes in non-commercial laboratories for open publication. They will be posted on the Internet.

Much valuable science has emerged from research undertaken for thoroughly practical reasons, but it is a truism that almost all of the important, radical and useful scientific advances have originated from unconstrained research: from work by scientists who were simply curious to know the whys and wherefores of something or other, and who believed it natural to disseminate their findings among their peers as soon as possible. Yes, I know; truisms are often flawed, but this one is sound: for example, in 1998 a survey of a million recent patent applications, undertaken for the USA's National Science Foundation, discovered that three out of four cited basic, publicly-funded research as the basis for their innovations. That is the kind of research that the taxpayer has to pay for, be it in Universities or in Government Research Institutes. And if it proves to be exploitable in practical ways, splendid; but let it always be for the public good. In the now extinct Government laboratory where I started my career, concessions to commercial secrecy were anathema. We had a wise and honourable rule that anything we discovered, and any practical advice we could give, was freely available to anyone, but we would not undertake research on behalf of an industry or business unless the problem bore upon our own research programme – in which case any new findings would be freely available to all and sundry. It worked very well, and many a small firm, as well as a few large ones, were grateful for our help.

To me one of the saddest developments of the past two decades, at least in the UK, has been pressure by Government, through its Research Councils, to urge research scientists to form themselves into businesses, or engage in 'networking' with industrialists, especially in the emergent biotechnology and electronics industries. The intention has been to promote the rapid practical application of fundamental discoveries, and indeed it has done this, because the emerging industries have found it both useful and economic to collaborate with academic laboratories and research institutes: much cheaper than employing their own research teams. But the longer term effect has been to impede basic research by promoting secrecy and inhibiting communication between scientists, by deflecting good and experienced researchers from basic science into management, sometimes into naïve financial operations, and by generating an unwholesome atmosphere in which the objective of research is money-making rather than either the public good or the advancement of knowledge. I have known laboratories funded by the taxpayer in which graduate students were forbidden to talk about their research to other students in the same room, let alone cooperate or help each other, because their respective research programmes were sponsored by different industrial concerns – a dreadful way to train young scientists, let alone to advance knowledge.

Even in laboratories which have escaped direct entanglement with business there is a trend away from basic research into practicality. 'Accountability' is the watchword: what is the economic justification for the last three months' research? Time and again I have watched good and productive experimentalists in middle life pressured by accountants and administrators out of the laboratory and into their offices, to devote their time to writing progress reports and convoluted economic justifications for research grants and funding.

Well, there has always been need for accountability and for keeping an eye on the practical fall-out of research, even during science's golden years during the 1950s and 1960s. Scientists can be starry-eyed about publishing for the benefit of mankind as a whole, and so lose patenting rights to quick-witted and grasping industries. But the pendulum seems now to have swung to the opposite extreme. A cynic might argue that science is progressing too fast for society, that to slow it up through paper-work, secrecy and under-funding is no bad thing. Well, it's a point of view. But meanwhile people starve, get ill and die unnecessarily; and the environment gets worse – and mysticism and credulity replace knowledge and understanding.

Enough! This chapter is becoming too political. Everyday life is fraught with risks, perceived, unsuspected and imagined. What with Advisory Committees, Standing Committees, Royal Commissions and Public Inquiries, the new hazards revealed by modern molecular microbiology, including its applications and its fall-out in biotechnology, are being watched over and regulated more closely than technological advances have ever been.

Is the wider public reassured? Of course not. Things scientific are still arcane mysteries to the vast majority of society, a situation abetted by an all but universal ignorance of even the most elementary science among those in journalism and broadcasting. For example, readers will by now be well aware that bacteria and viruses are about as different from each other as living things can be, in their biological aspects and in the measures needed to cope with them. Yet during the fuss in 1988–9 about salmonellae in eggs the British media referred to the bacterium *Salmonella* as 'a virus' quite half the time, and even when they got it right, they were as likely as not to call it 'a bacteria'. And things do not improve: during the late 1998 assault on Iraq's chemical and nuclear weaponry, the media as often as not referred to the anthrax agent, a bacterium called *Bacillus anthracis*, as a 'virus' or a 'toxin'.

I like to think that that figment of the media's collective imagination of half a century ago, the benevolent, absent-minded boffin producing scientific miracles, stemmed from a serious interest in, and some understanding of, matters scientific. But today he has become the crazed Professor gleefully generating sinister hazards: a sort of hybrid of young Victor Frankenstein and Dr Strangelove. No matter how carefully, simply and reassuringly scientists or officials may explain scientific or technical issues, that sort of fear will not go away until science, at least at an elementary level, is a part of everyday culture. As far as microbiology is concerned, children ought at least once to make some yoghurt; look at the microflora of their mouth under a microscope; see what grows up after a fly walks on a nutrient jelly or a hair brushes over it; see a sewage plant working and so on. The world of microbes ought to be as familiar to us as the world of plants and animals. Only then will there be any point in inviting public discussion of sophisticated matters such as genetic engineering or AIDS. If microbes were familiar to everyone, then microbiology would escape the anti-science posture once more – except, of course, in those aspects where it deserved antipathy. Only then can we hope that new projects and discoveries, with any risks they might entail, will generate a measured response from the lay public and not startle it with alarms which it cannot possibly assess.

So, back to the world of microbes. In the early chapters of this book, I looked at microbes from the point of view of an individual, even if that individual seemed over-preoccupied with health and food. Then for a few chapters I wrote of microbes from the point of view of society and its economic structure. Now I shall consider the relevance of microbes to man as a biological species, looking at the part they played in our evolution and in the evolution of other living things. Then, finally, I shall be able to say something of microbes and man in the space age and even make predictions, not all absurd, of what the future may have in store for us.

10 Microbes in evolution

In this chapter I shall consider the place of microbes in the evolutionary sequence of living things and attempt to assess what importance they had in influencing the directions which biological evolution has taken. Since I shall be dealing with questions that cannot usually be verified experimentally – for I shall be mainly concerned with events that took place in the darkest recesses of prehistory, even before recognizable fossils were formed – I must recall to readers the warning I gave about scientific fact early in this book. Even in everyday matters, laboratory science contains elements of uncertainty, particularly in interpretation of experimental findings. When one is concerned with retrospective deduction from today's knowledge about the state of our planet during its juvenile millennia, interpretation is so uncertain a process that it amounts to informed speculation. The surprising thing, really, is that one can say anything at all about the biology of those distant eras. Yet, as the reader will see, if one accepts geologists' views about the broad outlines of this planet's geological history, one can put together a coherent and reasonably plausible account of how the earliest living things developed. Whether it bears any relation to the truth is another matter, but it is a form of speculation that widens our understanding of life and its potentialities, as well as exercising the imagination. So, for this chapter, I shall relax scientific puritanism and see what sort of theoretical picture can be built up about the infancy of terrestrial life and the way in which today's microbes arose.

The accepted age of this planet, by which I mean the period for which geologists and cosmologists believe it has existed as an independent celestial body, has doubled during my lifetime. It is deduced by such scientists from the distribution of naturally radioactive materials, whose half-lives are known very precisely, and is now taken to be in the region of 4,500,000,000 years: four and a half thousand million years. This is an unimaginable figure, and could well be out by several hundreds of millions, but it is unlikely to be out by more than 10 per cent. If one accepts that the Earth was hot at the time of its formation – not all scientists do, but I shall follow the majority – there followed an immensely long period of cooling, involving intense volcanic activity, during which time the land remained too hot for liquid water to exist. Such free H_2O as there was took the form of steam, and much of it would have escaped into space. However, comets, most of which consist mainly of ice, were much more abundant than now and were crashing into the newly-formed planet some thousand times more often than they do today. They would have replenished the Earth's supplies of H_2O, even though much of it would have boiled off on the way in. Various chemical fractionations took place in this period; these I shall disregard and I shall only begin to show interest when, around 4 thousand million years ago, the Earth had cooled sufficiently for liquid water to exist permanently on its surface. What the atmosphere then consisted of is still very uncertain. For much of the twentieth century cosmochemists took the view that it contained principally methane, hydrogen and ammonia, with small amounts of water vapour, hydrogen sulphide, dinitrogen and the rare gases (helium, neon, argon, krypton and xenon). However, doubts were raised in the last three decades, partly as a result of the findings of space probes sent to Mars, the moon and Venus, and it now seems more likely that dinitrogen and carbon dioxide were the dominant gases, with the others present in only small amounts. But the one point which seems to be fairly certain is that no free oxygen was present: the chemical compositions of rock formations that were exposed at the time make it clear that any oxygen which might have been released by chemical reactions in the then turbulent atmosphere was consumed in other reactions as fast as it was formed. For there would have been almost constant thunderstorms, with consequent lightning and electrical disturbances; moreover, there was no ozone layer, which today protects the surface of the Earth from much of the ultraviolet radiation emitted by the sun, so the type of radiation received from the sun would have been quite different from today's.

A most suggestive set of experiments, performed by Dr S. L. Miller in Professor H. C. Urey's laboratory in the early 1950s and abundantly confirmed in other laboratories, showed that a wet mixture of methane, hydrogen and ammonia, exposed to an electrical discharge for a while, formed traces of organic compounds, including organic acids and amino-acids hitherto regarded as exclusive products of living things. Subsequent experiments, modifying the gas mixture, adding traces of hydrogen cyanide, hydrogen sulphide, phosphates and so on, have shown that all sorts of organic chemicals turn up in these conditions, many of them, such as the purines, being particularly characteristic of living things. Moreover, ultraviolet radiation is also a potent agent for causing the formation of organic matter from gas mixtures of the kind then thought likely to have existed in the primitive atmosphere. Though Miller and Urey chose a now unfashionable gas mixture with which to mimic pristine terrestrial conditions, chemically its most important feature was the absence of oxygen. It remains likely that, before life originated here, organic matter was being formed by electrochemical and photochemical reactions of this kind. But such chemical process were not the sole sources of pristine organic matter; another was, surprisingly, comets. Interstellar space is not a perfect vacuum, and among the molecules out there are simple organic compounds, very sparsely distributed. Those comets that replenished the Earth's water are likely to have swept up some of these compounds in their travels and brought them in as well. It is therefore likely that, before life originated here, the seas of this planet were dilute soups of organic matter, partly imported, partly formed in local electrochemical and photochemical processes. The seas had become the sort of environment in which many present-day anaerobic bacteria would have flourished. But there were no such bacteria, nor any other living things, so those materials accumulated. The seas, in fact, must have been a turmoil of photochemical products, with all sorts of organic compounds forming, interacting and breaking down. However, I must emphasize that the concentration of these materials was probably very low; the pristine organic molecules would only rarely have interacted in the open seas. These, believed to be about a third as salty as they are today, contained vastly more inorganic matter, salts and so on, than amino-acids and other organic chemicals. The only places where the concentration of organic matter would be high would be at the edge of drying pools, or adsorbed on the surfaces of materials such as silicate rocks and clay, which have a particular affinity for organic matter, and it is at these sites that chemical interactions between pristine organic molecules, to form more complex substances, were most likely to have taken place.

In Chapter 4, I was very emphatic about spontaneous generation being an event of astronomical improbability today. Four billion years ago, with the chemical turmoil that I have described taking place, it may have been less improbable. Bernal, Oparin, Haldane, Pirie and others have discussed how organic matter, concentrated by adsorption at rock or clay surfaces, might take on complexities analogous to present-day biochemical molecules; many examples have been demonstrated in the laboratory in which simple molecules such as amino-acids or components of nucleic acids join up to form a more complicated chain at a mineral surface. It is likely that a sort of chemical evolution took place, especially at such surfaces, with molecules forming and breaking down in all sorts of ways, until one emerged with the capacity to facilitate the synthesis of others like it. This property might never have been an attribute of one molecule – a fortuitous conjunction of molecules might have led to reproduction of the whole set. Such a molecule or molecular complex would have one of the basic properties of living things, self-reproduction. No doubt many such systems were formed and fizzled out before one became established, but once one *did* get established, it would tend to prevent the emergence of others by using up the available organic matter to form more of its own type.

These mechanistic views of the origin of life, which I have sketched very superficially, are popular today among those scientists who consider these questions, though they differ about the details. Some place emphasis on local volcanic heating as against radiation as the source of the chemical turmoil that allowed the evolution of pseudo-living molecules. Others insist on the importance of forming co-acervates, droplets of organic matter which form under special conditions in water and which divide in two when they exceed a certain size, rather like a living microbe. Yet others prefer to postulate the intervention of a divine agency. And the view that terrestrial living things were seeded by dormant living matter from elsewhere in space is not excluded, though it is at present unfashionable; it displaces, but does not answer, the question of how life originated. For present purposes, in this chapter, I shall accept that living creatures appeared somehow, in a slightly salty, watery environment containing all sorts of dissolved organic materials, some of which would become concentrated at the surfaces of solids such as rocks, clay or sand particles. The atmosphere contained no oxygen and the primitive organisms, though one has no idea what they looked like, behaved like microbes, as far as their chemistry was concerned, in that they formed more of themselves from available organic matter. In particular they performed the first step of an evolutionary process: they used up the available material and thus made the emergence of competing organisms less probable.

What kinds of microbes did they most closely resemble? Obviously the anaerobic bacteria have most properties in common with these primitive creatures. They can grow in the absence of air by breaking down organic materials and obtaining energy for growth from these reactions. Present-day anaerobes have quite complex structures – for microbes – and it is most unlikely that they include any representatives of the earliest living things, but several species of present-day anaerobes would have managed quite well in what one imagines the pre-biotic environment to have been like.

One important difference would be that these primitive microbes probably had very limited synthetic abilities – they themselves probably consisted of a relatively small number of complex molecules, and they made themselves from precursors that were almost as complex, available in the water around them. They needed to conduct very few chemical reactions to duplicate themselves. Yet they did not resemble viruses, as some people have been inclined to believe, because viruses need a complete living system to grow on: they programme another organism's enzymes to make more virus instead of normal products. Viruses depend on the existence of quite a complex biochemical system.

The primitive microbes would have had no such systems to work on, but they had a splendid reserve of food, at least to start with. Moreover, they were subject to constant ultraviolet irradiation, and ultraviolet light causes chemical transformations in a variety of molecules, even those composing our primitive microbes. Many such transformations probably killed the microbes, put a stop to their ability to multiply. But it is probable that, over millions upon millions of years, a few altered their chemistry appreciably without impairing their ability to reproduce themselves. Thus a new organism, inherently different from its predecessor, created itself, able to reproduce itself in the new form. This process is a crude example of what we know as a mutation: a change in the inheritable structure of a living thing, brought about by some accident, which leads to the formation of progeny having a different character from the parent. As every reader knows, mutation leads to evolution, because if a mutation confers an advantage on a mutant, that strain tends in the long run to outgrow and replace its predecessor. In this way, by the process called natural selection, the slow transformation of living species we call evolution has taken place over the millennia on this planet.

I wrote about mutation in Chapter 6. For present purposes,

however, I remind you that a very effective inducer of mutation in today's microbes is ultraviolet light. Therefore, if the primitive microbes bore any relationship to present-day organisms, their mutation rate was probably very high.

Since the emergence of living things in the primordial waters would lead to the removal of complex organic molecules, and their incorporation into primitive organisms, it follows that any mutation that enabled the organism to make do with less complex organic matter would give that mutant an enormous advantage over its neighbours. Thus there would be a strong selective pressure in favour of development of increased synthetic ability: biological evolution would take place in the direction of simpler and simpler nutritional requirements until, ultimately, the first autotrophs appeared. (To save flipping to the glossary, I remind readers that autotrophs are microbes that can use wholly inorganic materials for growth: sulphur, CO_2 and oxygen, for example, or sunlight, CO_2 and water.)

The idea that autotrophs developed after heterotrophs arises naturally from this picture of evolution, though this point has not always been accepted. Sixty years ago most microbiologists were inclined to regard the autotrophs as the most primitive of living things, simply because the majority of present-day microbes are heterotrophs and it is possible to plot plausible evolutionary pathways among existing microbes in the direction of heterotrophy. In this latter point they were quite correct, as I shall tell later, but it is difficult to regard autotrophs as representatives of the *most* primitive living things, simply because of the enormous number of separate enzymes they need to possess to be able to make up their bodies from inorganic matter. There is no doubt that autotrophs appeared early in evolution, but it is most logical to consider that they developed from even more primitive creatures that were heterotrophs.

What kinds of autotrophs exist today which might resemble the earliest kinds of autotroph? They would need to be anaerobes, so the coloured sulphur bacteria, which oxidize sulphides to sulphur and sulphate with the aid of sunlight, are candidates; there are certain cyanobacteria that can grow anaerobically, if sulphide is present, and also reduce CO_2 with sunlight; there are bacteria that can reduce carbonates to methane or acetic acid using hydrogen; one strain of bacterium has been reported that can reduce nitrates while oxidizing ferrous iron to ferric; another can oxidize sulphur with nitrates; others can oxidize hydrogen with sulphur or nitrates. Yet the choice among anaerobic autotrophs is not very wide: some of those I have mentioned are not very likely candiates, because, though there was probably plenty of sulphur and sulphide around, there was not, in the chemical conditions then obtaining, likely to be much nitrate.

One answer to this question arises from the discovery in 1960 of a nutritional group of microbes that seems halfway between autotrophs and heterotrophs. *Desulfovibrio*, a group of sulphate-reducing bacteria, can oxidize hydrogen with sulphate forming water and sulphide:

$$CaSO_4 + 4H_2 \rightarrow CaS + 4H_2O$$

and use the energy of this reaction to assimilate organic materials. There exists also a species of the sulphur bacteria *Thiobacillus*, called *T. intermedius*, which can couple the oxidation of sulphur to the assimilation of organic matter. There is also evidence that some hydrogen-oxidizing bacteria couple hydrogen oxidation to assimilations. Some of the photosynthetic sulphur bacteria certainly assimilate acetate as a result of photosynthesis and, as I told a moment ago, photosynthetic sulphur bacteria are among the best candidates for primitive status among present-day microbes.

Thus it seems likely that true autotrophy, though it probably developed at an early stage in evolution, was preceded by what has been called mixotrophy: the coupling of an inorganic reaction that yields energy to the assimilation of simple organic matter into the cell, and its use to form cell material. It is but a short evolutionary step from such assimilations to true autotrophy, the assimilation of CO_2, and among microbes there is an overlap between the two types of nutrition. Within the same groups of bacteria, as in the thiobacilli, one can today find types that conduct either or both processes. But the emergence of autotrophy was a vital step in evolution, because it provided the first reliable alternative to a primaeval photochemical turmoil for the accumulation of organic matter on this planet. Though the partial autotrophs I have described might utilize the components of the primaeval soup much more efficiently than their primitive predecessors, they still depended on it absolutely for their existence. They could grow no faster, nor more abundantly, than photochemical formation of organic matter permitted. The true autotrophs were the first creatures to become independent of spontaneous organic synthesis, and it is likely that the most effective ones were those that used solar radiation to do this: that is, the anaerobic precursors of green plants.

Let me pause at this stage and see if I can rustle up some facts bearing on this picture. I imagine a world with permanent seas which, though they were once rather weak, had probably become quite saline as storms and rains washed soluble salts out of the rocks, hills and mountains. No oxygen was present in its atmosphere, but there was plenty of sunshine with a strong component of ultraviolet light. (A photochemical reaction between UV-light and water vapour did generate a little oxygen, but it was rapidly mopped up by other chemical reactions.) A fair amount of free hydrogen sulphide was present in the seas, partly formed by microbial action, partly dissolved from volcanic emissions, and a population of primitive microbes existed, conducting the sulphur cycle (reducing sulphates, sulphur, and other inorganic sulphur compounds to sulphides; oxidizing these sulphides, via sulphur to sulphate) and assimilating photochemically-produced organic matter together with any organic matter produced by the emergent autotrophs. Though iron- and hydrogen-oxidizing microbes may have been present, they would have been strange by today's standards because their present-day representatives generally need oxygen or nitrates and these were probably rare. Thiobacilli were probably rare for a similar reason, but methane-producing microbes, though they are rarely autotrophic, were probably abundant, reducing CO_2 to methane while oxidizing any available organic matter. Likewise, organisms capable of reducing CO_2 to acetate could well have been plentiful and would have provided, in acetate, one of the best-known substrates for the mixotrophic assimilations I described a couple of paragraphs ago.

Facts, you remind me? Well, there is one set of experiments that has considerable bearing on this question. I remind you that the sulphate-reducing bacteria fractionate the isotopes of sulphur during sulphate reduction, a point that is of considerable importance in establishing the microbial origin of sulphur deposits. A comparable fractionating of isotopes takes place when carbon dioxide is taken up by autotrophs – bacteria or plants. Carbon consists almost entirely of atoms which are twelve times as heavy as hydrogen atoms, but it always includes a tiny proportion of atoms which are a mite heavier: thirteen times a hydrogen atom. (There is also a minute amount of a third isotope which I shall not bother with just now.) Biological uptake of carbon dioxide slightly favours the lighter isotope and, as with sulphur, the slight separation of isotopes that ensues can be detected and measured. Sedimentary rocks usually contain carbon which is slightly enriched with the light isotope, indicating that the carbon has undergone biological transformation, and in the mid-1990s fractionation of this kind was detected in rocks in Greenland which were around 3.8 billion years old. This implies that autotrophs of some kind were about even then. In a similar way, geochemists have examined the distribution of sulphur atoms in minerals of known geological ages and obtained clear evidence of microbial action by sulphur bacteria as far back as 800 million years ago, and fractionation in some samples about 2 billion years old.

A second line of evidence comes from the microscopic examination of pre-Cambrian rocks. In a formation called Gunflint Chert, which lies north of Lake Superior in Canada, various scientists reported, in the 1950s and 1960s, microscopic bodies which look very like traces of cyanobacteria; more recently such traces have been seen in Australian sedimentary rocks as old as 3.5 billion years.

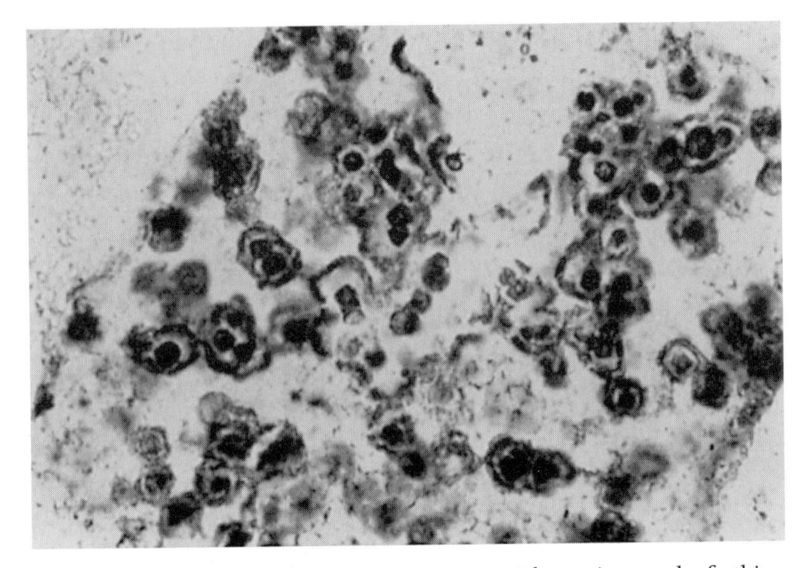

FOSSIL MICROBES IN ANCIENT ROCK. Photomicrograph of a thin section of more than 2 billion-year-old rock from the Gunflint Chert in Ontario, Canada. The spheres and filaments resemble present-day cyanobacteria. Magnification about 150-fold. (Sinclair Stammers/Science Photo Laboratory)

(Perhaps I should add here that the Gunflint Chert formations also look like the thermophilic flexibacteria I introduced briefly in Chapter 2. There is a certain logic in the view that the most primitive microbes were thermophilic, because the first liquid waters of this planet would have been hot. But for the present purposes I must leave that thought as just one of many speculations that are possible about the nature of primitive life.)

Microscopic traces in rocks of things that look like bacteria are perhaps rather dangerous pieces of evidence on which to base an evolutionary scheme; similar structures have been found in certain meteorites and have proved to be of non-biological origin. But isotope fractionation seems reasonably reliable and provides a couple of cogent facts. One can assert that, as early as 3.8 billion years ago, some kind of autotrophic carbon dioxide assimilation was taking place, and that between 2 billion and 800 million years ago abundant microbial sulphur metabolism was taking place. (Let me note in passing, to get the time-scale into perspective, that the first definite fossils of multicellular creatures appear in rocks of about 500 million years old.) During all this time this planet's atmosphere was anaerobic; the only living things would have been microbes, and dominant among them were the distant ancestors of the present-day sulphur bacteria. The main biochemical process on earth was the sulphur cycle. Or so it seems.

What happened between 800 million years ago, when there was virtually no oxygen, and 500 million years ago, when there was some, if not as much as today?

The coloured sulphur bacteria today contain chlorophyll, the green pigment of plants that is essential for photosynthesis. One can represent the chemistry of their photosynthesis very crudely this way:

$$2H_2S + CO_2 \xrightarrow[\text{chlorophyll}]{\text{sunlight}} 2S + [CH_2O] + H_2O$$

where $[CH_2O]$ represents carbohydrate. (For non-chemists this means they make carbohydrate from carbon dioxide, with the aid of sunlight, while splitting hydrogen sulphide to sulphur and water.)

Today there exists a group of coloured non-sulphur bacteria which conduct a photosynthesis using organic matter in place of H_2S. They are still anaerobes (at least when they photosynthesize) and they can be regarded as removing hydrogen from organic matter and using it to reduce CO_2. If I write H_2A as a formula for an organic molecule from which the bacteria can remove hydrogen, then their photosynthesis can be represented so:

$$2H_2A + CO_2 \xrightarrow[\text{chlorophyll}]{\text{sunlight}} 2A + [CH_2O] + H_2O$$

very like the mechanism in the sulphur bacteria.

By the wisdom of hindsight one can see that it was only a matter of time before sufficient mutations took place to enable organisms to do the whole exercise without H_2S or H_2A, using H_2O instead:

$$H_2O + CO_2 \xrightarrow[\text{chlorophyll}]{\text{sunlight}} [CH_2O] + O_2$$

splitting water to release oxygen.

This reaction, as it became widespread, would have had a profound effect on the whole planetary ecology, because H_2S and O_2 (hydrogen sulphide and oxygen) react with each other. They do so only slowly, but they cannot co-exist for long: they form sulphur and water. Thus, the emergence of microbes able to make oxygen from water would have a catastrophic effect on the sulphur cycle. It would remove H_2S, deplete the sulphide by oxidation and tend to put a stop to the whole process. And if the process stopped, it would mean that the organisms responsible for it would cease to flourish: most of them would die and the survivors would persist only in limited environments, in the sulfureta, for example, where special local conditions kept oxygen away.

In this way one can see a logical process leading to the emergence of photosynthetic autotrophs able to generate oxygen from water while converting CO_2 to organic matter. Slowly, because of the chemical reactivity of oxygen, gases such as ammonia and hydrogen sulphide would be removed from the atmosphere. Hydrogen would escape continuously into space – it is too light to be retained for long by a planet having the mass of the Earth. So the atmosphere would tend to consist of oxygen, dinitrogen and CO_2, possibly with residual methane as well, but most of the residual ammonia would be dissolved in the seas. The primitive anaerobic microbes would be finding conditions highly unsatisfactory in general: the environment would favour creatures able to develop some way of making biological use of the oxygen. The oxygen in the atmosphere would form ozone at the outer fringes of the atmosphere, as it does today, and this would screen out much of the ultraviolet light responsible for the early photochemical turmoil. Thus spontaneous generation would become an even less probable event and the average mutation rate of organisms would decrease. But this situation would favour the living things that were already established, for mutations would still occur, though less often. Heredity, like the environment, would become a more stable quality and species of a given type would persist for longer periods unchanged. One knows that oxygen-breathing creatures did develop; can one say anything about how?

Among the enzymes that present-day air-breathing organisms possess is a group called cytochromes. These are chemically related to the red haemoglobin of blood: in addition to the usual amino-acids they contain iron atoms bound in a special chemical grouping called a porphyrin. The porphyrin group, as classicists will guess, gives the molecules a red or purple colour. Cytochromes are concerned in the final reactions with oxygen that take place during respiration: they undergo reversible oxidations and reductions (the iron atom switching back and forth from the ferrous to the ferric state) and, by a fascinating process which must not distract us here, all air-breathing organisms, from humans to microbes, obtain much energy for their biological processes from these changes.

Fermentative anaerobes do not possess cytochromes. They have brown iron–containing enzymes called ferredoxins which undergo reversible oxidations and reductions, but as far as is known, these have no energy-providing function. To make efficient use of oxygen, it seems probable that the evolving microbes would need to develop the iron–porphyrin system and integrate it with an energy-generating process. We know, of course, that they did; but how? One suggestive point is the fact that among the anaerobes at least two groups, the methane-producing and the sulphate-reducing bacteria, contain cytochromes. There are other anaerobic bacteria that possess cytochromes – the photosynthetic sulphur bacteria – but their particular cytochromes seem to be concerned in photosynthesis and not in respiration. (Photosynthesis in green plants also involves cytochromes.) Thus the methane-forming and sulphate-reducing bacteria today contain representatives of the cytochromes universally encountered in aerobic organisms. Recollect that methane-producing bacteria actually reduce carbonate just as the sulphate-reducers reduce sulphate; if the primitive ancestors of either also contained such enzymes, then it was probably a fairly simple evolutionary step for organisms to develop the capability of reducing oxygen from the capability of reducing carbonates or sulphates. Simple, that is to say, compared with the evolution of the complex synthetic abilities involved in autotrophy.

Once an organism arose able to reduce oxygen to water, assuming evolution did proceed in this way, it would find a new world awaiting it. All those areas of the planet which the presence of free oxygen now rendered unsuitable for the anaerobes would be available to it and its progeny. It is likely, in fact, that air-breathing organisms evolved from several groups of primitive anaerobes and another promising ancestor might well have been found among the photosynthetic bacteria which, as I just mentioned, also contain cytochromes. I wrote in Chapter 2 that there is a link between some of the photosynthetic bacteria and the cyanobacteria. These organisms have a number of characteristics in common, and there exist borderline species that seem to span the bridge between photosynthetic bacteria that are anaerobes and those cyanobacteria that are aerobes. Some cyanobacteria are both: they can grow with air or metabolize sulphides. At the other extreme there are organisms on the borders of cyanobacteria and ordinary green algae, so one can see that, if the types one recognizes today are representative of creatures which evolved at the time when the atmosphere of this planet changed from a reducing to an oxidizing type, then there is a clear-cut evolutionary sequence through the photosynthetic bacteria to the cyanobacteria and then to the green algae. These events would have set life on the pathway to the whole plant Kingdom of today.

At this point I must interpolate a warning. I have written about ancient cyanobacteria, methane-producers, sulphur bacteria and so on as if they were just like those that exist today. I am quite confident that they would not have been. The planetary environment has changed and evolved dramatically since the pre-Cambrian era, and microbes, being creatures which not only evolve very rapidly but exchange genes readily, must have evolved and changed with it. For example, cyanobacteria split water by photosynthesis, breathe oxygen and fix nitrogen today. However, it is highly unlikely that they needed to do any of these things for the first couple of billion years of their existence: there was hydrogen sulphide to split, there was virtually no oxygen and there was adequate fixed nitrogen as ammonia. The properties of today's representatives of the groups of bacteria I have mentioned are no more than a guide to the sorts of microbial processes that one can imagine took place in those days.

Once oxygen-evolving, oxygen-consuming autotrophs were well established on this planet, especially when more complex algae and lower plants arrived, the biosphere as 'perceived' by microbes changed yet again. Plenty of CO_2 was now being fixed, so plenty of organic matter was becoming available to microbes when autotrophic organisms died. Selection pressure in favour of autotrophy was no longer over-riding; for a substantial portion of microbes it became advantageous to lose that property. And they did so. Most of the bacteria handled in laboratories today are not autotrophic and, in some groups, one can find examples that have increasingly complex nutritional needs. In my discussion of culture media I told how some bacteria will grow with a few simple chemicals whereas others require the most complex of brews and some, indeed, have not been cultured away from living tissue. One of the early contributions to bacteriology of the distinguished French scientist, Professor A. Lwoff, was the recognition that the trend of evolution among microbes, once autotrophs were well established on Earth, has been in the direction of *loss* of self-sufficiency. Microbes, particularly the pathogenic ones, have become more and more dependent on organic materials accumulated by plants, animals and more versatile microbes for their existence. Higher organisms or their detritus replaced the primaeval soup as the habitat of most microbes; autotrophy, or even highly developed synthetic abilities, conferred no evolutionary advantage, so that microbial evolution tended to go in the direction of loss of biochemical versatility.

It is easy to envisage loss of function in microbial evolution leading from nutritional dependence on the residua of autotrophs to more radical kinds of dependence. There exist in the soil the tiny bacterium-like creatures called *Bdellovibrio* which grow on organic matter and which, given the opportunity, infect true bacteria and parasitize them. There exists the tiny organisms, mollicutes, which are almost certainly like bacteria (though, because they contain sterols, a relationship to protozoa or fungi is also possible) but which lack their structural rigidity, and there exist large viruses which contain quite complex protein structures but no metabolic enzymes. These creatures form a sequence of increasingly refined parasites until one reaches the small viruses, which seem to consist of only two or three huge molecules capable of perverting the genetic apparatus and therefore the metabolism of more complex organisms to synthesize themselves, but unable to do anything whatever with non-living substrates. Though, *a priori*, one's instinct is to think of creatures so chemically simple as the small viruses as extremely primitive, it is in fact just as probable that they are elaborately degenerate, indeed fragmented, descendants of organisms that were at least as complex as bacteria.

Viruses are, in fact, almost perfect parasites: they do nothing for themselves until a host appears, whereupon they cause the host to form more virus. Even more refined parasitism is shown by the temperate bacteriophages, viruses which are parasitic on bacteria but which do them no apparent harm unless some stress affects the host.

Thus, despite the lack of concrete data, one can produce an analogical account of how the most primitive blobs of life might have evolved into organisms resembling present-day microbes. Circumstantial evidence makes one attribute crucial importance to the sulphur bacteria, particularly the sulphate-reducing bacteria, but perhaps this is only so because the one hard fact available which applies retrospectively, the fractionation of sulphur isotopes, involves such bacteria. But even if the apparent preponderance of sulphur bacteria in the early history of the biosphere is fortuitous, it is still true that the atmosphere of this planet was transformed, about

500 million years ago, through the activities of microbes, and the stage was set for the development of the air-breathing creatures we know (and are) today. Air-breathers inherited the Earth, but not without resistance. Even today, the catastrophic instances of natural pollution that occur, for example, in Walvis Bay can be seen as a sort of mindless take-over bid by the sulphur bacteria. Happily these outbursts are transient, and the surprising thing is that the bacteria responsible survive so successfully in what is, for them, a hostile environment. The persistence of sulphate-reducing bacteria, for example, throughout geological aeons of time undoubtedly depended on the fact that they grow best in an environment that is lethal to most present-day creatures. Successful evolutionary types not only develop characters which suit them to their environment, they also modify the environment to suit themselves.

The reader may care to reflect that this is as true of man as of microbes. Does man count as a successful species?

In Chapter 6 I discussed the self-transmissible plasmids and the Hfr genes which are chromosomal: both encode genetic information causing bacteria to conjugate and to pass genetic material from one to another. I pointed out that such gene transfer is a very primitive kind of sexuality. Indeed, it might be an evolutionary precursor of the sexual reproduction of higher organisms. The hereditary factors responsible for maleness in the Hfr strains have, in fact, many properties in common with the temperate bacteriophages just mentioned. The Hfr strains of bacteria possess DNA which is incorporated with the rest of the genetic material in the chromosome but which can excise itself and become transferred to a new host. Some temperate bacteriophages do the same thing. These facts lead to an interesting speculation on the evolution of sexuality: if bacterial sexuality originated as a mechanism for the transfer of a bacterial virus, has sexuality in higher creatures a similar origin? Is it a degenerate mechanism for transferring what was once, in an evolutionary sense, a parasite?

I also discussed the three major methods of gene transfer now known in microbes: conjugation, transduction and transformation. Transduction, as I told, involves infection of the recipient cell with a bacteriophage from the donor, and some of the hereditary characteristics of the previous host may then be carried into the new one. Generally speaking, the information co-transferred with the 'phage is carried on a rather small piece of DNA. Conjugation and transformation may, however, lead to the transfer of large packets of DNA, even in nature. A plasmid could easily transfer a hundred or so genes from one host to another and, if all that information is expressed, the recipient is virtually a new species of microbe.

These facts lead me to an important principle concerning the evolution of microbes, one which has been gradually revealed as molecular biology developed and which bears on all the evolutionary considerations I have presented so far. Plasmids, once thought to be rare, are now known to be very common among all sorts of microbes. They do not have to be self-transmissible to become transferred to new hosts: one can extract plasmid DNA in the laboratory and transform bacteria with it. Bacteriophages nip in and out of the bacterial chromosome, sometimes forming plasmid-like bodies in the cytoplasm, co-transferring genes and DNA as well as excising DNA from their first host. Some genes found on plasmids, such as the genes which specify resistance to the antibiotics penicillin or kanamycin, can transfer themselves from plasmid to plasmid. Such mobile genes are called transposons; they can also get into the chromosome and, when they do, they distort and usually obliterate the chromosomal genes in whose neighbourhood they have become inserted, a property which is very useful in modern molecular genetics. Furthermore, the two enzymes used by molecular geneticists, restriction enzymes and DNA-ligase, come from perfectly ordinary bacteria which use them for much the same purposes as do the scientists: to chop up and reassemble DNA. The essential message is that the sort of genetic shuffling that microbiologists and microbial geneticists do in laboratories today – making and isolating mutants, constructing plasmids, bacteriophages and new species, transforming cells and fusing genes – has been going on in nature for thousands of millennia. One thinks of humans, giraffes, trees, frogs and so on as having a fairly stable genetic background, and the whole of macroscopic biology confirms that this is so. But it is not true of the microbes – or, to be more precise, it is not true of the prokaryotes, to which the majority of microbes belong. The genetic fluidity of bacteria as it is now being revealed is amazing; their capacity for evolution and adaptation by exchange of genetic information in large and small packages would hardly have been credited four decades ago. It follows that, during their evolution, virtually all possible forms of microbe could have emerged, disappeared and re-emerged, perhaps many times. That is why, in my discussion of evolution, I have dealt with properties, not species. One can say something about the age and evolutionary relevance of sulphate reduction, photosynthesis or methane formation, which are microbiological processes, and still recognize that the bacteria which conduct these processes today may not be, indeed are unlikely to be, in any direct sense descended from those which did these things a couple of billion years ago.

A final point before I leave this topic. Biotechnologists who exploit molecular genetics today have successfully cloned animal, including human, and plant genes into bacteria. Is this the first time in the history of the biosphere that this sort of thing has happened? A moment's reflection and I am sure the reader will agree with me that the answer is a resounding 'no'! A dead animal or plant, being decomposed by bacteria which are themselves living, dying, becoming infected with bacteriophages, swapping plasmids, raw DNA and so on, is a marvellous environment for the restriction, ligation and transformation which are the everyday tools of the genetic engineer. Just as virus diseases result from the transfer of prokaryotic genes into our own genetic material so, I am sure, we and other eukaryotes have been passing out our hereditary material to microbes since we came into existence. Has it been of much use to them? I think not, but who really knows? Equally, how much of our present genetic complement has been collected relatively recently from microbes? There are still some fascinating evolutionary questions for modern molecular geneticists to answer, and the questions themselves open vistas for the future modification of genetic material, even that of man, by deliberate manipulation.

Evolution by accretion of genetic information, so that evolution proceeds in discontinuous jumps, is now a totally familiar concept to microbiologists (though not, I find, to macrobiologists, who have yet to come to terms with it). It is but a small step from the acquirement of a package of alien genes to the acquirement of a whole organism. An association of a microbe, particularly a symbiotic one, with a higher organism could lead to an interdependence so strict that the pair become essentially a single organism. Some authorities believe that the photosynthetic apparatus of certain protozoa, their chloroplast, is a vestige of what was once a symbiotic cyanobacterium. For example, the protozoon *Euglena*, which I introduced in Chapter 2, is a single-celled animalcule which lies half-way between animals and plants. It possesses a chloroplast and can photosynthesise its needs like a plant, but it can also assimilate pre-formed food like an animal. If it is cultured in the dark it tends to lose its chloroplast, and after several generations its progeny lose chloroplasts completely and become wholly animal-like. They do not regain them when returned to light, they have come to resemble another species called *Astasia*. In the 1940s the conversion of *Euglena* into a pseudo-*Astasia* was regarded as a model for the evolution of animals: motile algae, by virtue of their ability to move around, might have found it more efficient to seek pre-formed organic matter and to assimilate it, rather than to make it for themselves, thus becoming dependent on those types that had remained autotrophic. Some would lose their chloroplasts altogether and become protozoa; diversification would ensue, multiple aggregates of such cells – metazoa – would emerge and the evolution of animals would be under way. A nice story; but the converse may contain more truth: a protozoon-like creature may have ingested, but not digested, a cyanobacterium, acquiring it as an internal symbiont which multiplied along with its host. That host found the situation advantageous, gave up foraging for food and became wholly dependent on its symbiont: thus it became the ancestral plant. There is certainly good evidence that the chloroplasts of higher plants originated from symbiotic cyanobacteria within a primitive cell, as evidenced by the fact that chloroplasts have DNA separate from that of the rest of the plant's hereditary apparatus and that their own is read in a distinctive way, rather as in bacteria. Another organelle of eukaryotic cells, called the mitochondrion, also has its own DNA, the base sequence of which suggests that it evolved from a symbiotic bacterium related to today's *Rickettsia*. As a model for such a process we have met the protozoon *Crithidia oncopelti*, which contains symbiotic bacteria in its protoplasm that aid its nutrition. Microbes, it seems, not only set the stage for the emergence of plants and animals in the terrestrial biosphere but contributed substantially to the internal anatomy of their cells.

Those are the considerations which have made it so difficult to set up a phylogenetic classification of bacteria. But they raise another question. If most of the genetic properties of prokaryotes are interchangeable, are there any at all that have remained stable? Are there any genes of which one can say 'this is the stable genetic background of this kind of microbe'? Well, there seem to be. All living things must synthesize themselves, which means that they must make protein. Their machinery for making protein, the ribosomes which I mentioned briefly in Chapter 6, are specified by genes which change very little indeed from organism to organism, for the simple reason that all living things make protein in much the same way. Ribosomes contain a special kind of ribonucleic acid called r-RNA, and analyses of the chemical composition of r-RNA from ribosomes of over 50 species of bacteria have provided a sort of catalogue which seems to have evolutionary significance: one can array the species according to the relatedness of their r-RNA, and prepare a family tree which broadly matches what bacteriologists had earlier felt, usually on rather weak scientific bases, were evolutionary relationships among bacteria. A simple version of the r-RNA family tree, looked at from the microbial end, is sketched in the illustration opposite. But the tree has not merely confirmed, or overthrown as the case may be, a few earlier views and prejudices, it has also added impressive new insights; here are three examples.

First, the 'Gram stain' reaction which bacteriologists used, without clearly understanding why, for about a century to divide bacteria into two large groups, proves to be a valid test: the r-RNA compositions of Gram-positive bacteria form a branched cluster in the catalogue separate from the Gram-negatives.

Second, the idea which I mentioned a few paragraphs ago that the chloroplasts of higher plants evolved from endosymbiotic cyanobacteria, which colonized the evolutionary precursors of plant cells, has been gratifyingly confirmed: the r-RNA from plant chloroplasts is indeed closely related to cyanobacterial r-RNA. (My sketch of the tree does not take into account accretions of this kind, of course.)

Third, perhaps the most spectacular new insight has been the discovery of a third great class of living things: the Archaeobacteria. I introduced them in Chapter 2 and they have cropped up now and again since then. As r-RNA specimens from more and more bacteria were examined, it became clear that they fell into two groups which seemed hardly to be related at all. On the one hand were the 'regular' bacteria, those most usually encountered in medicine, soil and water; on the other hand were a cluster of rather exotic bacteria, including very strict anaerobes capable of forming methane (the methanogens, which have featured often in these pages), certain sulphur bacteria able to grow in very hot and acid environments such as hot springs (called thermoacidophiles), and bacteria which inhabit strongly saline environments such as salt pans (the halobacteria). In fact, these exotica had already revealed biochemistries very different from ordinary bacteria – possessing unique enzymes or cell walls, even, among the halobacteria, a unique mode of photosynthesis – so

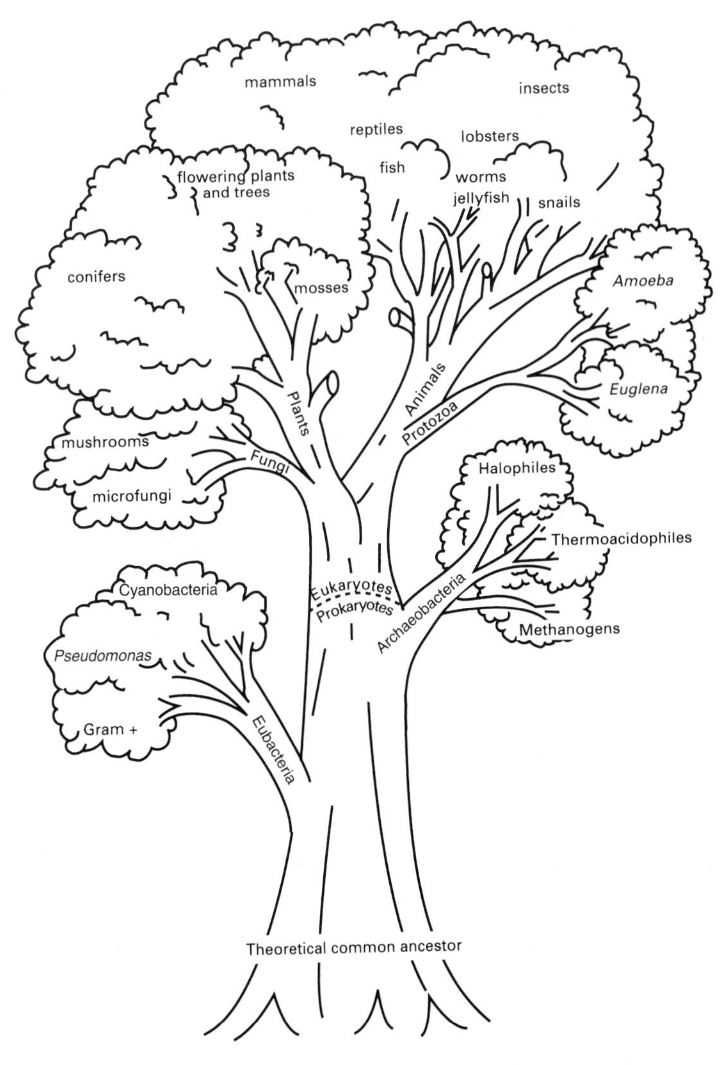

A MICROBE'S-EYE VIEW OF ITS FAMILY TREE.

the discovery that they have very distinctive r-RNAs made sense. The microbes in the new group proved to be more like each other than they are like regular bacteria (eubacteria) or like eukaryotes (protozoa, plants, animals, etc.). They were given the name Archaeobacteria for reasons I shall shortly discuss. The r-RNA catalogue has thus revealed that all living things fall into three very distinct groups, which are of higher rank than the five Kingdoms I mentioned in Chapter 2. To distinguish these new groups from Kingdoms, which of course they encompass, their discoverers called them Domains, and their formal names are now the Archaea (earlier Archaeobacteria), Eucarya (earlier Eukaryotes) and Bacteria (earlier Eubacteria). One can look upon the domains as three grand branches of a classical evolutionary tree with, at its root, an even more primitive ancestor which is no longer represented at all among living things.

The group of microbes whose discovery brought about this new insight was called Archaeobacteria because microbial evolutionists thought at the time that they were ancestral to the eubacteria and eukaryotes. But as the r-RNA catalogue has been extended, improved, and linked up with other biochemical features, it has come to seem likely that it is the Domain Bacteria which represents the most archaic of living things. This is a rapidly developing area of study which constantly raises new questions. Was the common ancestor of all three Domains a cellular creature? (Probably not.) Should the Domain Archaea be divided further, because there is considerable diversity within it, or simply sub-divided? (The current consensus is to stick to a single Domain.) Some of the organelles within the cells of higher Eucarya are clearly derived from Bacteria; was the Eucaryal cell itself of Archaeal origin? (Possibly; and yes, we ourselves just might be complex Archaea which have symbionts which were once Bacteria within our cells.) As might be expected, the scope for fascinating speculation is abundant.

However, these name changes are troublesome for students and non-specialists, and further changes are not unlikely. At present most microbiologists, while accepting the validity of the Domains, and delighting in the revolution they have brought to taxonomy, still find it useful to use older terms such as prokaryotes, archaeobacteria and eubacteria in their teaching, writing and discussions – as I have done in this book.

Where does all this leave the viruses? You may well ask. They are indeed composed of DNA, or sometimes RNA, and they certainly evolve. But they have no ribosomes nor anything else of long-term genetic stability; in fact, the medical problems they present often stem from the rapidity with which they can mutate and so change their genomes. Moreover, many viruses pick up DNA from their hosts. Analyses of their DNA or RNA have been used successfully to establish evolutionary relationships within groups of viruses, but their origins and relationships to other organisms remain obscure.

Ribosomal RNA, however, represents a basic, relatively stable framework within the bacteria of today, one from which we can deduce their evolutionary relationships, but superimposed on this is the remarkable genetic flexibility which I emphasized earlier. Because of this genetic fluidity, bacteria can adjust themselves to the wide variety of environmental conditions already discussed – which is one reason why they have persisted so successfully throughout geological time. They provide an indication of the intrinsic versatility of living things. Though the larger denizens of this planet are oxygen-breathing creatures living in a temperate environment, this is just a freak of evolution. The existence of anaerobic bacteria, sulphate- and nitrate-reducing bacteria, makes it clear that oxygen is a prerequisite neither of life nor of evolution. The waters of this planet are about neutral, neither particularly acid nor particularly alkaline, but the existence of thiobacilli and their associated acid-tolerant flora tells us that life could have developed and evolved on a much more acid planet. Water is sufficiently abundant here to be fresh or only weakly saline, but the existence of halophiles shows us that, had water been much more restricted and, therefore, had the few seas and lakes been highly saline, living things would nevertheless have managed. Barophiles tell us that high pressure would have been no obstacle; psychrophiles indicate that a temperature constantly near freezing would have been acceptable. Spore formation shows that life could have adjusted to periods of considerable heat and desiccation – as have some desert plants; thermophiles tell us that life could have developed at temperatures over 90 °C. Even here one must make the proviso that this limit is set by the boiling point of water in Yellowstone Park, USA. Under high pressure, water boils at much higher temperatures, and microbes have been grown in such conditions; there seems to be no upper limit to the temperature of terrestrial-type life provided liquid water persists.

The Earth's ordinary flora and fauna, then, today represent only a limited aspect of the biochemistry of which terrestrial life is inherently capable: our communal biochemistry became dominant about five hundred thousand millennia ago and only among microbes are there now representatives of what might have been. But it makes one think. How might carbon-based life have fared elsewhere in the universe? Will halophilic psychrophiles be found beneath the arid, cold wastes of Mars? Or CO_2-fixing, acid-loving thermophiles on Venus? Or anaerobic heterotrophs in the icy mush of the Jovian satellite Europa? If inter-stellar travel is forever closed to mankind, as relativists would seem to have us believe, may we nevertheless hope one day to receive television pictures of the sulphate-reducing equivalent of *Homo sapiens* from its anaerobic home in a distant solar system? I opened this chapter with a warning that much of what I should write would be of a speculative character; perhaps I should now separate off the really wild speculation by opening a new chapter.

11 Microbes in the future

One statement can be made with as great certainty as any other in this book: short of some cosmic catastrophe, such as the sun becoming a nova, microbes on this planet have a future. This is more than can be said for many animals. It is highly probable, for example, that the days of the mountain gorilla and bonobo chimpanzee are numbered. Though attempts are regularly made to control hunting and poaching, it is unlikely that they will succeed while war and civil strife cause whole populations to hunger for 'jungle meat'. Likewise the rhinoceros, the pangolin, the osprey and at least a hundred and fifty other large mammals and birds are destined to disappear from this planet unless they are successfully preserved in zoos or game reserves; only whales have attracted sufficient international attention for their anthropogenic decline to have been halted – if only briefly. Immensely greater numbers of lesser-known species of animals and plants are likely to vanish, unnoticed except by biologists, as mankind colonizes and modifies the more remote habitats of this planet; the environmentalist Paul R. Ehrlich has asserted that one in ten of the known plant species is threatened. But it was always so: throughout biological evolution, natural selection has involved an unceasing succession of extinctions and emergences of species; mankind has merely distorted the process, in recent centuries rather drastically. Is the startling variety of dogs, brassicas and the like, which our unnatural selection has generated, a sort of compensation, I wonder? Forgive the side-issue. On a more positive note, genetic manipulation promises the means to recover a lost species, if a specimen of its DNA can be found – provided, of course, that our present high-tech civilization lasts.

For even man's own future is in doubt. As the twentieth century approached its end, it became obvious to all educated people that atomic armoury had powers of universal destruction that even writers of science fiction had not imagined fifty years ago. Militarists consider quite seriously devastation and radioactivity spread over hundreds of square miles, such that no visible living thing survives. It is not beyond the powers of a war-based technology to sterilize this planet of plant and animal life; it is fairly easy to calculate the number of nuclear weapons that would need to be exploded to do this and the figures have been published.

In these circumstances, microbes would still survive, for they flourish in conditions of squalor, disease and human deprivation. Even in the extreme scenario in which the whole surface of the planet were lethally radioactive to higher organisms, microbes would survive and evolve. For example, *Deinococcus radiodurans* is a remarkably radiation-resistant organism that tolerates some hundreds of times the γ-radiation of ordinary cells, and other microbes are known that tolerate considerable amounts of radioactivity. They appear to be able to repair the damage caused by radiation very effectively – a good example of the adaptability of micro-organisms. To produce a level of radiation sufficient to eliminate such microbes from this planet would require an almost inconceivable number of atom bombs.

Of course, it is now almost inconceivable that any nation would be so stupid. Campaigners against nuclear weapons, as well as scientists, have at least alerted people and their political leaders to the certainty that unrestricted nuclear war would abolish the macroscopic component of the biosphere. Today most political leaders think of atomic war more as a threat than as a matter of practical action, and their military advisers think in terms of limited or strategic nuclear warfare. Though the threat of nuclear war fluctuates in intensity, and though there are religious or nationalist fanatics on the sidelines who cannot wait to get their hands on these marvellous God-given weapons, it is nevertheless fair to say that the prospect of global catastrophe receded over the two decades from 1970 to 1990. The reason why must be obvious to all thinking people: it is because the standard of living went up in those countries which offered the major threat. Mankind cares more about survival the more it has to lose.

This is not the place to tangle with the debate about nuclear weapons. Anyway, I take the view that the threat of war, nuclear or conventional, is a symptom of a more radical problem. For, in my opinion – everything in this chapter is my opinion, of course – mankind's control over microbes has generated a greater threat to its own future than has control of the atom. And my reason is quite unconnected with biological warfare and its possibilities, disgusting as these may be. The reason is simply this. By the control and prevention of disease, civilized communities have prolonged the lives of their own individuals, increased their potential fertility and decreased infant and child mortality. They have also, and quite rightly, introduced such medical benefits to backward and underdeveloped countries. Therefore we have the population explosion. Professor P. M. Hauser of the University of Chicago once quoted a simple calculation to the effect that, if the population of the world keeps increasing at its present rate, by about the year 2600 there will be one person for every square foot of the planet's land, poles, deserts and mountains included. This sort of calculation is good for coffee-table conversation but is, of course, meaningless, because it will not happen. But the serious information underlying such calculations is this: even if all present birth-control programmes proceed smoothly, the world's population (assuming there is no unforeseen global catastrophe) will double by the early decades of the twenty-first century. To give real numbers, the world's human population, which was around 1,500,000,000 at the beginning of the twentieth century, will be around four times that number by the time this book is published (it passed 5,500,000,000 towards the end of 1996) and will approach 8,000,000,000 by about 2020. How can one be so dogmatic? Quite easily. Population censuses are now available for most countries in the world and, although they may be inaccurate in detail, they are not wildly wrong. So demographers need only look at them to discover that something like a third of the world's population is under child-bearing age. These are the children who will grow up, mate and produce the new offspring, and most of them will do so before their parents and grandparents die. This is why one can be reasonably certain about population trends for a generation ahead; things get fuzzy after that, but up to twenty-five years ahead the pattern is quite clear. Official estimates of world population figures in the further future are no longer quite as drastic as they were in the 1970s and 1980s, largely because most national governments, international agencies and charities have heard the message and taken heed. There are some lamentable exceptions, but in most countries education and the availability of contraceptives, together with a degree of emancipation of women, have had a discernible effect and, in consequence, though the global population is still growing alarmingly, its *rate* of growth began to fall in the 1990s. However, the numbers are so vast that it will make no serious difference over the next few decades if censuses have been out by the odd million here and there, or if the overall growth rate drops little. Overpopulation is already with us and it will get worse during the foreseeable future.

The population explosion has been underpinned, as I said, by advances in medicine, in hygiene, in health care, in nutrition and in food production, to all of which areas microbiology has made enormous, indeed critical, contributions. Does the population explosion matter? Many people seem not to think so, especially the millions who adhere to political or religious dogmas which encourage fecundity, but I am utterly convinced that it presents a major threat to society. It threatens all sorts of societies: capitalist, communist, Roman Catholic, Islamic, nomadic, tribal, alternative and so on, but not for the reason given by so many benevolent persons and organizations (including me, in early editions of this book). The conventional reason for anxiety about the world's huge future population centres on food production. In the 1960s it seemed that these teeming billions could not possibly be fed, and in the 1980s there were appalling famines in East Africa. Despite the fact that European harvests are regularly in surplus, today a tenth of the world's nearly six billion people are clinically undernourished (some authorities put the figure much higher) and a tenth of these will die (or have died) of starvation or malnutrition. Yet I now accept the views of agronomists that this is not a technical, agricultural problem. It is a political problem, a problem of distribution. Enough is now known about the use of chemical fertilizers, about irrigation, intensive farming, plant breeding, control of weeds and plant disease, and about the exploitation of microbial processes such as biological nitrogen fixation, composting and recycling of animal wastes, protection of foods from pests and deterioration, for the world's agricultural soils to support nearly double its present population at a reasonably high level of nutrition. As an example, India, plagued by famine and malnutrition in the mid-twentieth century, upgraded its agriculture faster than its population grew, so as to become a net exporter of food in 1980. At the level of agricultural technology, ways are already available to improve the world's present level of nutrition and ways to feed the millions of the future are in sight. All that is needed is the political will to do it.

I excuse myself from prescribing how to generate that political will.

I accept, also, the view of most economic forecasters that energy and raw material shortages, of the kind I discussed in some detail at the beginning of Chapter 6, will prove manageable, at least for a couple of centuries and, if fusion reactors are developed, forever. But I agree, too, with those environmentalists who point out that all their pet problems – global warming, acid rain, depletion of the ozone layer, smog and atmospheric pollution, marine pollution, pesticide residues and comparable nasties in drinking water, you name the rest – arise from overpopulation and will therefore get worse. Of course, there are remedies for most, if not all, of these troubles, remedies which are usually simple – at least in theory. But progress with the few that are in hand does not give one confidence that we possess the will to put remedies into practice vigorously.

Yet even if mankind does muddle through, coping with the physical and environmental threats which arise from the population explosion, there is a more immediate, if rarely discussed, danger, which will be familiar to many biologists (but not necessarily to microbiologists, actually). It arises from the fact that overcrowding of any kind, in mammals, generates conflict. In mankind, it generates neurotic, irrational and criminal behaviour and, the more people there are, the more social deviants one has. Of course, one also has more people of kindness, altruism and benevolence, but which of these types, the saints or the sinners, is more destructive to society? To put the point succinctly, population expansion increases the number of social deviants, and therefore the number of foci of social breakdown, in any society. This phenomenon is blatantly evident in the West today in the form of street crime, violence and vandalism; at a communal level, and in other societies, it appears as political or religious extremism and terrorism. The social diseases of the present century are nationalism, terrorism, racism, fundamentalism and extremism; they flourish in conditions of crowding and competition for resources, and the demagogues who are willing to canalize them into anything from local intimidation to multinational disaster become ever more numerous. Considerations of this kind should be familiar to every thinking person and I apologize for the apparent digression from microbes; for the purposes of this book war is an extreme case of neurotic, irrational and criminal behaviour. It is largely our control of microbes in sickness that keeps the threat with us.

What is the answer? It is not, as some think, a deliberate reintroduction of disease, a sort of controlled biological warfare. Nor, obviously, could anyone of humanity withhold the benefits of medicine from communities simply because they then breed too rapidly. Obviously – an again every thinking person accepts this, unless some religious or political dogma prohibits the thought – births must be reduced and food and consumer goods must be increased. Which means more contraceptives and less dogmatism, more food and the goodies of civilized life, fewer weapons. All so easy, is it not? Forgive me once more if I do not here explain how to arrange these things.

If I have taken a gloomy view of the future, I have at least justified a preoccupation with the good things of life. What goodies have microbes, or rather, has applied microbiology, in store for us?

It is possible that new and more effective antibiotics will be discovered, though, as I mentioned earlier, penicillin was the first to be discovered and remains the best when it can be used at all. I have already discussed the problems presented by resistant strains; one can be fairly confident that more antibiotic-resistant strains will develop but that these will be kept in bounds by the discovery of new antibiotics, or the deliberate modification of existing ones. It is likely, generally speaking, that new patterns of disease will develop as existing pathogens become eliminated and, indeed, one can see this happening already. Despite setbacks such as the unexpected appearance of legionellosis in the late 1970s, and despite the emergence of antibiotic-resistant but otherwise familiar pathogens, bacterial diseases are generally manageable in civilized communities today. The troublesome diseases are caused by viruses. One class of ailment, classified under the general name of cancer, has no obvious microbial origin (except for one or two types which are definitely caused by viruses). Nevertheless, the manner in which cancer develops has much in common with the consequences of certain types of viral infection, and the reasons why the disease sometimes regresses seem to be much involved with the general topic of immunity and antibody formation. Thus a furtherance of knowledge of virus infection and of the processes involved in immunity, both originally microbiological topics, will probably lead to the most practical of medical advances.

At the end of Chapter 6 I discussed the prospects of biotechnology and molecular genetics, noting that the developments closest to practical application lay in the area of medicine. It seems likely that genetic disorders will come under control, for example, and the production of substances such as the interferons and factors to control blood clotting from cloned human (or animal) genes are likely to revolutionize parts of medicine. Social medicine, hygiene in particular, will advance – but a truly hygienic society loses its immunity to quite ordinary infections, and the clinical microbiologist will need to be increasingly alert for resurgent epidemics of diseases we had almost forgotten.

To turn to production process. Today the use of microbes to produce heavy chemicals, such as alcohol and industrial solvents, is obsolescent. Generally speaking, microbes can only be used economically to produce substances that are too difficult to synthesize chemically on a factory scale. But their role in the manufacture of more expensive materials will clearly increase. Their use in the production of steroids, where the industrial chemist uses them rather as a chemical reagent, turned up in Chapter 6. There is a touch of poetic justice about the thought that the systemic contraceptives are steroids: microbes, whose control in medicine made the development of systemic contraceptives a matter of social urgency, help in their manufacture.

The most complex chemical mixture mankind needs is food. Food, one trusts, will remain outside the province of the synthetic chemist for many centuries. No doubt minor pickling or fermentation processes using microbes will be developed in the future, but the main importance of microbes that one can foresee is as a bulk food themselves. I have touched on this question before: a microbial crop such as *Chlorella* would be independent of the weather and require far less space than conventional agriculture. (A culture space twenty-six yards square would supply the protein needs of a family of five or six people if the productivity of pilot experiments is any guide.) Likewise food yeast, and bacterial food from methane, will probably be made use of and will thus make waste materials palatable and nourishing. A process was developed in the 1970s in the USA for growing mushroom mycelium on meat residues (apparently about three-quarters of the material handled by a modern slaughterhouse is thrown away), but I do not know what became of it. Mushroom soup? It was said to be nutritious and tasted good. The question of taste and palatability is all-important in this kind of discussion, for it is no good producing nourishing foods if they disgust people. In fact, techniques of flavouring and fortifying foods have now developed to such an extent that today the real problem is to ensure that these abilities are used sensibly, to improve the quality and quantity of food, not, as so often in the past, to defraud the customer.

That people will change their eating habits to match these developments goes without saying. The transformation of the English diet in my lifetime has been spectacular: from roast and 2 veg. to scampi and ratatouille. Yeast extract has been part of my daily diet, and of my family's, since I was a boy – it appears daily on the breakfast table with the marmalade and such, and this is true of millions of English families. Sixty years ago eating yeast extract was an eccentric practice of vegetarians and food faddists; at the turn of the century it was unheard of. No doubt chlorella cookies and methano-burgers will one day be a delicious meal that one will take for granted; as one reconstitutes one's dehydrated *Château Latour* (esters specially blended to reproduce that greatest of great years, 1937), one may wonder at the barbarian habits of one's ancestors who grew large animals, killed them and actually ate their flesh . . .

Perhaps the most significant development in pharmacy in recent decades, one which society has not wholly come to terms with, is the arrival of psychomimetic drugs. These are the tranquillizers, antidepressants and hallucinogenic drugs that have revolutionized the practice of psychiatry. It has been said that one-third of the population of a civilized community is neurotic. This sort of statement depends on how eccentric speakers allow their neighbours' (but rarely their own) activities to become before they regard them as neurotic, but it has a certain substance. As soon as life becomes sufficiently comfortable for us to consider the question, we realize that we are knotted-up and illogical in a variety of responses, and that these responses get worse the more complex, stressful and crowded daily life becomes. (And let us be quite clear, parenthetically, that noble savages, carefree nomads, sturdy peasants and such paragons are equally subject to neurosis, anxiety and obsession – it is just that in their way of life it does not show.) These disorders are not new, nor are they particularly a product of modern civilization; they have been part of everyday life for centuries and the major advance made in the last few decades has been to recognize them and treat them. Drug abuse is one of the most appalling and destructive features of society, a practice which is increasingly difficult to control as younger and younger people become trapped. Yet the drug user, in a crude and self-defeating way, is pointing to a road mankind may well have to take. Mentally, we live with a heritage of reflexes left over from millennia of savagery: aggressiveness, dread and gregariousness, which lead people to wars, race riots, child beating, murder, religious manias and the rest of the social diseases I have already ranted about. Fleetingly one may recognize the irrationality of these responses, but they are generally outside an individual's control. For the first time pharmacy has developed drugs that enable one to stand back from, to reflect upon and even control this mental lumber. At least some of these drugs are of microbial origin, derivatives of fungi. As their constitution and action become better understood it is likely that microbiological processes will be involved in the manufacture of the acceptable varieties. At the simplest level, tranquillizers have already removed burdens of wholly unnecessary misery from the lives of millions of struggling citizens; if people can use like materials, without abuse, to improve the rationality of their social structure and behaviour, then once more microbes will have made a transcendent contribution to the human condition.

Without abuse, I said. Lysergic acid derivatives have been proposed as weapons in chemical warfare, the idea being that enemies will become so depressed and introverted that they cannot be bothered to fight. They would work, and so would peace drugs, derivatives of tranquillizers that render enemies too peacefully inclined to struggle. Such weapons would certainly humanize warfare, but one fears for the state of mind of the victors, given such power. Perhaps good, old-fashioned atom bombs are to be preferred? Or a really virulent microbe for biological warfare? I offer no opinion; I merely remark once more that scientific advance has always been subject to abuse.

Leaving our minds alone now, let me return to the more mundane aspects of this planet's economy. Nitrogen-fixing bacteria bring 50 to 400 pounds of nitrogen to each acre of soil, and today this is not nearly enough. Already one-third of the world depends on artificial nitrogen fertilizers for its food, and one could foresee the day when the population became so large that all the world's shipping facilities would have to be devoted to carting nitrogenous fertilizers about – the year 2000, according to one calculation I have seen. The arithmetic may have been wrong, but the message is clear: nitrogen-fixing bacteria must replace chemical fertilizer wherever possible. One consequence of this use of artificial nitrogenous fertilizers is that nutrients other than nitrogen are running out in certain types of soil. In the 1950s, sulphur-deficient soils were a rarity; the only examples I am aware of were found in East Africa. By late 1965 they had been detected in Australasia, Western Europe, India and Sri Lanka, both North and South Americas, West as well as East Africa. In other areas cobalt and copper deficiencies have been found. Soils deficient in phosphates have been known for years. Tropical and subtropical soils known as laterites are remarkably lacking in minerals, because they are regularly washed by the equatorial rains. As mankind learns to add nitrogen to the soil, other defects in the local soil composition become exposed.

Though these deficiencies can often be remedied by the use of chemicals, particularly in advanced communities, it is not easy to see how this could be done in practice on a global scale. The subtropical savannah, for instance, is a zone that is almost totally non-productive in terms of human food, though it is warm, wet and has lots of sunshine. The sheer mechanical problems of making such an area productive by chemical means are dispiriting; the solution is far more likely to arise from an understanding of the microbes involved in the local nitrogen, sulphur and phosphorus cycles. Understanding and control of microbes in agriculture, together with their deliberate use for the disposal and recycling of complex products, seem to me to be an obvious large-scale contribution applied microbiology has to make to this planet's economy.

Another pedestrian but important area arises from what I wrote in Chapter 8. Already disposal and recycling of the detritus of 5,800,000,000 people presents drastic problems – imagine what it will be like with another couple of billions! The microbes will, I know, rise to this smelly and disagreeable challenge as mankind struggles to keep its environment wholesome.

Mundane, plodding fields of advance, you may say? Perhaps, but even the most romantic research is mundane and plodding in its day-to-day reality. Let me, however, indulge my romanticism and look outside this planet. What of microbes in the space era?

One consideration arises at once. If man goes into space, microbes go too. You cannot produce a germ-free human; moreover, even if you could, he or she would probably die of obscure kinds of malnutrition. Anything people handle, indeed anything that emerges from the biosphere, is contaminated by microbes. For this reason both Russian and American space agencies have been at pains to sterilize equipment sent up outside this planet's atmosphere. But a space probe in fact crashed accidentally on Venus and the question of how efficiently it was sterilized was a matter of considerable concern to microbiologists until the Venera probes of 1975 confirmed that Venus is hotter than the hottest autoclave at its surface (about 480 °C), so no terrestrial life could have survived. But it would be a tragedy if the Moon and Mars became contaminated by terrestrial microbes before a proper evaluation of indigenous biological conditions there could be made. For terrestrial microbes might swamp, and conceivably eliminate, indigenous populations before space travel became sufficiently advanced to permit the detection of alien life. Then one could never be sure that the microbes one found had not arrived with the early Moon or Mars probes. For we can be certain of one thing: the cold airlessness of deep space provides little obstacle to the survival of bacterial spores. Radiation in deep space may be lethal, we do not know; but the average bacterial spore, within the shell of a space vehicle, would have no difficulty in remaining viable through an interplanetary journey, provided it survived the initial heating-up which occurs as the projectile traverses the Earth's atmosphere.

Some have thought that the Earth has been scattering creatures the size of bacteria throughout space during the period since life originated, but this is not likely. The Earth's gravitational field is so strong that the probability that a particle with the mass of a bacterium could reach escape velocity is infinitesimally small. Even effects of high-speed atmospheric winds and volcanic explosions do not significantly increase this probability. Viruses, being one or two orders of magnitude smaller, could reach escape velocity somewhat more readily than bacteria and so, presumably, could the tiny microbial parasites called *Bdellovibrio*, but the likelihood that actual living material has escaped from the Earth, except via the space probes of the last few years, remains infinitesimal. The notion that the surface of the moon will be peppered with spores of *Bacillus subtilis* (a common aerial bacterium), though attractive to scientists with a taste for anticlimax, is not likely to be correct. Hence any planetary exobiology that space exploration may encounter can be expected to have evolved independently of terrestrial life.

Can one say anything about what extraterrestrial organisms might be found? One had best set aside the more extravagant ideas of science fiction writers and assume that extraterrestrial life would resemble terrestrial life in principle. By that I mean that it would be based on compounds of carbon, would conduct its life processes in liquid water, and would generate the necessary energy by exploiting the sorts of biochemistry that I have alluded to in this book. Let me remind you that terrestrial life depends on the sun, not only to keep the planet's surface warm enough for water to be liquid, but also to provide, through microbial and plant photosynthesis, a steady supply of carbon compounds for the inhabitants of the biosphere to exploit as energy sources. In addition, photosynthesis today supplies the oxygen which enables higher organisms to exploit these energy sources efficiently. And air-breathing creatures that live in the sunlight are by no means the only creatures that rely on solar energy for their existence; so do organisms which live in the dark, in deep marine sediments or in soil, because they feed on the fall-out from photosynthesis: the organic dust, sediment and detritus that falls their way – and, usually, they depend on the oxygen photosynthesis provides. Anaerobes, though they may eschew the oxygen, also need the organic matter which ultimately derives from photosynthesis.

Making the assumption, then, that one is looking for terrestrial-type life, one can make a couple of informed guesses about the possibilities within the solar system. The first is that, unless they have another source of heat to keep them wet, the outer planets, their satellites, and the asteroids lie too far from the sun for its radiation to keep liquid such water reserves as they may possess. Any water would be frozen solid, and life cannot establish or sustain itself without liquid water. The second is that the inner planets can be written off: Venus is too hot – all its free water is in the form of steam – and Mercury is too hot and dry on one side, too freezing cold on the other. However, the Moon and Mars are worth taking seriously in the first instance as celestial bodies potentially fit for habitation.

On any inhabited planet it is likely that there will be microbes, because it is unlikely that living things would evolve without a microbial stage, and it is equally unlikely that microbes, once evolved, would be eliminated. The Moon is a dry body with no atmosphere, subject to meteoric bombardment at the surface and a large temperature gradient between the insolated face that we normally see and its dark rear. The moon rocks brought back by the Apollo missions were dry and sterile. If any liquid water exists on the moon it probably rests beneath the surface as saturated salt solution, protected from temperature extremes. Life depends on cyclical transformation of biological elements such as in the nitrogen, carbon and phosphorus cycles. Though one could imagine salt-tolerant sulphate-reducing bacteria surviving beneath the surface of the Moon in, for example, a saturated magnesium chloride solution, they would require a source of carbon to use. But it is difficult to envisage a carbon cycle, for what microbial process could one envisage that would then return CO_2 to an organic form? An anaerobic oxidation of iron, perhaps? One would need to know more about the chemistry of the Moon to reach a useful conclusion, but the primary moral is obvious: to seek life on the Moon, dig, and look for halophilic, chemo-autotrophic, anaerobic microbes.

Mars is a rather more promising candidate for our kind of life. Though it is cold and has a most tenuous atmosphere, it does seem to have water, and near its equator this is probably liquid for some of the Martian year. Its microbes would need to be anaerobes, but several terrestrial types might survive there. The psychrophiles would be tolerant of the low temperature; the water would be pretty briny, there being little of it, so again one would expect halophiles. But they could exist on the surface and be reached by sunlight – so a carbon cycle based on photosynthesis is feasible. Anaerobic iron bacteria might have developed; sulphate-reducing bacteria and sulphide-oxidizing bacteria could be expected; there seems to be reasonable scope on that planet for several combinations of chemotrophy and phototrophy.

As many readers will know, there is slight evidence for seasonal colour changes on Mars, which were earlier taken as indicating some analogue of terrestrial plant life. So it was a disappointment when the Viking lander of 1976 found not the slightest trace of life (despite some excitement to the contrary in the first few days). Yet it may have landed in an unsuitable place. Early in 1998 a NASA spacecraft, Mars Global Surveyor, sent back convincing photographs of a dried-up river valley fed by tributaries. It has been named Nanedi Vallis, and it lies in a Martian plain. Its existence supports earlier evidence that liquid water was once relatively abundant on Mars: there are other Martian features which resemble the shorelines of dried-out seas or

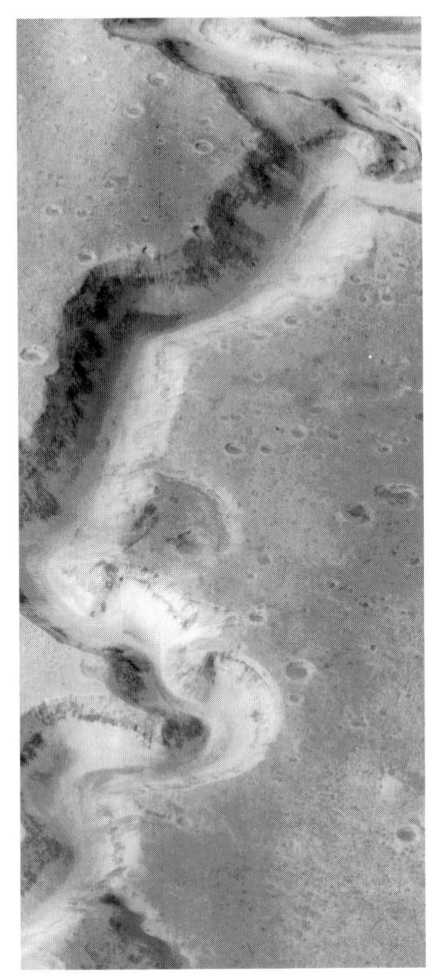

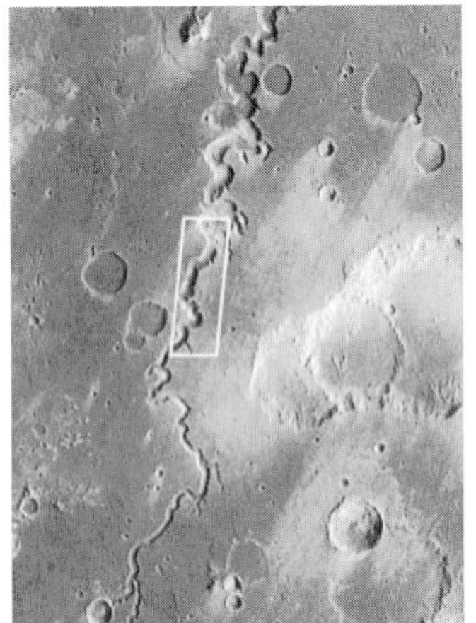

NANEDI VALLIS, a putative dried-up watercourse in the Xanthe Terra region of Mars. Taken by the Mars Orbiter Camera of the Mars Global Surveyor on June 8, 1988. The pock-mark indentations on the plain are impact craters. The right-hand picture is a close-up of the boxed area on the left, showing meanders and rocky outcrops; the canyon is about 2.5 km wide. (NASA image)

lakes. The current consensus among planetary scientists is that Mars was warm and wet some 3 billion or more years ago and at that time had a thicker atmosphere. If that is so, Mars may now be in the terminal stages of habitation: a place where life once flourished but where, as the atmosphere became increasingly tenuous and water more scarce, only the toughest organisms survived. Microbes are the toughest of living things: space scientists have placed terrestrial microbes in simulated Martian conditions and they survived and multiplied. At the terminal stages of evolution, as at its commencement, one would expect microbes to predominate; there may yet be survivors of Mars's erstwhile microbial inhabitants in wet places beneath its dry and dusty surface. The next generation of landers will be designed to dig for samples and examine them for signs of life.

For a brief period in the summer of 1996 the evidence for life having existed on Mars seemed stronger than I have just indicated. Meteorites have been found in the Antarctic which, from analyses of their mineral composition, are believed to have come from Mars, catapulted off that planet over 13,000 years ago by a huge impact with a small asteroid. Scientists working for NASA announced that one of these, which contained carbon compounds, also contained tiny tubular structures which they thought were microfossils of organisms of about the size of terrestrial ultramicrobacteria. Much excitement was generated, but the idea did not survive critical examination, and the consensus seems to be that the tubules were wholly inorganic formations. Nevertheless, NASA's enthusiasm was understandable; and Mars is certainly the planet a dedicated microbiologist would most wish to visit.

The outer planets and their satellites are, as I said, mostly too cold for water to exist in liquid form. However, the Voyager spacecraft, as it flew past Jupiter in 1979, revealed an exception. Early pictures of the surface of Jupiter's satellite Europa showed it to be unexpectedly smooth, and as more detailed pictures were received (slowly, because spacecraft return details of their photographs only gradually), evidence accumulated that it consists of a sheet of ice. This conclusion was later backed up by ground-based spectroscopy. Details of its surface features, as well as measurements on the satellite's gravity, suggest strongly that Europa's icy crust is shallow, some 150 kilometres thick, and floats on a sea of liquid water or ice–water mush, all surrounding a rocky silicate core. Under the influence of Jupiter's gravitational pull, tidal eruptions of water and ice regularly renew and smooth out its surface; tidal heating and compression also keep the watery layer fluid.

More recently the Galileo space probe, launched in 1989, has sent back evidence that sodium carbonate and magnesium sulphate are present in Europa's icy crust. It is more than likely that these salts, together with organic matter and other biological elements, are present in Europa's ocean, though at what dilution is anyone's guess. It is also probable that tidal effects will have led to their being regularly perturbed and unevenly distributed. As with the juvenile Earth, one can imagine scenarios in which organic matter became sufficiently concentrated at mineral surfaces for appropriate chemical interactions to take place and carbon-based life to originate. If so, could tidal heating sustain such life in the absence of a contribution from sunlight? Yes. There is at least one terrestrial model for life which seems truly independent of photosynthesis: deep in soil near the Columbia River in the north-west of the USA a microbial community has been discovered whose primary source of carbon is provided by methane-producing bacteria – anaerobic autotrophs which make carbohydrate plus methane from carbon dioxide by reducing it with hydrogen. That bacterial process in itself is nothing unusual, of course, and in the vast majority of microbial comunities the hydrogen originates from photosynthetically formed organic matter. Its ultimate source is sunlight, albeit perhaps at one or two removes. The feature which makes the Columbia district's subterranean community different is that sunlight is in no way involved: the hydrogen comes from a chemical reaction between iron silicates in the local rocks and water, and it is geothermal heating that keeps the environment wet and warm.

Well, the persistence of that community demonstrates that terrestrial-type life, once it has been initiated, can persist without sunlight. It will be a long time before Europa can be visited by astronauts, and before the situation thereon, or more correctly therein, can be assessed directly. It will be a subject for speculation for many years, periodically refuelled, I hope, by reports from new space probes. So let me return to Mars, where a manned visit is already a serious possibility.

On such a visit, the biologist would have problems with his or her own microbes. A space ship with a few astronauts taking a year-long trip to Mars would be a physically isolated community, and a peculiar thing happens to the commensal microbes of people in such communities. One type of microbe tends to become dominant, from mouth to anus, and if this germ happens to be pathogenic the situation can be dangerous. Likewise, immunity to infection by ordinary microbes tends to be lost. It seems probable that astronauts will have to keep cultures of the varieties of microbes they started out with, and will need deliberately to re-infect themselves at intervals. On the other hand, astronauts will have considerable disposal problems: getting rid of urine and faeces; removing carbon dioxide exhaled and regenerating oxygen. A most pleasing microbial system has been proposed to help in these processes, which seems quite feasible. A solar cell on the space ship would generate electricity which would be used to electrolyse water. Oxygen and hydrogen would thus be formed, which would be used to grow the bacterium *Pseudomonas facilis*, a chemotroph that fixes CO_2 while forming water from hydrogen and oxygen. Thus, with no net waste of water, CO_2 would be removed from the atmosphere. But these creatures require a nitrogen source, for which the urea of urine will do very well. Thus one would grow microbes at the expense of urine and CO_2; these, once a well-trained astonaut got used to the idea, would be a useful protein food. *Chlorella*, the alga, could also be used to form oxygen and yield food, since sunlight would be available; a methane fermentation of faeces would dispose of waste and, aided by methane-oxidizing bacteria, produce food. In all, there would be a curious fulfilling of the biblical threat of Rabshakeh (2 Kings 18:27). To generalize, it will be almost impossible to transport the bulk food, water and disposal requirements of astronauts for long space trips: little microbiological microcosms will have to be set up to recycle the chemical environment which the astronauts inhabit, and here an understanding of terrestrial microbial ecology will be critically important.

Can one say anything of life outside the solar system? Some cosmologists believe that there must be a vast number of planets suitable for terrestrial life in the universe, and it is likely, if our views about the origin of life on Earth are correct, that life will have developed on them. Dr H. Shapley's estimate of 100,000 habitable planets among the 100 billion stars of our own galaxy, the Milky Way, is often quoted. The prospects of exploring and visiting such planets seem remote indeed; unless our theories of the cosmos are completely awry, journeys lasting not only centuries but millennia would be needed. But communication with such systems by radio is feasible, even if conversation would have its one-sided aspects. (Dr F. D. Drake calculated that planets on which life has evolved to a level of being able (and willing) to communicate across space are likely to be separated, on an average, by a distance of 1,000 light years. When one has to wait several centuries for a reply to one's opening remark, the give-and-take of day-to-day intercourse tends to be lost.) Though such systems will undoubtedly have representatives of the microbes, communication will necessarily be with creatures of advanced intelligence. Hence, though the sulphate-reducing equivalent of *Homo sapiens* is an interesting entity to speculate on, it would be a macrobe, not a microbe, and thus outside the scope of this book.

This chapter is about microbes and the future. One can predict, as I have done, that benefits will arise from further development of economic microbiology, that there will be advances – and retreats – in medicine, health, environmental conservation, social behaviour and, indeed, sanity. One can point to roles for microbes in space exploration and food production. Once a grandiose scheme was proposed to melt the polar ice caps by seeding them with red algae, thus increasing their absorption of solar heat, an expedient which, I am told, would flood many of the lowlands of Europe and Asia. (Some argue that, by promoting global warming, we have inadvertently embarked upon that project already.) But when all is said and done the real importance of microbes will prove to be, and this I assert with complete confidence, in the advance of knowledge. I told in Chapter 6 how modern molecular biology has arisen from microbiology; how microbial genetics kicked biology violently into the twentieth century. Biology is today in something of an ecstatic state: information derived from the bacterium *Escherichia coli* has proved, speaking very broadly, to be of universal validity. The application of the principles it has engendered has revolutionized the study of cells of higher organisms. Fortunately, understanding that there are microbes other than *E. coli* in due course penetrated to the less obsessed molecular biologists and, with such experimentally amenable material available for laboratory use, it is certain that microbes will continue to be exploited to further our knowledge, not only of themselves but of all living things. The techniques of microbiology are used in tissue culture; microbes can be hybridized and transformed; DNA can be passed from one type of microbe to an unrelated one and what amount to wholly new species can be created. DNA from higher organisms can be cloned in bacteria, or yeasts, analysed and either returned, if necessary altered, to its origin or installed in a different macrobe. Certain subcellular structures, such as plant chloroplasts and the mitochondria of higher organisms, are evolutionary relics of symbiotic associations. Microbes, it seems, show a remarkable range of capacity for association: the almost casual commensalism of intestinal bacteria, which is essential to the nutrition of many animals; the more obligatory association of root nodule bacteria with leguminous plants; the intimacy of the symbiont of *Crithidia*, which actually lives inside the cell protoplasm and reproduces with it; the complete parasitism of a temperate bacteriophage; finally, the total loss of individuality that must occur when the symbiont or parasite becomes an organelle. Evolution is not necessarily divergent: it seems probable that cooperative associations of increasing intimacy have developed during evolutionary time and have led to the emergence of new creatures by what amounts to accretion. The evolutionary chart may well be a network rather than a family tree. If such associations occur spontaneously, why cannot we induce them deliberately? For example, would it not be convenient if, as I have earlier speculated, one could confer nitrogen-fixing properties on wheat and thus bypass the use of chemical fertilizer or the ploughing-in of leguminous crops? And plants are not the only creatures that can be so manipulated; in principle, from a study of microbes, scientists now have a real prospect of altering our own heredity.

'Never!', you might cry, 'That would be eugenics!'. Yes, in a way it would. Eugenics has become a terrible word: a term which recalls dreadful memories of the enforced sterilization of individuals, of secret experiments on prisoners and ethnic groups, of pogroms, ethnic cleansing and above all of the Nazi holocaust. Yet to Sir Francis Galton, who coined the word, eugenics meant the use of our emerging understanding of genetics to improve the lot of mankind. Historically that understanding was inadequate, and in the hands of maniacal enthusiasts the results were horrific. Our understanding of human genetics is still inadequate for a rational and beneficient eugenics – make no mistake about that. Yet would it be so terrible, so unethical even, to manipulate, when it becomes possible, the germ lines within a human family so as to cure its descendants for ever of an hereditary disease such as porphyria, cystic fibrosis, or whatever genetically inherited disease you care to name? I am no ethicist, but a still, small voice within me says that, when it can be done safely and reliably, it would be unethical to withhold such a cure.

Understanding of microbes has opened new vistas for the futures of biology and medicine, and will continue to do so. And it will equally play a part in some sticky ethical problems. We shall have to learn to live with the possibility of regenerating organs for transplants, even whole individuals, from tissue cultures of genetically manipulated cells, sometimes from embryos; of preparing cloned animals, such as pigs, whose genomes have been 'humanized' so that their organs may be transplanted into people. Perhaps we shall learn how to upgrade the intelligence of animals and alter their characters, deliberately to alter the heredity of strains even of mankind, so as to adapt them to space travel or life on inhospitable planets. In a few centuries it may become desirable, indeed necessary, to cool down the planet Venus and make it habitable; I am sure the first colonizers will be judiciously introduced consortia of microbes. In millennia to come, it is conceivable that a creature that once was man could meet an intelligent sulphate-reducing organism on its own ground, having survived several centuries of space travel to be there. To such prospects the study of microbes will have made a major contribution, and herein lies their most profound importance for the future of mankind. But such concepts offer horrifying prospects for abuse – let us hope that man will have escaped from the infantilism so apparent today long before these prospects become a reality. Science, perhaps unfortunately, is morally and ethically neutral; it is also irreversible. Its consequences are what mankind makes of it and this, particularly to a scientist, is its most terrifying – if exciting – aspect.

Further reading

The subject of microbiology is now very well documented, at elementary, intermediate and advanced levels, and encyclopaedic tomes are available for the finer detail. Such material is regularly up-dated, in the form of revised editions or new works, so I shall discontinue my practice in earlier editions, in which I offered a selection of texts on various branches of the subject: general microbiology, medical microbiology, environmental microbiology, biotechnology and so on. In a similar way, elementary and more advanced texts are available which deal with those facets of chemistry and biochemistry to which I have alluded. Books which deal more fully than I have with such matters ought to be available, in your nearest university or technical college library – though ordinary public libraries are unlikely to be well provided with such material.

The following books amplify some of the more peripheral topics to which I have alluded.

HISTORICAL

De Kruif, P. P. (1954) *Microbe Hunters*. New York: Harcourt Brace. This is a classic study of the founders of microbiology.

Schierbeek, A. (1959). *Measuring the Invisible World: the life and work of Antoni van Leeuwenhoek*. London & New York: Abelard-Schuman. This is a readable survey of the observations of Antoni van Leeuwenhoek, amazing for his day, which he communicated in letters to a doubting, but ultimately convinced, Royal Society of London over the couple of decades following its foundation in 1660.

THE ORIGIN OF LIFE

Orgel, L. E. (1973) *The Origins of Life*. London: Chapman & Hall. Presents the older orthodoxy crisply.

Mason, S. F. (1991) *Chemical Evolution*. Oxford: Clarendon Press. This is an impressive survey of modern ideas on the origins of stars, planets, the chemical elements etc., which deals *inter alia* with recent views on the early terrestrial environment and their impact on ideas about the origin of life.

ENVIRONMENTAL AND POPULATION PROBLEMS

This is an emotive, sometimes politicized, area of contemporary discussion and, despite a large bibliography, it is difficult to point to an unbiassed and readable survey. However, if you can set aside some embarrassing mysticism about Gaia, the Earth Mother, there is a lot of common sense in:

Myers, N. (ed.) (1985) *The Gaia Atlas of Planet Management*. London & Sydney: Pan Books.

THE NATURE OF SCIENTIFIC RESEARCH

Watson, J. D. & Tooze, J. (1981) *The DNA Story: a documentary history of gene cloning*. San Francisco: W. H. Freeman. This account of the turmoil and absurdities surrounding the introduction of recombinant DNA technology is a sometimes hilarious documentation of the fears, posturing and general turbulence which can be generated by scaremongering. There is an ironic touch in that Watson was one of those who signed the published letter which started it all off.

Crick, F. (1989) *What Mad Pursuit; a personal view of scientific discovery*. London: Weidenfeld and Nicholson (paperback: Penguin Books, 1990). A revealing and idiosyncratic account of the thinking, as much as the actual experimental work, which ultimately disclosed the genetic code and the way it is read, written by one of the father figures of molecular biology.

Glossary

Aerobe An organism which, like you and me, respires by consuming the oxygen of air (contrast **Anaerobe**).

Aerosol A suspension of droplets in air so fine that it settles extremely slowly.

Anaerobe An organism that does not use the oxygen of air for its respiration (contrast **Aerobe**).

Antibodies Proteins, formed in response to foreign, usually infectious, materials entering the bodies of higher organisms, that react with such foreign matter, coagulating it and making it easier for the body to dispose of.

Antigen A substance such as a bacterial toxin (*q.v.*), that provokes the formation of antibodies in the blood or tissues of higher organisms (see also **Vaccine**).

Archaeobacteria A distinct class of bacteria which grow in extreme conditions; members of the Domain Archaea.

Autotroph An organism capable of growing at the expense of wholly inorganic substrates (*q.v.*) (contrast **Heterotroph**).

Bacteriophage A virus parasitic on bacteria.

Barophile A microbe capable of growing at very high pressures.

Biosphere The skin of the planet inhabited by living creatures.

Chemotherapy The science of curing disease with the aid of chemicals.

Chloroplast An organelle (*q.v.*) in microbes and higher organisms that conducts photosynthesis (*q.v.*).

Commensalism The property of living in harmless but independent association with a second organism (contrast **Symbiosis**).

Continuous culture A culture of microbes that is fed slowly but continuously with medium (*q.v.*) so that the microbes multiply continuously.

Cyanobacteria A class of bacteria which photosynthesize (*q.v.*) and evolve oxygen, like plants do. Earlier called blue-green algae.

DNA Deoxyribonucleic acid: a natural polymer that carries the genetic information determining the character of an organism (see also **PCR**, **Plasmid**, **RNA**, **Mutation**, **Transposon**).

Enzyme A protein which, without itself undergoing permanent change, accelerates a biochemical reaction that would otherwise scarcely take place at all.

Eukaryote The class of living things which consist of cells with nuclei (plants, animals, fungi, etc.) (contrast **Prokaryote**) ; members of the Domain Eucarya.

Halophile A microbe capable of growing in solution containing concentrations of salt (sodium chloride) in excess of about 3 per cent; such concentrations are toxic to fresh-water microbes.

Heterotroph An organism requiring pre-formed organic matter for growth (contrast **Autotroph**).

Ion An atom or molecule carrying an electric charge.

Medium (pl. media) The environment in which a microbe grows.

Metazoa Multicellular organisms.

Mitochondrion (pl. mitochondria) An organelle (*q.v.*) present in eukaryotic cells concerned primarily with respiration and energy generation.

Motility (adj. motile) The property of being able to move deliberately.

Mutation A chemical change in the DNA (*q.v.*) leading to a change in the genetic character which is inherited unless the mutation is lethal. An organism that has undergone such a change is a **mutant**.

Mycelium The thread-like ramifications of a fungus.

Organelles Subcellular structures having functions comparable to the organs of metazoa (*q.v.*).

Pathogenic Capable of causing disease.

PCR Polymerase Chain Reaction: a laboratory means of making multiple copies of a stretch of DNA (*q.v.*).

Permafrost The zones of arctic and antarctic soil which do not thaw in summer.

Photochemical Pertaining to chemical reactions brought about by light.

Photosynthesis The property of forming organic matter from carbon dioxide using radiant energy from light; the basic growth process of green plants.

Plasmid A genetic element, like a small chromosome, frequently found in bacteria. Often confers ability to resist antibacterial substances.

Prion An aberrant protein thought to cause spongy brain diseases.

Prokaryote The class of living things consisting of cells without nuclei (contrast **Eukaryote**); covers both regular bacteria and archaeobacteria.

Protoplasm The living contents of cells.

Psychrophile A microbe capable of most rapid growth at temperatures below 20 °C (contrast **Thermophile**).

Psychrosphere That zone of the sea below the thermocline (*q.v.*) where the temperature is low and not subject to seasonal variation (see also **Thermosphere**).

RNA Ribonucleic acid: a natural substance concerned with the transfer and interpretation of genetic information (see also **DNA**).

Rumen The first stomach of a ruminant mammal.

Serum The colourless, fluid component of blood.

Spore A dormant form of a microbe capable of enhanced resistance to heat, drying and disinfection.

Sublime, to To distil from the solid state without melting.

Substrates The components of a medium used by the microbes for growth; also the chemicals used by enzymes for their action.

Sulfuretum A microcosm involving the main bacteria of the sulphur cycle.

Symbiosis An association of two different organisms involving some degree of interdependence (contrast **Commensalism**). The partners in such an association are **symbionts**.

TCDD 2,3,7,8-Tetrachlorodibenzo-*p*-dioxin, a toxic industrial chemical used in making a herbicide.

Teratogen A substance which causes deformity of the foetus.

Thermocline That zone of the sea separating the psychrosphere (*q.v.*) from the thermosphere (*q.v.*)

Thermophile A microbe capable of growing at temperatures above the 45 to 50 °C lethal to ordinary organisms.

Thermosphere The (upper) zone of the sea subject to seasonal fluctuations of temperature (contrast **Psychrosphere**, see also **Thermocline**).

Toxin A toxic protein, usually of microbial origin.

Transposon A length of **DNA** which is able to move from one place in a **DNA** chain to another.

Vaccine An antigen (*q.v.*) from a pathogenic (*q.v.*) microbe which, virtually harmless itself, provokes immunity to the pathogen when introduced into the blood stream.

Vitamin An organic substance essential in small amounts for the growth and health of an organism.

Index

(First mentions of chemical formulae quoted in this book are indexed under 'Formula; italicized page numbers signify illustrations)

abattoir effluents 274
Abd-el Malek, Y. 181
Aberdeen, typhoid outbreak 80–81
acetic acid 153, 162–163, 192, 274
Acetobacter 162, 203
  suboxydans 164
Acetomonas 162
acetone 192
acidophiles 40, 49
acquired immune deficiency syndrome (AIDS) 57, 71–73, 111, 233, 307
acriflavin 52, 103, 300
actinomycin 202
ADA (adenosine deaminase) 227
adaptability of microbes 32–33, 52
adenovirus 68, 227
aerobe, defined 49, 360
aflatoxin 249
African horse sickness 143
African swine fever 143
Agave 157
Agrobacterium tumefaciens 217, 229, 236
Ain-ez-Zauia 176–179, 177
Alcaligenes eutrophus 194
alcohol, fuel 193
alder 149
algae, as group 20–21
alkalophiles 40, 49
allergy 79
Alternaria 36
Alvin, submersible 43
amodiaquine 109
Amoeba 21, 28
ampicillin 296
amylase 205–206
Anabaena azollae 150
anaerobes
  ancestral role 320, 321–322
  classified 44–46, 49
  enriched 115
  extraterrestrial? 349, 352–353
  opportunist pathogens 75–76
  primitive? 312–313, 314, 319, 320
Angkor Wat 261
anthracite 186
anthrax 94, 109, 306
antibiotics, in animal feed 138–139
anti-depressants 343
Apollo missions 348
Archaea 31, 332
Archaeobacteria 38, 43, 46
  as group 31, 330–332
Archaeoglobus 46
arsenic in paint 252
ascorbic acid – *see* Vitamin C
asepsis 129
Ashbya gossypii 164
Aspergillus 22, 194, 200, 204, 248, 252
  flavus 249
  fumigatus 249
  glaucus 250
  niger 165, 194
  oryzae 156, 163, 206
  restrictus 250
asphalt, spoilage 256
assimilation, defined 133
Astasia 21, 328
athlete's foot 92
atmosphere, microbes in 1–2, 97, 245, 347
aureomycin 201

autotrophs
  defined 21–22, 49
  in evolution 314–316, 323–324
avian 'flu 69
azobacterin 150
Azolla 150, *151*
Azotobacter *29*, 149–150

Bacillus *29*, 144, 200, 204, 206
  anthracis 306
  cereus 146
  israeliensis 146
  megaterium 165
  subtilis 347
  thuringiensis 146–147, 237–239, 295
bacitracin 200
bacon 247–248
bacteria, as group 28–31
  L-forms 30
bacteriophages 24, 101
  as parasites 324
  in transduction 220, 325, 326–327
Bacteroides 45, 60, 137
badgers 139
baking 161–162, 206
barophiles 38–39, 48–49, 333
Bath spa 260
BCG vaccination 78
Bdellovibrio 30, 101, 324, 347
Beecham Research Laboratories 197–199
beer 133, 151–153, 156, 163, 206
beijerinckia 149
beri-beri 137
Bernal, J. D. 311
biocontrol 145–147
biodegradable substances 253–254
biogas 192, 193
biological warfare 94–96, 145, 224, 248, 340, 345
bioplastic 194
bioremediation 280, 284–285
biosensors 207–208
biosorption 284
biosphere, defined 1
biotechnology 53, 208–9, 231, 341
Black Death 64
black fly 146
Black Sea 264
Blakeslea trispora 165
blooms 265–266
blue ear disease 143
blue-green algae 20–21, 30, 43
bog iron 183
bog myrtle 149
Bordeaux mixture 131, 144
Bordetella 58
  pertussis 56, 58
Borrelia burgdorferi 64, 90, 210
Botrytis cinerea 154
bottle feeding 130–131
botulism 76, 79, 248
bovine mastitis 157
bovine spongiform encephalopathy (BSE) 139–142, *140*, 299, 302–303
bovine tuberculosis 139, 157
brandy 155
bread 151, 161–162, 164, 170, 248
Brenner, S. 293
Brock, T. D. 37
Brucella 75, 157
  abortus 64, 73–75
brucellosis 143
bubonic plague 64, 94
building land, reclamation 288–289
butanol 192, 193
Butlin, K. R. 177, 191, 267, 282–283
butter 254–255

calcivirus 145–146
Campylobacter 56, 89, 90, 91
cancer 55, 70, 76, 202, 225, 249
Candida 57
  albicans 73
  utilis 167
carbolic acid 128
carbonate reduction 48
carbon cycle 10–12, *11*, 348–349
caries 99–100
Carnegie Institute of Washington 168
carotenes 43, 96, 165
cassava 163, 193
cauliflower mosaic virus 239

cellulase 206
cellulolytic bacteria 133, 136, 251
cephalosporins 200
Cephalosporium 200
Cetus Corporation 231
Chain, E. B. 107
chalk 181, 184
Chakrabarty, Dr 278
Champagnat, Dr 167
Champagne 155
  elderflower 157
cheese 157, 159–161, 248
chemo-autotrophs 40, 42, 44, 349
chemotherapy 103, 107, 109, 360
Chlamydia 63, 93
Chlamydomonas 20
chloramphenicol 199, 202
Chlorella 20, *21*, 168–169, 342, 353
chlorine 130–131
chlorophyll 20, 43, 96
chloroplast 19, 20, 219, 328–329, 330, 355, 360
chloroquine 109
chlortetracycline 201
chocolate 163, 206
cholera 56, 61, 65, 70, 86, 87, 109, 219, 271
cholesterol 205
Chromatium 178, 265
chromium-plating 281
cider 156, 163
citric acid 165–166, 185, 193–194, 283
Citrobacter 284
citrus stubborn disease 144
Cladosporium 36
Claviceps 203
cloning of DNA 222, 327, 342
Clostridium 45, 115, 137
  acetobutylicum 192
  botulinum 76, 94, 248
  butyricum 76
  pasteurianum 149
  tetani 76
  thermosaccharolyticum 248
  welchii 75–76
co-acervates 312
coal 12, 172, 186
  formation 186, 252
  microbial attack on 51, 256–257
cobalamide – *see* Vitamin B_{12}
cobalt 138, 185, 345
cocoa 163
cold, common 24, 57, 67, 68, 83–84, 119, 121–122
Columbus, C. 67
compost 185, 191, 252, 287–289
concrete, corrosion 261–262
conjugation 215–217, 220, 326
continuous culture, explained 123–124
Cook, Captain 67
copper 181, 182, 184, 255, 284, 345
  strip test 255
copper sulphate, growth of microbes in 52
Corrosion, microbial 257–262, *259*
Corynebacterium 204
  glutamicum 164
  renale 75
cowpox 78, 81–82, *82*
Coxsackie virus 68
cream 254–255
Creutzfeldt–Jakob disease 26, 141–142, 299, 302
Crithidia oncopelti 136, 329, 355
Cryptosporidium 57, 88
curing, of meat 247–248
cutting emulsion, spoilage 256
cyanide, microbial decomposition 51
cyanobacteria, defined 30, 37
  as ancestral organelles 328–330
  as food 169
  in lichens 23–24, 149
  nitrogen-fixing 149, 150–151
  primitive 43–44, 315, 322
  spoil paintings 262
  toxic blooms of 87, 265–266
cyclodextrins 194
cycloserine 202
cystic fibrosis 226–227, 356
cysts 36
cytochromes 321

2, 4-D 279
Dalton, J. 214

damping-off disease 145
dandruff 98
Deinococcus radiodurans 127, 336
Delwiche, C. C. 9
denitrifying bacteria 8, 13, 45–46, 116, 276, 285, 333
deoxyribonucleic acid (DNA) 209–220
absence from prions 25
amplification 231–235
fingerprints 231, *232*
ligase 221–222, 326
manipulation 220–224
mutations in 214–215
of choroplasts 231–235
polymerase 206, 231
profiling 233–234
sequencing 223
transfer 257, 281
vaccines 230, 231
Derxia gummosa *117*
Desulfobacter 46, 47
Desulfobacterium oleovorans 255
Desulfonema 46, *47*
Desulfotomaculum 46, 47
nigrificans 248, 259
Desulfovibrio 46, 47, 258, 315
detergents 51, 131, 281–282
dewatering, of sewage sludge 274, 283
dextrans 204–205
diabetes 225
diaminopimelic acid 164
dinitrogen (defined) 7
diosgenin 204
Domagk, G. 104
domains, of living things 332
dough 161–162
Drake, F. D. 354
Dunaliella 38, 165
Dutch elm disease 101, *102*
dysentery 56, 61, 86, 111

Ebola 70–71
echovirus 68
ecology 2, 354
economic microbiology 53
Ehrlich, P. 103, 109
Ehrlich, P. R. 335
'Ehrlich 606' 103
elderflower champagne 157
electric discharge, lethal effect 97
enrichment culture 115–116, 119
Entamoeba gingivalis 98
enterovirus 68
Eremothecium ashbyii 164
ergosterol 165, 203
Erwinia 144
erysipelas 56
erythritol 75
erythromycin 119
Escherichia coli
and knowledge 355
as cloning vehicle 225, 298
chromosone and DNA 204–210
gene into potatoes 239
habitat 60, 137
in lysine production 164
multiplication rate 4, 119
mutation rate 215
pathogenicity 56, 61–62, 294
plasmids in 217, 220–222, 240–241
Eubacteria 31, 331
eugenics 356
Euglena 21, 328
eukaryotes 28, 31, 331, 332
Europa 334, 351–353
extra-terrestrial life 334, 347–353, 354, 357
Exxon Valdez disaster 277

ferredoxin 321
fertilizer 5, 9, 115, 140, 265, 345, 356
Fildes, Sir Paul 105
first plague 265
fish farming 168, 287
flagellum 20
flamingoes 265
flatus 137
'Flavr Savr' tomato 239, 296
flax, retting 205
Fleming, Sir Alexander 107, 197
Flexibacterium 37, 318
Flor 155
fluoracetamide 51, 278

fluoride 100
food yeast 167, 342
foot and mouth disease 65, 139, 143
Foraminifera 184
formula 4
ammonia 8
benzene 5
calcium sulphate 13
calcium sulphide 13
carbon dioxide 10
carbon monoxide 12
chalk 181
copper 184
copper sulphate 184
glucose 13
hydrogen 257
hydrogen sulphide 181
iron chlorides 14
iron hydroxides 183, 257
iron pyrites 182
methane 5
nitrate ion 6
nitrite ion 8
nitrogen (dinitrogen) 7
oxygen 10
p-amino benzoic acid (*p*-AB) 105
penicillanic acid 199
penicillin 199
prontosil 104
salvarsan 103
sodium 4
sulphanilamide 104
sulphate ion 6
sulphide ion 14
sulphonamides 104
sulphur 14
water 10
fossil fuels 172, 174
fossils 182, 308, 319
fowl pest 139
Francisella tularensis 64
Frankia 149
freeze-drying, of microbes 125–126
fumaric acid 194
Fungi, as group 22
predaceous 146
Fusarium graminarum 168
oxysporum 143

Galapagos Rift 40–42, *41*
gallery vegetation 286
Gallo, R. 71
Galton Sir Francis 356
gamma rays, lethal effect 118, 126, 127–128, 129, 197, 336
gas gangrene 65, 75
genes
defined 112, 211, 213–214
general account 211–224
in alien organisms 224, 229–230, 235–340, 356
in therapy 226–229, 230, 356
synthetic 240–241
transfer 215–220, 292, 326–328, 329
genetic code 213, 240
Genetic Manipulation Advisory Group 295
genital herpes 93
geomicrobiology, defined 174
Geotrichum 160
German measles 79, 93–94
Gibberella fujikuroi 165
gibberellin 152, 165
globulin 79
gluconic acid 194
glutamic acid 166
gold, bacterial release 185
gondola 267
gonorrhoea 56, 92, 108
grain spoilage 248
Gram, C. 122
stain 122, 130
gramicidin 200
Great Plague 64
green monkey disease 7
griseofulvin 202
Gunflint Chert 317–318, *318*
gyppy tummy 61
Gypsum 174, 191

Haber process 297
haemophilia 55, 229
Haldane, J. B. S. 311

hallucinogens 343
halophiles 38, 48, 49, 330, 331, 333, 349
Hata, Dr 109
Hauser, P. M. 337
Helicobacter 62–63, 90
hepatitis B 228, 230
herpes 25, 112
heterotroph 20, 21, 23, 29, 40, 49, 312–314
hoatzin 134, *135*
Hoja blanca 144
Hopwood, Sir David 201
human immunodeficiency virus (HIV) 57, 71–73, 111–112, 223
Hutner, S. 168, 272
hyaluronic acid 194
hydrocarbon-oxidizing microbes 254–256, 280
hydrothermal vents 40, *41*
hydrotroilite 182

'ice-minus' pseudomonas 235–236, 290
immunity 77–79, 86, 112, 341–342
Imperial Chemical Industries 106
influenza 53, 68–69, 79, 83
insects, biological control 146–147
insulin 225
interferons 77, 112, 225, 342
invertase 206
ions 5–6, 13, 42, 44, 45
iron bacteria 42, 44, 116, 183, 262, 315
iron, ferrous and ferric 13–14
iron pyrites, formation of 182
itaconic acid 194

Jeffreys, Sir Alex 231, 233

kanamycin 296, 326
Kelly, Mrs B. 157
keratin 50, 252
Khorama, H. G. 240
kidney disease 75
kingdoms, of living things 31, 332
Klebsiella 26, 149
  aerogenes 164
Kluyver, A. 120
Knolles, A. S. 289
Koch, R. 121
Koch's postulates 121–122
Krakatoa 149
kuru 26

Lactic acid 153, 154, 158, 162, 192
Lactobacillus 98, 152, 156, *158*, 159, 160, 162, 192
  bulgaricus 130, 166
  casei 166
La Rivière, J. 253
Lascaux, spoilage of paintings 262
Lassa fever 71
laterites 185, 345
leaching, of metal ores 184–185
leather, microbial spoilage 50, 250
leben 159
Leeuwenhoek, A. van 18, 21, 358
Legionella 56, 66–67, 90, 119
legionellosis 56, 66, 341
leguminous plants 147, *148*, 229, 355
leprosy 56, 109, 119
Leptospira buccalis 98
Leptothrix 183
Leucaena 135–136
leucocytes 77
Leuconostoc 159
  mesenteroides 204
lichens 23–24
lignite 186
liposome 227
Listeria 56, 89, 90, 235
  monocytogenes 90
London Underground 97
luminous bacteria 208, 266–267
Lwoff, A. 323
Lyme disease 64
lysergic acid 344
lysine 136, 163–164, 239
lysozyme 77

macrophage 77
malaria 57, 64, 73, 108–109, 111
malignant ulcers 58
malo-lactic fermentation 154
Marburg disease 71

Mars 309, 334, 346, 348–351, *350*
measles 57, 78, 79
Medical Research Council 167
medium, defined 115
meningitis 55, 59
meningococcus 58
Mendel, G. J. 214
Mercury 348
mesophiles 40, 49
metal sulphide ores 182
Metarhizium flavoviride 147
methane 5
  as fuel 187, 188, 191–192, 288
  in early atmopshere 309, 310
methane bacteria, in carbon cycle 10–11
methane-forming bacteria (methanogens) 31, 48, 133, 137, 187, 290, 315, 316–317, 330
  in Carboniefererous era 185–186, 252
  in evolution 315, 316–317, 321
  in intestines 133, 137
  in land in-fills 288
  in sewage treatment 191, 273–274, 282–283
  see also Will-o'-the-Wisp
methane ice 187–88
methane-oxidizing bacteris 42, 48, 188, 254
  as protein 168, 353–354
Methanomonas 42
methicillin 129
methylene blue 104
microbicides 130, 131–132
microbiological assay 124–125
Micrococcus 36, 98, 100, 127
  radiodurans 127
Microsporon 57, 92
Miller, S. L. 310
Milton 131
mitochondrion 219, 329, 355, 361
mitomycin 202
mixotrophy 315
Mogden Sewage Works 272
molasses 248
mollicutes 30, 144, 324
Monsanto Company 237, 275
Montagnier, L. 71
Montagu, M. van 237
Moon, the 309, 346, 348–349
morbillivirus 57
Moses 265
MRSA (methicillin-resistant *Staphylococus aureus*) 129
Mucor 23, 252
multiplication of microbes 4, 246–247
mumps 55, 57, 78, 79, 94
muscular dystrophy 226
must 153
mutant 108, 197, 313, 314, 326
mutation 28, 197, 211–212, 214–215, 313–314
myalgic encephalomyelitis (ME) 73
Mycobacterium leprae 56, 119
  tuberculosis 56, 119
mycoplasma, *see* mollicutes
Myrothecium verrucaria 251
myxomatosis 145
myxovirus 57, 68, 78

nanobacteria 30
nappy rash 100
National Collection of Industrial and Marine Bacteria 17
necrotizing fasciitis 58, 59
Neisseria 92
  gonorrhoeae 56, 92
  meningitidis 58–59, *59*
neomycin 201
neurospora 22, 211–212
Nicol, H. 151
nisin 202
nitrates in drinking water 276
nitrifying bacteria 9
Nitrobacter 42
nitrogen cycle 7–10, *8*, 46, 346, 348–349
nitrogen-fixing bacteria 9, 32, 115–116, 135, 147–150, *148*, *151*, *345*, 356
  genetics of, 217–218, 240, 355
nitrogen oxides 77, 97
Nitrosomonas 42
Norwalk virus 230
novobiocin 201
nuclear energy 172, 173, 190
nystatin 201

ochre 184
Oidium 144, 154
oil 172, 175, 190, 262
  formation 188–190
  marine pollution 276–278
olivomycin 202
Oparin, A. I. 311
Opisthocomus hoatzin 134
opportunist pathogens 65
optical activity 194–196
organelle 211, 238, 256
origin of life 311–312
oxidation, defined 12–14
oxytetracycline 201
ozone 97

Pacific trenches 40–41
paint, microbial spoilage 51, 252, 262
paludrine 106
Panama disease 143
pandemic, defined 69
paper, spoilage 50, 250–251
papillomavirus 70
para-amino benzoic acid 105–106
Paracoccus denitrificans 43, 45
  halodenitrificans 247–248
Paramecium 21, 22
parasitism 55–57
Pasteur, L. 18
Pasteur Institute, Paris 261
Pasteurella pestis 64
  tularensis 64
pasteurization 127
PCR, *see* Polymerase Chain Reaction
peat, formation 186
pectinases 205
penicillin 107, 136, 138–139, 197, 200, 201, 341
  resistance 32, 52, 122, 222, 326
Penicillium 23, 160, 197, 200, 252
  chrysogenum 197
  griseofulvum 202
pernicious anaemia 138
perry 156
Petri dishes 117, 127
phagocytes 77–78
PHB (poly-β-hydroxy butyric acid) 194
phenol 51–52, 128, 131
phocine distemper virus 73, 74
phosphorus cycle 12, 346, 348–349
photo-autotrophs 20, 49, 264
photosynthesis, formulations 319–320
Phytophthora palmivora 144
picoplankton 11
picornavirus 68
Pirie, N. W. 311
pituitary growth hormone 225
Pityrosporum ovale 98
plague 56, 64, 79, 94, 109, 265
plankton 11
plasmids 108, 215–219, 220–222, 236–237, 241, 278, 325–327
Plasmodium 57, 64
plastics 51, 243–244, 253–254, 289
Pneumocystis carinii 57
pneumonia 55, 56, 71, 105, 109
pneumonic plague 79
Pochon, J. 261
polar ice caps, to melt 354–355
poliomyelitis 24, 67, 78, 83
Polymerase Chain Reaction (PCR) 232–233, 234
polymyxin 200
polythene 51
population explosion 337–339
porphyrins 189, 321
post-viral fatigue syndrome 73
potato eel-worms 146
prions 26–28, 142, 209
progesterone 204
prokaryotes 28, 30, 327, 329, 331
Prontosil 104
Propionibacterium 160
proteinases 206
protein engineering 241
Protozoa, as group 21–22
Prusiner, S. 27
Pseudomonas 115
  facilis 42, 353
  syringae 235
psychomimetic drugs 343
psychrophiles 38, 40, 247, 330, 333, 349

psychrosphere 38
ptomaines 45, 246–247
puerperal fever 105
pulque 156
pyrites 182, 184–185

quinine 108, 205
Quorn 168

rabies 65, 228
Rabshakeh, biblical threat 354
recalcitrant substances 253, 274, 281
Red Sea 265
reduction, defined 12–14
rennin 159–160
restriction enzymes 221–222, 326
rhinovirus 57, 68
Rhizobium 147
Rhizopus 23, 163, 194, 203, 204
  arrhizus 204
Rhodospirillum 29
riboflavin – *see* Vitamin B_2
ribonucleic acid (RNA) 27, 209, 212–213, 233, 235, 329–332
ribosome 212, 329
Rickettsia 30
rifampicin 119, 202
rinderpest 143
River blindness 146
rubber, spoilage 253
rumen 45, 133–136
rusting of iron 257–258

Sabin vaccine 78
Saccharomyces beticus 155
  carlsbergensis 152
  cerevisiae 152, 153, 155
  ellipsoideus 153
Salmonella 56, 89, 90, 122, 217, 228, 306
  enteritidis 89
  typhi 80, 121
salmonellosis 88–89
salvarsan 103, 109
saki 156
sauerkraut 162
sausages 161
scarlet fever 56
Scenedesmus 20, 168
scent 194
schistosomiasis 55
scrapie 26, 139–140, 142
scrub typhus 30
scurvy 164
seal disease 73, 74
Seveso 278
sewage 38, 125, 191–192, 266, 270–275, 280–284, 285, 287
sewerage 270–271, 283
sexuality, evolution 325
Shapley, H. 354
Shell Petroleum Company 168
Shepherdia 149
shipworms 136
silage 162
sleeping sickness 57
sludge, activated 272–273, 285
  digested 273–274, 285
  settled 272–273, 283
smallpox 57, 70, 78, 81–83, 94
Smarden, Kent 51, 278
Smith, H. 75
Smith, J. Maynard 28
Sneath, P. 32
soda bread 162
soda deposits 173, 180–181
solera 155
somatotropin 298
sourdough 162
soy 163
Sphaerotilus 183
Spiroplasma citri 144
Spirulina 169
spoilage of food 245–249
spores 1, 35–37, 45, 75, 97, 116, 118, 122, 126, 248, 333, 347
Sporobolomyces roseus 267–268
Staphylococcus 79, 129, 246–247
  albus 56, 60
  aureus 56, 60, 79–80, 129–130
  piscifermentans 161
sterilization 126–128
steroids 50, 203–204, 342

sterols 324
Stinky cowbird (Stinky pheasant) 134
stranglervine 144
Streptococcus 56, *130*, 137
  agalactiae 157
  lactis 130, 202
  mutans 100
  pneumoniae 56, 58
  pyogenes 56, 58
  thermophilus 158
Streptomyces 201–203, 235
  aureofaciens 201
  fradiae 201
  griseus 202
  mediterranei 202
  olivaceus 165
  rimosus 201
  venezuelae 199
streptomycin 201
sulfuretum 15, 181, 182, 255, 320
sulphanilamide 104
sulphate-reducing bacteria 46–48, *47*, 116, 290, 357
  as primitive microbes 315, 317, 319, 325
  cytochromes in 321
  effect on environment 14–15, 48–49
  in corrosion of metals 257–260
  in effluent treatment 273, 275, 282–283
  in food spoilage 248
  in intestines 134, 137
  in mineral formation 176–180, 180–182, 191
  in oil technology 188–190, 262
  in petroleum spoilage 252–256
  in water pollution *263*, 264, 266–267
  isotope fractionation by 178–179, 317
  on moon? 349
sulphide-oxidizing bacteria 14, 41, 42, 43, *178*, 180, 264–265, 314, 319–320
sulphonamides 52, 104–106, 107, 108, 109, 122
sulphur bacteria 39–40, 41, 42–43, 116, 144, 314, 319, 321, 324
sulphur cycle 14–15, *14*, 175, 181, 264, 319, 320, 346
  farming 191
  formation 175–180
  industrial use 144, 174–175, 192
  isotopes 178–179, 317
  world shortage 190–191
sulphur-deficient soil 260, 345
sulphur stinker spoilage 248
sunlight, lethal effect 95, 96–97
Swakopmund 267
sweat 98, 131
symbiosis 147–149, 328–329, 355
Synergistes jonesii 137
syphilis 56, 67, 92–93, 103, 109

2, 4, 5-T 279
Tamiya, H. 279
TCDD (2,3,7,8-tetrachlorodibenzo-*p*-dioxin) 278–279
teeth 98–100, *99*
tempe 163
temperate phages 24, 324
temperature range of microbes 36–37, 333–334
tequila 156
teratogen 94, 279
termites 136
terramycin 201
tetanus 65, 76, 79, 205
tetracycline 200, 322
Thames, pollution of 266, 271
thermoacidophiles 330
thermocline 38
thermophiles 36–37, 39, 46, 49, 116, 233, 248, 288, 310, 333
Thermoproteus 43
thermosphere 38
Thermus aquaticus 206, 233
Thiobacillus 39, 40, 42, 144, 179, 185, 261–262, 315, 316, 333
  denitrificans 42, 45
  ferro-oxidans 42, 184–185
  intermedius 315
  thio-oxidans 184, 253
Thiovulum *42*
Thode, H. 178
Ti plasmid 236–237
toilet paper 85

Tokagawa Institute of Japan 168
tonsillitis 56, 58
Torrey Canyon disaster 277
toxic shock syndrome 79–80
toxin 73, 76–77, 78, 238
tranquillizer 343–344
transduction 220, 326
transformation 219, 220, 326, 327
transgenic organisms 224
transposon 326
trench fever 30
Treponema pallidum 56, 92
Trichodesmium 265
Tristan da Cunha 67
Trypanosoma 57, 64
trypanosomiasis 103, 109, 143
Tsetse fly 64
tuberculosis 55, 56, 63, 73, 78, 79, 109–111, 139, 143, 301
tularaemia 64
Tutankhamun 35
Tyndall, J. 114
typhoid 63–64, 70, 80–81, 86, 88, 109, 271
Typhoid Mary 63–64

UHT (ultra-high temperature) 127
ultramicrobacteria 30, 351
ultraviolet light, lethal effect 127
undulant fever 64, 73
Upjohn Company 204
uranium ores 182, 185
Urey, H. C. 310
urine 75, 85–86, 100, 229, 353

vaccination 65, 78–79, 82–83, 227–228, 230–231
vancomycin 129
Variola 82
Venera Venus probe 346
venereal diseases 67, 92–93
Venice, 267
Venus 309, 334, 346, 348
Vibrio cholerae 65, 217, 219
  fischerii 267
Viking Mars probe 349
vinegar 162–163, 192
viruses, as group 24–26
  living? 26, 28
Vitamin A 165, 201
  B group 56, 137, 163, 166
  B_2 (riboflavin) 106, 164–165
  B_{12} (cobalamide) 124–125, 137–138, 165, 283
  C (ascorbic acid) 50, 164, 203
  D 165, 203
  E 163

Wadi Natrun, Egypt 181, 265
Walvis Bay 267, 325
war gases, disposal of 279–280
warts 69–70
waste stabilization ponds 287
water calamity 262
Wells, H. G. 291
whooping cough 56, 58
Will-o'-the-wisp 186, 187, 257
wine 133, 153–157, 162
wood-rot 251
Woods, D. D. 97, 105
wool, spoilage 252
Work, E. 164
World Health Organization (WHO) 82–83, 111

yeast – *see* Fungi, *Saccharomyces*, food yeast, *Candida*
Yellow fever 70
Yellowstone National Park, USA 36–37, 37, 334
Yersinia pestis 56, 64, 94
yoghurt 158–159, 158, 307

ZoBell, C. E. 39, 189